Manufact

A unified approach ... ...gy, production
management, and industrial economics

Manufacturing Systems Engineering

A unified approach to manufacturing technology, production management, and industrial economics

SECOND EDITION

KATSUNDO HITOMI

*Ryukoku University and
Nanjing University*

UK   Taylor & Francis Ltd, 1 Gunpowder Square, London EC4A 3DE
USA Taylor & Francis Inc., 1900 Frost Road, Suite 101, Bristol, PA 19007

First published 1975 in Japanese by Kyoritsu Shuppan Co Ltd
Translated into English by Taylor & Francis Ltd, 1979

Copyright © Katsundo Hitomi 1996

All rights reserved. No part of this publication may be reproduced, stored in a retrieval system, or transmitted, in any form or by any means, electronic, electrostatic, magnetic tape, mechanical, photocopying, recording or otherwise, without the prior permission of the copyright owner.

British Library Cataloguing in Publication Data

A catalogue record for this book is available from the British Library

ISBN 0-/484-0323-X (cased)
ISBN 0-7484-0324-8 (paperback)

Library of Congress Cataloguing in Publication Data are available

Cover design by Grand Union Design

Contents

Foreword	*page* xii
Preface	xiii
About the author	xix

PART ONE Essentials of Manufacturing Systems   1

1 Fundamentals of Manufacturing   3
  1.1 Definitions of production and manufacturing   3
  1.2 Principles of manufacturing   5
  1.3 Resources of production—inputs for production   8
  1.4 Goods produced—outputs of production   11
  1.5 Production processes—transformation of inputs into outputs   14
  1.6 Production organisation   22
  Notes   22
  References   23
  Supplementary reading   23

2 Fundamentals of Systems   24
  2.1 Basic concepts of systems and chaos   24
  2.2 Definition of systems   25
  2.3 Basic problems concerning systems   29
  2.4 Systems design   30
  2.5 Decision-making procedures   38
  Notes   45
  References   45
  Supplementary reading   46

3 Fundamentals of Manufacturing Systems   47
  3.1 Meaning of the term 'manufacturing systems'   47
  3.2 Structural aspect of manufacturing systems   48
  3.3 Transformational aspect of manufacturing systems   48

3.4	Procedural aspect of manufacturing systems	50
3.5	Integrated manufacturing systems (IMS)	55
3.6	Manufacturing systems engineering—an academic discipline	56
	Notes	58
	References	58
	Supplementary reading	59

4 Modes of Production — 61

4.1	Types of production	61
4.2	Mass production	62
4.3	Multi-product, small-batch production	65
4.4	Product diversification	72
	Notes	73
	References	73
	Supplementary reading	73

5 Integrated Manufacturing and Management Systems — 74

5.1	Basic functions and structures of management systems	74
5.2	Basic framework of integrated manufacturing management systems	76
5.3	Framework of an integrated manufacturing system	79
	References	82
	Supplementary reading	82

PART TWO Process Systems for Manufacturing — 83

6 Material and Technological Information Flows in Manufacturing Systems — 85

6.1	Logistic systems	85
6.2	Material flow	85
6.3	Technological information flow	88
	References	88
	Supplementary reading	89

7 Product Planning and Design — 90

7.1	Product planning	90
7.2	Product design	93
7.3	Product structure and explosion	103
	Notes	106
	References	106
	Supplementary reading	106

8 Process Planning and Design — 107

8.1	Scope and problems of process planning	107
8.2	Process design	107
8.3	Operation design	114
8.4	Optimum routing analysis	125
8.5	Line balancing	130
	References	134
	Supplementary reading	134

9 Layout Planning and Design — 135
- 9.1 Scope and problems of layout planning — 135
- 9.2 Systematic layout planning (SLP) — 138
- 9.3 Mathematical layout design — 143
- 9.4 Production flow analysis — 145
- References — 146
- Supplementary reading — 146

10 Logistic Planning and Design — 147
- 10.1 Transportation problems — 147
- 10.2 Distribution problems — 152
- Notes — 153
- References — 153
- Supplementary reading — 153

11 Manufacturing Optimisation — 154
- 11.1 Evaluation criteria for manufacturing optimisation — 154
- 11.2 Optimisation of single-stage manufacturing — 156
- 11.3 Optimisation of multistage manufacturing systems — 174
- Notes — 183
- References — 183
- Supplementary reading — 184

PART THREE Management Systems for Manufacturing — 185

12 Managerial Information Flow in Manufacturing Systems — 187
- 12.1 Managerial information flow — 187
- 12.2 Decision problems in managerial information flow — 188
- References — 189

13 Aggregate Production Planning — 190
- 13.1 Production planning defined — 190
- 13.2 Short-term production planning — 190
- 13.3 Multiple-objective production planning — 205
- 13.4 Product mix analysis — 209
- 13.5 Lot-size analysis — 213
- 13.6 Material requirements planning (MRP) and machine loading — 217
- 13.7 Long-term production planning — 225
- 13.8 Production forecasting — 228
- Notes — 233
- References — 233
- Supplementary reading — 234

14 Production Scheduling — 235
- 14.1 Scope of production scheduling — 235
- 14.2 Operations scheduling — 237
- 14.3 Project scheduling—PERT and CPM — 258
- Notes — 268
- References — 268
- Supplementary reading — 269

15 Inventory Management — 270
- 15.1 Inventory function in manufacturing — 270
- 15.2 Fundamentals of inventory analysis — 271
- 15.3 Inventory systems — 271
- 15.4 Multiple-product inventory management — 277
- 15.5 Probabilistic inventory models — 279
- References — 281
- Supplementary reading — 281

16 Production Control — 282
- 16.1 Scope and problems of production control — 282
- 16.2 Process control — 283
- 16.3 Just-in-time (JIT) production — 285
- 16.4 Productive maintenance — 288
- 16.5 Replacement — 294
- References — 301
- Supplementary reading — 301

17 Quality Engineering — 302
- 17.1 Quality control (QC) — 302
- 17.2 Quality function deployment (QFD) — 307
- 17.3 Quality engineering — 309
- Notes — 311
- References — 311
- Supplementary reading — 311

PART FOUR Value Systems for Manufacturing — 313

18 Value and Cost Flows in Manufacturing Systems — 315
- 18.1 Value/cost flow in manufacturing systems — 315
- 18.2 Concepts of cost and time-series value of money — 317
- References — 318

19 Manufacturing Cost and Product Cost Structure — 319
- 19.1 Classification of costs — 319
- 19.2 Product cost structure — 319
- 19.3 Selling price — 320
- 19.4 Computing the manufacturing cost — 322
- References — 324
- Supplementary reading — 324

20 Profit Planning and Break-even Analysis — 325
- 20.1 Profit planning — 325
- 20.2 Break-even analysis — 328
- Notes — 333
- References — 333
- Supplementary reading — 333

21 Capital Investment for Manufacturing — 334
- 21.1 Investment for manufacturing automation — 334

	21.2	Evaluation methods of capital investment	334
		Notes	338
		References	338
		Supplementary reading	338

PART FIVE Automation Systems for Manufacturing — 339

22 Industrial Automation — 341
- 22.1 Towards automation — 341
- 22.2 Meanings of automation — 342
- 22.3 Kinds of automation — 343
- 22.4 Development of automatic manufacturing — 344
- Notes — 352
- References — 352
- Supplementary reading — 352

23 Principles of Computer-integrated Manufacturing (CIM) — 353
- 23.1 Essentials of computer-integrated manufacturing systems — 353
- 23.2 Definition of CIM — 354
- 23.3 Effectiveness of CIM — 355
- Notes — 357
- References — 357
- Supplementary reading — 357

24 Computer-aided Design (CAD) — 358
- 24.1 Brief history of computer-aided design (CAD) — 358
- 24.2 Computer-aided design/drawing (CADD) — 359
- 24.3 Computer-automated process planning (CAPP) — 364
- 24.4 Automatic operation planning—autoprogramming system — 370
- 24.5 Computerised layout planning — 373
- References — 379
- Supplementary reading — 379

25 Factory Automation (FA), Computer-aided Manufacturing (CAM) and Computer-integrated Manufacturing (CIM) Systems — 381
- 25.1 Automatic machine tools for mass production — 381
- 25.2 Numerically controlled (NC) machine tools — 384
- 25.3 Computer-controlled manufacturing systems — 389
- 25.4 Flexible manufacturing system (FMS) — 390
- 25.5 Automated assembly — 395
- 25.6 Automatic materials handling — 399
- 25.7 Automatic inspection and testing — 404
- 25.8 Computer-integrated automation system—unmanned factory — 406
- References — 408
- Supplementary reading — 408

PART SIX Information Systems for Manufacturing — 409

26 Fundamentals of Information Technology — 411
- 26.1 Concept of information — 411

	26.2	Information systems	413
	26.3	Management information system (MIS) and strategic information system (SIS)	415
	26.4	Information networking	417
		Notes	419
		References	419
		Supplementary reading	419

27 Parts-oriented Production Information Systems — 420
27.1 Concept of parts-oriented production information systems — 420
27.2 Structure of parts-oriented production information systems — 421
27.3 Advantages of parts-oriented production information systems — 423
References — 424
Supplementary reading — 424

28 Computerised Production Scheduling — 425
28.1 Interactive group scheduling technique — 425
28.2 Computer-aided line balancing (CALB) — 430
References — 431
Supplementary reading — 431

29 On-line Production Control Systems — 432
29.1 Concept of on-line production control — 432
29.2 Structure of on-line production control — 433
29.3 Scheduling and control of on-line production — 434
29.4 Optimum-seeking method for on-line production control — 436
References — 439
Supplementary reading — 439

30 Computer-based Production Management Systems — 440
30.1 Computerised production management — 440
30.2 Computerised manufacturing information systems — 441
References — 449
Supplementary reading — 449

PART SEVEN Social Systems for Manufacturing — 451

31 Social Production Structure — 453
31.1 Social manufacturing systems — 453
31.2 Social production modes — 455
31.3 Establishment of management systems — 456
Notes — 457
References — 458
Supplementary reading — 458

32 Manufacturing Strategy — 459
32.1 Strategy and tactics — 459
32.2 Corporate strategy — 460
32.3 Manufacturing strategy — 460

	32.4	Computer-integrated manufacturing as a corporate strategy	461
		References	462
		Supplementary reading	462
33	**Global Manufacturing**		464
	33.1	Movement towards globalisation	464
	33.2	International production	465
	33.3	Export and import of industrial products	466
		References	468
		Supplementary reading	468
34	**Industrial Structure and Manufacturing Efficiency**		469
	34.1	Industrial structure	469
	34.2	Japan's manufacturing and industry	470
	34.3	Industrial efficiency in advanced countries	479
	34.4	Industrial efficiency in developing countries	484
	34.5	International comparison of manufacturing efficiency	485
		Notes	489
		References	489
		Supplementary reading	490
35	**Industrial Input–Output Relations**		491
	35.1	Relationships between industrial activities	491
	35.2	Industrial input–output analysis	492
	35.3	Requirements for input–output analysis	495
		Notes	495
		References	495
		Supplementary reading	496
36	**Manufacturing Excellence for Future Production Perspectives**		497
	36.1	Importance and dilemma of today's manufacturing	497
	36.2	Concept of manufacturing excellence	499
	36.3	Approaches to manufacturing excellence	500
	36.4	Green production	501
	36.5	Socially appropriate production as ultimate manufacturing excellence	507
		Notes	508
		References	508
		Supplementary reading	509
	Concluding Remarks		511
	Review Questions and Problems		515
	Index		529

Foreword

It is twenty-one years since Professor Hitomi promoted manufacturing systems engineering, with publication of the first edition of his book. Since then there has been continuous evolution of the concepts of management, organisation, control and the technologies of manufacturing, and revolution in some aspects fuelled by the explosion of computer applications. Although much of the first edition remains fully pertinent as the basis for design of manufacturing systems, he has now brought the book thoroughly up to date, drawing on research and industrial practices world-wide. The reader will find substantial lists of references and recommended readings throughout the text, and the book is littered with practical examples and anecdotes drawn from years of active involvement in manufacturing research and practice. A major strength of the book is that it is firmly solutions-oriented—not simply a collection of tools and methodologies. Following an introduction to the fundamentals of manufacturing systems, the book considers six aspects of manufacturing: Process Systems, Management Systems, Value Systems, Automation Systems, Information Systems, Social Implications. Each of these is presented logically, identifying industrial needs and problems and developing the tools and solution methodologies to enable design of effective, efficient, profitable and socially appropriate manufacturing systems. The inextricable linking or integration of these six aspects is also fully considered.

This new edition remains an ideal text for students of production engineering, production management and manufacturing systems. Through its structure and helpful indexing it will also provide a valuable roadmap for practising engineers and managers addressing manufacturing issues from either a local or a global standpoint.

JOHN MIDDLE
Editor: *International Journal of Production Engineering*

Preface

This is the second edition of *Manufacturing Systems Engineering*. This book is written to provide a modern introductory text for mechanical, industrial, and production engineering students and a reference book for mechanical, industrial, and production engineers and managers who are concerned with manufacturing technology and production management in industry.

This book is intended to integrate the following three aspects:

- manufacturing (or production) technology,
- production management, and
- industrial economy (or economics).

'Manufacturing technology' is concerned with the 'flow of materials' from raw material acquisition, through conversion in the workshop, to the shipment of finished goods to the customers. 'Production management' (or 'management technology') deals mainly with the flow of information, so as to manage the flow of materials efficiently by planning and control. 'Industrial economy' (or 'production economics') treats the flow of costs in order to reduce the product[ion] cost so as to set the price reasonably. (*Note*: Throughout this book [ ] is used to read in two ways; in this case we read 'product' and 'production'.)

The majority of the study of manufacturing technology has been concerned with manufacturing processes, machine tools, etc., in the areas of mechanical and production engineering, whereas management technology, such as production management, information technology for manufacturing, etc., has been within the fields of industrial engineering and business administration. As to industrial or production economics, this has been investigated and taught in the economics department, though it is an important subject in manufacturing studies.

'Manufacturing' is the production of tangible goods (products); it is a basic historical process which has lasted for several thousand years. Manufacturing matters, in that it is a basic means of human existence, it creates the wealth of nations, and it contributes towards human happiness and world peace. In 1991 the National Academy of Engineering/Science in Washington, D.C. rated manufacturing as one of the three important subjects necessary for America's economic growth and national security, the others being science and technology. In Japan the importance of these three subjects was pointed out as early as 1935.

These days 'manufacturing' is taken to include software and its serviceability. In this respect manufacturing is to be considered not only from the technological viewpoint, but also from the wide standpoints such as management, economy, social sciences, philosophy, ... In other words, study of manufacturing/production must include both hard and soft technologies. Such an integrated, systematic study of manufacturing is termed 'manufacturing systems engineering'.

Based upon the foregoing ideas and objectives, this volume contains seven parts with 36 chapters as follows.

Part I, Essentials of Manufacturing Systems, describes basic concepts and principles of manufacturing systems and manufacturing systems engineering. Principles of production and manufacturing (Chapter 1) and systems (Chapter 2) are discussed, then manufacturing systems are defined within three aspects— structure (plant layout), transformation (production process), and procedure (manufacturing management) (Chapter 3). In addition, two important modes of production—mass production and multi-product, small-batch production—are mentioned (Chapter 4). Fundamental frameworks of integrated manufacturing [and management] systems are displayed (Chapter 5).

Part II, Process Systems for Manufacturing, describes basic principles of production process technology—'material flow' as well as 'technological information flow' (Chapter 6), which is the most essential activity in production/manufacturing. Fundamentals of product planning and design for new product development (Chapter 7), process planning and design dealing with the effective conversion of raw materials into finished products (Chapter 8), and layout planning and design concerning spatial allocation of production facilities (Chapter 9) are mentioned. Logistic planning and design for solving the transportation and the travelling salesperson problems (Chapter 10) and manufacturing optimisation for deciding optimum machining conditions based upon the minimum time, the minimum cost, and the maximum profit rate (Chapter 11) are also discussed.

Part III, Management Systems for Manufacturing, describes basic principles of production management technology—'managerial information flow'. The meaning of this flow and the functions and decision problems of strategic production planning and operational production management are introduced (Chapter 12). Basic theories and solution algorithms of production planning including product mix, economic lot size, MRP, and production forecasting (Chapter 13), production scheduling including PERT/CPM (Chapter 14), inventory management which is 'stock' in the material flow (Chapter 15), production control including just-in-time (JIT) production, productive maintenance, and replacement (Chapter 16), and quality engineering including quality control (QC), quality function deployment (QFD), and the Taguchi method (Chapter 17) are discussed.

Part IV, Value Systems for Manufacturing, describes basic principles of production economy or economical production—'flow of value/costs' against the 'flow of materials' and the 'flow of information', which are mainly concerned with technical production. The concept of cost and the time-series value of money (Chapter 18), the product cost structure including price-setting methods and typical cost accounting and control procedures (Chapter 19), profit planning and break-even analysis determining the optimum plant size and production scale (Chapter 20), and effective capital investment with typical evaluation methods for manufacturing automation (Chapter 21) are introduced.

Part V, Automation Systems for Manufacturing, describes basic principles and the present state of factory automation (FA) and computer-integrated manufacturing (CIM) for automated 'flows of materials and technological information', in connection with Part II. Industrial automation (Chapter 22) introduces the development and kinds of automation, and CIM (Chapter 23) is defined as a system of integration of computer aids of design, production and management functions. Computer-aided design (CAD) including computer-automated process planning (CAPP), autoprogramming systems, and computerised layout planning (Chapter 24) and computer-aided manufacturing (CAM) including numerical control (NC), flexible manufacturing systems (FMS), flexible assembly systems (FAS), industrial robots, automated warehouse, and a real-life example of an unmanned factory (Chapter 25) are explained.

Part VI, Information Systems for Manufacturing, describes basic principles and software of manufacturing information processing for automated 'flow of managerial information', in connection with Part III and as an extension of Part V. Fundamentals of information technology including [management/strategic] information systems (MIS/SIS) and information network techniques are introduced (Chapter 26). Two effective methods for multi-product, small-batch production with computers are introduced—parts-oriented production information systems for order entry by customers (Chapter 27) and on-line production control and information systems for an efficiently running shop floor (Chapter 29). Computerised production scheduling introducing interactive group scheduling techniques and computer-aided line balancing (Chapter 28), and computer-based production management and manufacturing information systems are discussed. The communications-oriented production information and control system (COPICS) and the business and manufacturing control system (BAMCS) (Chapter 30) are explained.

Part VII, Social Systems for Manufacturing, describes basic principles of the social aspect of production and manufacturing systems. Social production structure in connection with consumption and inventory with the historical development of production modes (Chapter 31) is discussed. Essentials of manufacturing strategy stressing the strategic side of CIM (Chapter 32) and global manufacturing including theories of globalisation and export/import of industrial products (Chapter 33) are introduced.

The industrial pattern shifts from primary industry (agriculture) to secondary (manufacturing), and further to tertiary (service), as the economies of a nation grow. This industrial development and manufacturing efficiency (Chapter 34) and the input–output analysis expressing the influences among industrial sectors (Chapter 35) are mentioned.

In the last chapter of this book (Chapter 36) a concept of 'manufacturing excellence' is proposed for future production perspectives to enhance the elegance of goods production, and as its ultimate realisation 'socially appropriate manufacturing' is introduced to save our only earth from destruction caused by today's excess production and consumption accompanied by huge waste of useful resources. This is also stressed in *Concluding Remarks* with other issues such as human problems, the concept of philoso-technology, etc.

Each chapter contains references [and supplementary reading]. *Review Questions and Problems* are listed for the students at the end of this book.

In describing the concepts, fundamentals and principles of manufacturing system analysis/design and production management, as mentioned in the above, the following six approaches are vital.

1. To clarify the concept of manufacturing systems and their structure and functions; i.e. systems analysis and design for manufacturing, particularly, the 'flow of materials' (Parts I and II).
2. To optimise planning, implementation and control of manufacturing systems; i.e. optimum decision-making for production (Part III).
3. To recognise economy for manufacturing systems; i.e. cost engineering/management and profit planning, particularly, the 'flow of cost' (Part IV).
4. To automate manufacturing systems; i.e. factory automation (FA) and computer-integrated manufacturing (CIM) (Part V).
5. To process manufacturing information systems; i.e. information technology in production management, and timely 'flow of information' (Part VI).
6. To understand the social aspect of manufacturing, i.e. strategic and global manufacturing, industrial structure and input−ouput analysis, proposing 'manufacturing excellence' for future production perspectives (the 'flow of value') (Part VII).

Concepts, basic theories, algorithms and software technologies are emphasised with respect to manufacturing systems analysis/design/engineering/technology/management/economics. Hardware technologies for production facilities, machines, materials, jigs and tools are not explained in such detail.

This book may be used as a textbook of production engineering, production management, manufacturing systems [engineering], and other related courses. The flow chart for teaching as well as learning is indicated in Figure 1. For one quarter Parts I–IV should be treated, and Parts V–VII may be handled in the following quarter.

Practising engineers and managers can consult this book as a sort of 'map' of manufacturing by finding their present situation and issues from the global standpoint. The *Index* will help them to understand the meaning of technical, managerial, and industrial keywords.

The first book entitled *Manufacturing Systems Engineering* was published in 1975 in Japanese by Kyoritsu Publishing, Tokyo. This was translated into English and the English version with enlargement was published in 1979 by Taylor & Francis, London. The second edition of the Japanese book was published in 1990. These Japanese and English books were translated into Korean and into Chinese.

In 1979 the concept of manufacturing systems engineering was recognised as one of the disciplines concerning industrial engineering and management (T. Furukawa, *Journal of Japan Industrial Management Association*, **34** (2)), and several book reviews appeared (e.g. D.A. Dornfeld (1981), *ASME Journal of Dynamic Systems, Measurement, and Control*, **103**(4)).

It should be mentioned that in 1982 the International Business Machines (IBM) Corporation developed the Manufacturing Systems Engineering (MSE) curricula for Master's courses with five universities in the United States, and contributed to the advancement of this academic subject (*IE News* (Manufacturing Systems), **18** (3)). It is noted that this world-class giant company, who are accustomed to guarding their own intellectual property with great care, paid no attention to the earlier advocacy of this academic discipline, nor to the term manufacturing systems engineering, as

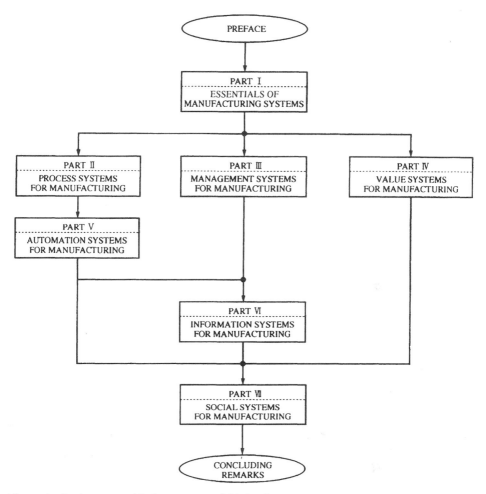

Figure 1 Explanatory guide for contents of this book.

developed by the author.

The subject is now recognised as having an impact potentially as great as a second Industrial Revolution (K.P. White and C.M. Mitchell (1989), *IEEE Transactions on Systems, Man, and Cybernetics*, **19** (2)), and quite a few departments, chairs and lectures bear this name. The University of Wisconsin-Madison established a program for this research; the director of its program recently mentioned that the term 'manufacturing systems engineering' was coined by me and has since become global (R. Suri (1993), *Journal of Manufacturing Systems*, **12** (3)). It was also stated that I proposed the concept of manufacturing systems engineering as 'integrated manufacturing unifying material flow (manufacturing processes) and information flow (production management)' (G.J. Colquhoun, R.W. Baines and R. Crossley (1993), *International Journal of Computer Integrated Manufacturing*, **6** (4)). Chongqing University in China established the Institute of Manufacturing Systems Engineering; the director of this institute also mentioned that the concept of 'manufacturing systems engineering' was first proposed by me in the 1970s (F. Liu *et al.* (eds.) (1995), *Manufacturing Systems Engineering* (in Chinese), Beijing:

Defence Industry Publishing, p. 2). It was a great pleasure and honour to me to be called by Professor J.T. Black of Auburn University, Director of the Advanced Manufacturing Technology Center, 'Mr Manufacturing Systems Engineering' in a dedication of his book (J.T. Black (1991), *The Design of the Factory with a Future*, New York: McGraw-Hill).

For a brief history and for future perspectives of manufacturing systems, refer to my recent article (1994): 'Manufacturing systems: past, present and for the future', (*International Journal of Manufacturing System Design*, **1** (1)).

The author wishes to thank the many persons who have provided valuable material used in this book. Special appreciation is expressed to John E. Middle, MSc, CEng, at the Department of Manufacturing Engineering at Loughborough University of Technology, and Editor of *International Journal of Production Research*, who kindly reviewed and commented on the original draft of this second edition.

Thanks are due for contributions to the translation of the earlier versions into Korean and into Chinese to Dr Inyong Ham, Distinguished Emeritus Professor at Pennsylvania State University, Dr Kyu-Kab Cho, Professor and Director at Pusan National University, Korea, and Zhenye Wu, Professor and Dean at Southwestern Jiaotong University, China.

The author is deeply indebted to Mr Richard Steele, the Publisher and Becky Norris, the Production Editor, at Taylor & Francis, for the publication of this second edition. Other notable persons who are acknowledged are Dr Masaru Nakajima, Associate Professor at Kyoto Institute of Technology, Dr Kenji Yura, Associate Professor at The University of Electro-communications, and Dr Hiroki Ishikura, Research Associate at Kyoto University, Japan, for their great efforts in preparing the figures and tables used in this book, and Noriko Hitomi for her patient word-processing effort.

KATSUNDO HITOMI
August 1995
(celebrating the twentieth birthday of 'manufacturing systems engineering')

About the author

Dr Katsundo Hitomi is a registered professional engineer in Japan and a certified professional economist in China. He is currently Professor of Business Administration at Ryukoku University (Japan) and Concurrent Professor at the International Business School of Nanjing University (China). His areas of interest are manufacturing systems engineering, industrial management, and production economics. He is the author of over 300 technical papers and 15 books. He is a fellow of the American Society of Mechanical Engineers, a member of the Trustees' Academy of Ohio University, an advisory member of the Nanjing Economist and Management Association (China), and serves on the editorial board of seven technical journals. Dr Hitomi is the recipient of the Japan Society of Mechanical Engineers Prize, the Japan Industrial Management Association Prize, and the Machine Tool Promotion Association (Japan) Prize. He is listed in *Jinjikoshinroku* (Who's Who in Japan), *Who's Who in Engineering* (USA), and *Marquis Who's Who in the World* (USA).

PART ONE

Essentials of Manufacturing Systems

This part describes basic concepts and principles of manufacturing systems and manufacturing systems engineering. Fundamentals of production and manufacturing (Chapter 1) and systems (Chapter 2) are discussed, then manufacturing systems are defined within three aspects—structure (plant layout), transformation (production process), and procedure (manufacturing management) (Chapter 3). In addition, two important modes of production—mass production and multi-product, small-batch production—are mentioned (Chapter 4). Fundamental frameworks of integrated manufacturing [and management] systems are displayed (Chapter 5).

Three major flows in manufacturing pointed out in Chapter 1—the flow of materials, the flow of information (technical, managerial), and the flow of value (or costs)—are related to Parts II, III, and IV.

CHAPTER ONE

Fundamentals of Manufacturing

1.1 Definitions of Production and Manufacturing

1.1.1 Transition of Production Mode

What is Production?

Production is the making of something new—either tangible ('products') or intangible ('services' that disappear in the very act of their creation). Today intangible 'ideas' are also included under the heading of production. Production is one of the most basic and important functions of human activities in modern industrial societies and is now viewed as a cultural activity. This English term appeared in 1483,[1] stemming from *producere* (Latin/lead forward).

Three Modes of Production

The mode of production has changed with time in the following three ways (*Encyclopedia Americana*, 1965).

(1) In ancient times nature was the sole source of wealth;[2] that is, agriculture, hunting, fishing, mining, and their like were the basic productive activities. This category is now known as *primary industry*.
(2) About two hundred years ago, pioneers of economics such as Adam Smith (1723–90), David Ricard (1772–1823), and John Stuart Mill (1806–73) included 'manufacture' as an element in the creation of wealth, introducing the concept of 'vendability: production for the market'. The term 'manufacture' originally appeared in 1622, stemming from *manu factum* (Latin/made by hand). Production in this sense puts special emphasis on making things which are tangible ('products'). This category is now known as *secondary industry*, which includes manufacturing, construction, and public utility generation.
(3) Towards the latter part of the nineteenth century, the concept of 'utility' came to be used by the marginal utility economists such as William Stanley Jevons (1831–82), Karl Menger (1840–1921), and Marie Esprit Léon Walras

(1834–1910). Economically, utility is an index expressing the degree of satisfaction of a human want. This term originally appeared in 1440, stemming from *utilis* (Latin/useful). With the introduction of this concept, the meaning of production was widened: 'production is a creation of utility'. Hence it further includes 'services', that is, accompanying transportation, sales, trade, and other service activities. Now this category is known as *tertiary industry*.

Free/Economic Goods

As above there is no clear distinction between material (tangible) production and immaterial (intangible) production[3] from the standpoint of economics; the only significant distinction has been between the following two goods:

- *free goods* available in unlimited quantities at no cost, such as air and river water, and hence, need not be produced;
- *economic goods* limited to a quantity sufficient to satisfy human wants completely.

Economic goods need to be produced at a required time and place with expenditure. In this sense, 'scarcity' specifies economic goods.

1.1.2 What are Production and Manufacturing?

Production and Manufacture[ing] Defined

In a narrow sense production is understood to be 'the transformation of raw materials into products by a series of energy applications, each of which affects well defined changes in the physical or chemical characteristics of the materials' (Danø, 1966).

Since this definition applies only to producing tangible goods (products) such as in the manufacturing and process industries, it is termed *manufacturing* (or *manufacture*). A place which executes manufacturing is a 'factory' or a 'workshop'. Firms concerned with manufacturing are 'manufacturing firms' or 'manufacturers'.

Today's Meaning of Manufacturing/Production

The original meaning of manufacture was to make things by hand, as mentioned previously. However, the present meaning has quite widened: 'Manufacturing' is 'the conversion of a design into a finished product', and 'production' has a narrower sense, namely the physical act of making the product (Young and Mayer, 1984). In 1983 CIRP (International Conference on Production Research) defined *manufacturing* as 'a series of interrelated activities and operations involving the design, materials selection, planning, manufacturing production, quality assurance, management and marketing of the products of the manufacturing industries'. Note how this definition includes the inextricable linking of design into the manufacturing system.

'Manufacturing' should be recognised as a series of productive activities: planning, design, procurement, production, inventory, marketing, distribution, sales, management.

1.1.3 Why is Manufacturing Important?

Manufacturing Matters in Three Features

Manufacturing—the production of tangible goods or products—has a history extending several thousand years and contains the following three important features (Hitomi, 1994).

(1) Providing *basic means of human existence*. Without the manufacture or production of goods a human being is unable to live, and this is increasingly so in modern society.
(2) *Creation of wealth of nations*. The wealth of a country or a nation is created by manufacturing. A country where manufacturing has been exhausted becomes poor and weakened.
(3) *Steps toward human happiness and the world's peace*. An affluent and prosperous country can provide security, welfare and happiness to its people; such a country no longer needs to invade other countries/nations, or to make war, so the manufacture of weapons (public 'bads') stops, resulting in world peace.

In 1991 the National Academy of Engineering/Science in Washington, D.C. rated 'manufacturing' as one of three important subjects necessary for America's economic growth and national security, the others being 'science' and 'technology'. In Japan the importance of these three subjects was pointed out as early as 1935 (H. Aikawa).

1.2 Principles of Manufacturing

1.2.1 Three Kinds of Flows Supporting Effective Manufacturing

Three Flows in Manufacturing

For efficient manufacturing activities, unification and harmonisation of the following three flows[4] are vital (see Figure 1.1) (Hitomi, 1991):

(1) *flow of materials* (or *material flow*)—conversion of raw materials into products, i.e. technical production;
(2) *flow of information* (or *information flow*)—planning and control of production;
(3) *flow of cost* (or *cost flow*)—economical production.

Flow of Materials

Production of goods is basically the utilisation of resources of production (manpower, materials, machines, money, and information) particularly raw materials in the manufacture of finished product. This input–output system is referred to as a 'production process (or production technology)', which is simply called the 'flow of materials' (*technical production*).

From a wide viewpoint, this flow constitutes a serial chain of functions: procurement, production, distribution, inventory and sales, as depicted in Figure 1.2.

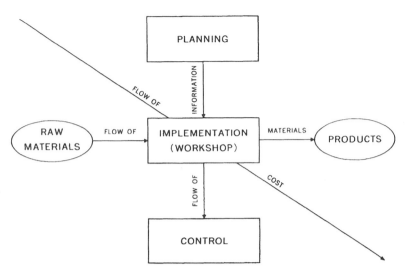

Figure 1.1 Three flows concerning manufacturing: flow of materials, flow of information, and flow of cost.

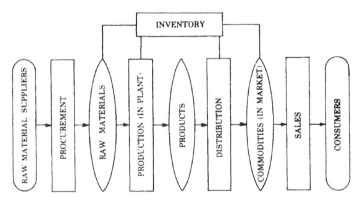

Figure 1.2 Flow of materials in manufacturing systems: procurement–production–distribution–inventory–sales.

Throughout this process raw materials are procured from outside suppliers, processed and assembled at workshops and stored in warehouses as inventories, and finished products are delivered to consumers as commodities throughout the distribution stages. Also involved will be the means of physically handling materials through and between these functions.

Flow of Information

In the consumer-driven age of 'market-in' (meeting a variety of market needs) rather than 'product-out' (just producing the products),[5] it is essential to grasp exact market needs and reflect those needs in the production processes. This is the 'management function (or management technology)' which conducts planning and control; it is called the 'flow of information' (see also Figure 1.1).

FUNDAMENTALS OF MANUFACTURING

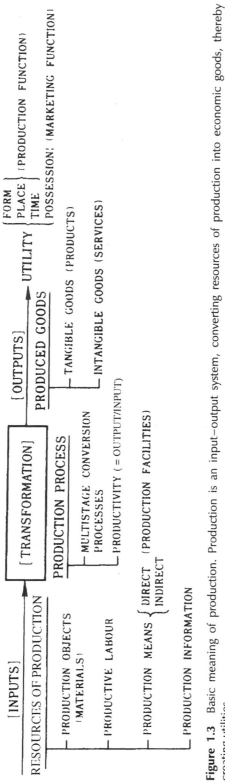

Figure 1.3 Basic meaning of production. Production is an input–output system, converting resources of production into economic goods, thereby creating utilities.

Planning and Control

Planning is selection of the future course of action. According to this plan (schedule) production implementation executes practical activities to make products in a workshop. *Control*[*ling*][6] is the measurement and correction of the performance of the activities in order to make sure that the management objectives and plans are being accomplished (Koontz and Weihrich, 1990).

Flow of Cost

Through the production processes raw materials are converted into finished products with value added. This added-value process is the 'flow of value' (*economical production*); it is more specifically called the 'flow of cost[s] (or cost flow)', as depicted in Figure 1.1. Through the production processes costs occur and are accumulated with successive production activities.

Importance of Information Flow

The 'flow of materials' is a basic indispensable function in manufacturing. This flow is accompanied by the 'flow of costs' as the raw materials are converted into products with high value. The driving force of this function is the 'flow of information'.

The flow of materials which generates the flow of costs proceeds according to the instructions issued from the flow of information based upon market needs. This synergistic action realises high quality, low costs and just-in-time (or quick-response) delivery. Realisation of this ideal production mode is now observed in automated factories or those concerned with computer-integrated manufacturing (CIM).

1.2.2 Production/Manufacturing Defined

Definition of Production/Manufacturing

From the above discussion we now define *production/manufacturing* as 'the process of producing economic goods, including tangible goods and intangible services, from resources of production (production process/technical production), thus creating utility by increasing value added (added-value creating process/economical production) (Figure 1.3) (Hitomi, 1972).

As Figure 1.3 shows, production can well be viewed as an 'input–output system'.[7]

1.3 Resources of Production—Inputs for Production

1.3.1 What are Resources of Production?

Resources of Production Defined

Resources of production are inputs procured (purchased) from the outside and used in the production processes for the making of goods.

The classical classification for resources of production from the economist's viewpoint is into the following three groups:

(1) *land*—natural resources;
(2) *labour*—human effort (physical and mental);
(3) *capital*—economic goods for reproduction (tools, machinery, factory, building, raw materials, etc.).

Occasionally 'management' (skill of managing and operating the firm) is included to differentiate this from the human effort within the production processes.

Resources of Production for Manufacturing Systems

The above broad classification for resources of production is not so suitable for a detailed analysis of a practical production process at the microscopic level. Hence, in this book resources of production are classified into the following four categories, which play essential roles in the manufacturing systems (Hitomi, 1975):

(1) *Production objects*—materials on which activities of production are performed. They consist of:
   (a) *primary materials*—converted into products through the production processes, such as raw materials, parts, etc. that construct the products;
   (b) *auxiliary materials*—added to the primary materials—e.g. paints—or supplementary to their production—e.g. electricity and lubrication oils consumed in the process of production, light and air-conditioning which support the productive labour, and others.
(2) *Production means*—media by which the raw materials are converted into products. They include:
   (a) *direct production means* or *production facilities* which directly work on raw materials—e.g. machines, equipment, apparatus, jigs and tools, materials-handling equipment, etc.;
   (b) *indirect production means* which do not directly run productive activities, such as land, roads, buildings, warehouses, etc.

   Production objects change and are consumed during production, while production means can be utilised repeatedly during a certain specified length of time (machine durable years).
(3) *Productive labour*—human ability, including the physical, spiritual, and mental ability of an individual worker, with which production activities are performed. An *organisation* where two or more persons cooperate for a common purpose has particular importance in most manufacturing systems.
(4) *Production information*—knowledge/intelligence/know-how to implement effectively, i.e. efficiently and economically, the productive processes for manufacturing. It includes *production methods*—technical procedures of implementing the productive processes. These methods include:
   (a) *production technology*, which follows objective engineering laws including empirical rules;
   (b) *production techniques*, which are subjective skills gained by training of individuals wherein experiences and intuition are highly relied upon;
   (c) *production knowledge-bases*, which are expert systems based upon production rules.

10   ESSENTIALS OF MANUFACTURING SYSTEMS

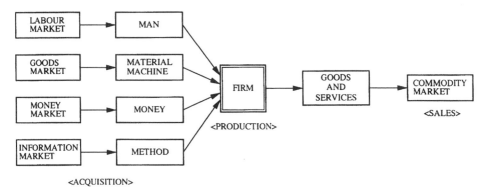

Figure 1.4 A firm acquires personnel, materials, machines, money, and methods from the labour market, the goods market, the money market, and the information market, then produced goods/services are sold as commodities at the commodity markets.

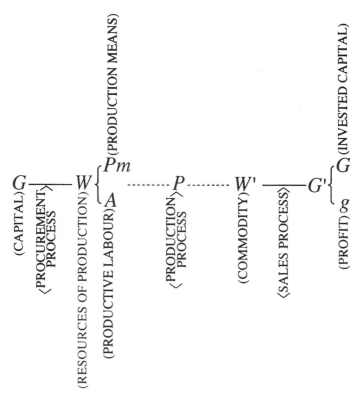

Figure 1.5 Karl Marx's capital circulation: the capital (G) is employed to acquire production means (P_m) and productive labour (A), both of which produce commodities (W'). Through sales of W' the capital is accumulated to be G' including the profit (g).

Production information is software and controls the other three resources of production that are all tangible (hardware). This resource of production has become increasingly important through the use of computers as a major means of information flow.

1.3.2 Firm's Activity and Capital Circulation

Manufacturing Firm's Activity

Funds are necessary for acquiring the above four resources of production. Hence, from the viewpoint of industrial engineering (IE) the inputs for production are said to be *man, material, machine, money* and *method*. The manufacturing firm (enterprise) provides these from the outside markets, and the produced commodities are sold (distributed and exchanged) at the markets and consumed by customers,[8] as depicted in Figure 1.4.

Capital Circulation Creating Profits

Through the above manufacturing activity the capitalistic mode of production increases the initial fund; the capital, G, is used to acquire the resources of production, W, production means/objects, P_m, and the labour force, A. W is converted into the commodities W' through the production processes, P, then the sales function changes W' to G'—the initially invested capital G with the surplus (profit) g, as depicted in Figure 1.5.

1.4 Goods Produced—Outputs of Production

1.4.1 Creation of Utilities

Kinds of Utilities

As mentioned previously, outputs produced through production are tangible goods (products)/intangible goods (services). The production function (the supply side) creates utility. From the viewpoint of production economics, it provides *form, time*, and *place utilities* for the produced goods.

—— EXAMPLE 1.1 ——

Creation of form utility as provided by a car, a TV set, etc.; creation of time utility as provided by communication with telephone; and creation of place utility as provided by using an aircraft.[9]   □

—— EXAMPLE 1.2 ——

Manufacturing provides the form utility, while physical distribution provides the time and place utilities.   □

Disutility Generated

Utilities created by the production function should be of 'positive' value. However, various by-products may be produced during this process when the raw material state is changed to a finished-product state through successive production stages. Some of these by-products are industrial wastes/air-polluting gases. The products also may generate air-polluting gases, and are disposed of as waste after use.

--- **EXAMPLE 1.3** ---

Figure 1.6 shows generation of wastes and air-polluting gases through production, use, and disposal of a car (Ohara, 1994). □

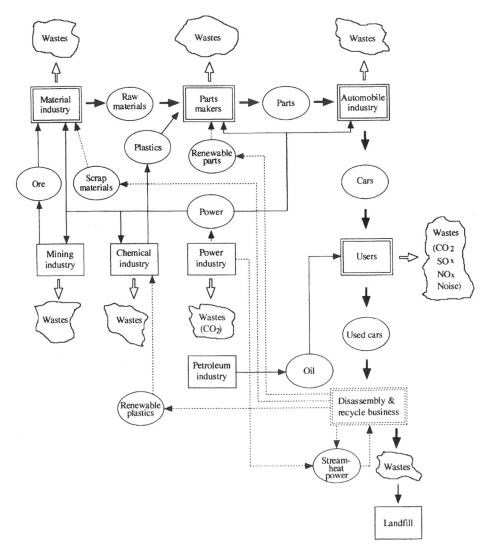

Figure 1.6 Many wastes and air-polluting gases are generated through production, use, and disposal of cars (Ohara, 1994).

The industrial wastes and air pollution definitely destroy the natural environment and bring about 'negative' utilities or 'dis'utilities[10]—public 'bads' or hazards.

It is of the utmost importance today to prevent this by establishing a closed-type or recycle-oriented production system as well as developing a new production technology for preventing creation of public hazards. This is 'green manufacturing', which will be discussed in the last chapter.

Possession Utility Provided by the Marketing Function

The *possession utility* is the 'use value' beyond the exchange value (price) or the total satisfaction which the consumer derives from using the product purchased—*customer satisfaction*. ('The customer is King!')

Creation of possession utility is the direct concern of the marketing or sales function that is based on the demand side (Timms and Pohlen, 1970).

Marketing

Marketing is defined by the American Marketing Association as 'a process to plan and implement concept building of ideas, commodities and services, pricing, sales promotion and distribution so as to create the "exchange value" of an individual/organization'.

1.4.2 Firm's Objectives

Profit

Subtracting the production cost from the product value (price/revenue) gives the value added created by production. This is the *profit* which the manufacturing firm obtains through production activities:

$$\text{profit} = \text{price} - \text{cost}. \tag{1.1}$$

The firm's objectives can be divided into two categories:

(1) *profit objective*—to be maximised by the management;
(2) *social objective*—to be used to contribute to the welfare of society (company's philanthropy).

1.4.3 Product's Value and Effectiveness of Production

QCD

From the practical standpoint the product's value and/or the effectiveness of production can be evaluated from the following three aspects:[11]

(1) function and quality—'Q';
(2) production cost and price—'C';
(3) production quantity and on-time delivery—'D'.

Thus, it is a primary objective of production to make products with desired functions quickly (just-in-time (JIT)—quick-response delivery) at lowest cost. The function which provides the plan/schedule for this purpose is *production management*.

—— EXAMPLE 1.4 ——

(Japan's Q) Japan introduced quality control (QC) from the United States after World War II and developed QC circles (or quality circles), which are practised in almost every workplace area, workers spontaneously find inferior articles and improve production methods. The concept of QC has been applied also to the management under the name of total quality control/management (TQC/TQM) or company-wide quality control and continuous improvement.

These activities have enhanced the quality of Japan-made products. ☐

—— EXAMPLE 1.5 ——

(Japan's C) The Japanese–English term 'cost-down' (cost reduction/cutting) is very popular in Japan; it is made by industrial engineering, value engineering, or *kaizen* (small improvements). However, the principal Japanese policy is toward big sales in spite of small profit margins; hence the return on sales was 1.5% for Hitachi, 1.3% for Matsushita Electric, and 0.9% for Toshiba, compared with 7.3% for General Electric in 1994 (data: *Fortune*, 7 August 1995). ☐

—— EXAMPLE 1.6 ——

(Japan's D) Japan seems to be excellent in JIT delivery. *JIT* (*just-in-time*) refers to the production and supply of the required number of parts/products when needed. Toyota Automobile Company put this principle into practice by the use of a *kanban* (instruction card) for information processing, so it is often called the 'Toyota production system' or the 'kanban system'.

Comparison concerning delivery times (leadtime between receipt of order and shipment of product) between the US and Japan shows 5 to 6 months in the US, compared with 1 to 2 months in Japan (*Business Week*, 6 June 1988). ☐

1.5 Production Processes—Transformation of Inputs into Outputs

1.5.1 What is the Production Process?

Production Process

The process of conversion of resources of production, in particular that of raw materials into tangible goods or products, is called the *production process[es]*. The production process is generally made up of successive multiple production stages, on which a series of operations—the work of producing the output—are performed successively on workstations (centres) or production facilities.

FUNDAMENTALS OF MANUFACTURING

---- **EXAMPLE 1.7** ----

A straight-line production process [system] consisting of part fabrication and product assembly is depicted in Figure 1.7. ☐

---- **EXAMPLE 1.8** ----

An information system inputs raw data, transforms them by use of computer facilities, and outputs useful information, as depicted in Figure 1.8. ☐

Material Flow with Cost Flow

The provision of form, time, and place utilities for the product takes place in increments throughout the transformation of the input into the output, and is accompanied by added value. This is the 'material flow' with 'cost flow' in the manufacturing firm, as mentioned in Section 1.2.

1.5.2 Productivity and Production Function

Productivity Defined

It is of upmost importance to convert effectively (economically and efficiently) resources of production into produced goods. A measure of the effectiveness of this transformation process is normally termed *productivity*, which is abstractly defined as the ratio between input and output:

$$\text{productivity} = \frac{\text{output}}{\text{input}}. \tag{1.2}$$

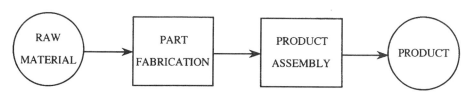

Figure 1.7 A typical production process consisting of part fabrication and product assembly, which transforms the raw materials into mechanical products.

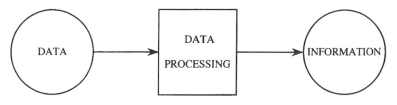

Figure 1.8 An information system converts the raw data into useful information through data-processing activities.

Production Function

To calculate production properly it is necessary to know the quantitative relationship which the production processes impose upon the simultaneous variations in the quantities of inputs and outputs. This is expressed as a *production function*, or a *production model*.

The production model for a single-output process expresses the technological relationship between inputs, $x_1, x_2, \ldots, x_n$, and output x_0:

$$x_0 = f(x_1, x_2, \ldots, x_n). \tag{1.3}$$

Production models are used to estimate, calculate, and evaluate a national or a firm's productivity. Gross national product (GNP) Y is often expressed in terms of capital K and labour-force L:

$$Y = f(K, L). \tag{1.4}$$

Typical Production Functions

Typical models are described below (Hitomi, 1995).

(1) *Fixed-coefficient* (or *Leontief*) *model*. If k units of capital and l units of labour are required to produce a unit of product,

$$Y = \min(K/k, L/l). \tag{1.5}$$

(2) *Cobb–Douglas* (or *Wicksell*) *model*. If the technological level is a, and elasticities of capital and labour are α and β, respectively,

$$Y = aK^\alpha L^\beta, \tag{1.6}$$

where a is concerned with technological progress with time t; it is expressed as $a_0 e^{ct}$ where a_0 and c are constants. The relation $\alpha + \beta = 1$ is often assumed for real-life examples.

(3) *CES (Constant Elasticity of Substitution) model*. If the elasticity of substitution is constant, σ,

$$Y = [\kappa K^{(\sigma-1)/\sigma} + \lambda L^{(\sigma-1)/\sigma}]^{\sigma/(\sigma-1)}, \tag{1.7}$$

where κ and λ are parameters showing the technological conditions.

Average/Marginal Productivity

Two kinds of productivity are calculated with the above models.

(1) *Average productivity*: $x_0/x_1, x_0/x_2, \ldots, x_0/x_n$.
(2) *Marginal productivity*—the ability of one more unit of a variable input to increase the total output: $\partial x_0/\partial x_1, \partial x_0/\partial x_2, \ldots, \partial x_0/\partial x_n$.

--- EXAMPLE 1.9 ---

The marginal capital productivity of the Cobb–Douglas function is

$$\frac{\partial Y}{\partial K} = a\alpha K^{\alpha-1} L^\beta = \alpha \frac{Y}{K}, \tag{1.8}$$

i.e. it is the product of the constant α and the average capital productivity, which is the product per unit capital, Y/K. □

Kinds of Productivity

The contents and dimensions of the inputs and the output specify various kinds of productivity as follows:

(1) *physical productivity*, where the outputs are measured in units;
(2) *value productivity*, where the outputs are measured in monetary values;
(3) *factor productivity*, such as labour productivity, capital productivity, land productivity, raw-material productivity, etc., which are related to each of the resources of production;
(4) *total productivity*, concerned with the total of the resources of production — an overall measure expressing the contribution of the resources of production to the efficiency attained by a firm.

Increase in productivity is a basic proposition for continuous economic growth in the modern world, especially in capitalist societies. Among various kinds of productivity labour productivity is usually considered of the utmost importance; if the output is measured in physical units, that productivity is 'physical labour productivity'.

—— **EXAMPLE 1.10** ——————————————————————

The trends of labour productivity over recent years for the manufacturing firms in various countries are shown in Figure 1.9. □

—— **EXAMPLE 1.11** ——————————————————————

Japan's physical productivity is high; that is, as shown in Figure 1.10, setting it as a standard index of 100, the index for the USA is 86 and that for Germany, 60. □

1.5.3 Value Added and its Productivity

Value Added [Productivity]

Using 'value added (or added value)' we obtain *added-value productivity*, which is the most important criterion of industrial productivity. *Value added* is a value created purely during the process of production; it is calculated as follows:

<subtraction procedure> value added = total production
  − purchasing expense from outside   (1.9)

<addition procedure> [pure] value added = profit
  + personnel expenditure + interest + land rent + tax   (1.10)

  gross value added = pure value added + depreciation expense   (1.11)

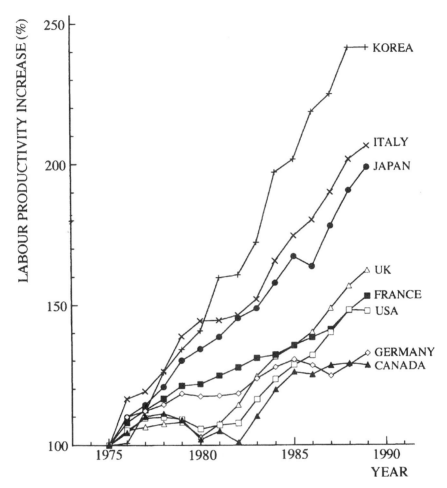

Figure 1.9 Increase of labour productivity: international comparison (1975: 100) (data: Japan Productivity Centre, September 1992).

GDP

Total value added created by a country in a year is *gross domestic product* (GDP); it indicates the economic power of a country.

—— EXAMPLE 1.12 ——

Some GDPs in 1993 were: 1. USA—$6260 billion, 2. Japan—$4207 billion, 3. Germany—$1911 billion, ..., 8. UK—$941 billion. □

—— EXAMPLE 1.13 ——

(International comparison of productivity) As discussed in Example 1.11, the physical labour productivity of the Japanese manufacturing industry is superior to those of American and European countries. However, the added-value productivity per capita per hour from the

FUNDAMENTALS OF MANUFACTURING

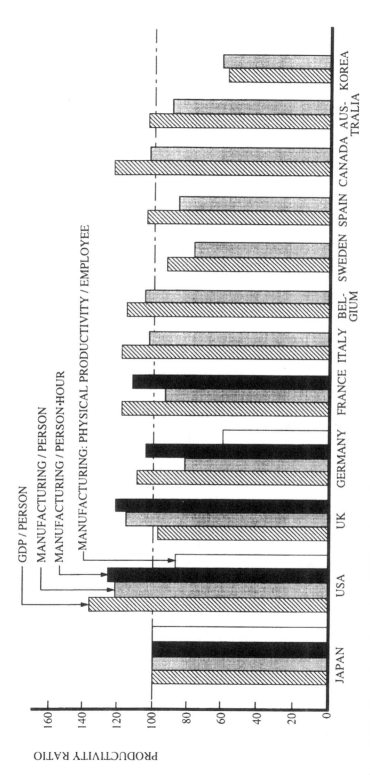

Figure 1.10 Gross domestic product (GDP) and manufacturing productivity: international comparison (Japan: 100) (data: Japan Productivity Centre, September 1995).

viewpoint of the purchasing power parity[12] is low in Japan, as represented in Figure 1.10, since yearly work hours are large in Japan compared to other developed countries.

In 1992, with Japan's productivity as standard of 100, the United States was 125 and Britain 121, Germany 1.04, and France 112 (data: Japan Productivity Centre, September 1995). □

—— EXAMPLE 1.14 ——

(Productivity comparison between the US and Japan) With American added-value productivity as standard of 100, the trends of the Japanese ratio over 1975 to 1992 for the whole manufacturing industry, the iron and steel industry, and the automobile industry are depicted in Figure 1.11. In 1975 Japan's manufacturing productivity was 45% of the United States and after closing the gap every year it stood at 71% in 1992.

Japanese productivity for iron and steel production is now larger than that of the United States, but the productivity for cars, which is representative of Japan's industries, is still low: 77% of that of the United States in 1992 (data: Japan Productivity Centre, September 1995). □

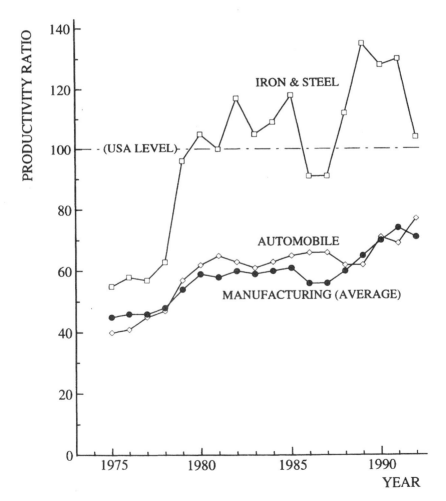

Figure 1.11 Trends of Japanese manufacturing productivity—iron and steel, car, and the whole manufacturing industry (USA: 100) (data: Japan Productivity Centre, September 1995).

FUNDAMENTALS OF MANUFACTURING

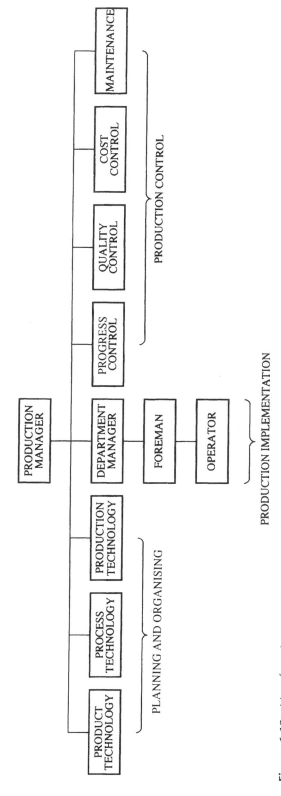

Figure 1.12 Manufacturing organisation consisting of line relationships: production manager–department managers–foremen–operators; and staff relationships for planning, organising, and controlling.

1.5.4 Manufacturing Effiency

Efficiency

In manufacturing industry a criterion called 'efficiency' is often used. An industrial engineering pioneer, H. Emerson discussed this and established 12 principles in 1916.

Efficiency is a capacity of performing a given task within the specified standard time;[13] it is usually expressed:

$$\text{efficiency} = \frac{\text{standard time}}{\text{actual time}} \times 100(\%). \qquad (1.12)$$

If this figure exceeds 100, the manufacturing activity is efficient.

Raising the efficiency depends upon two factors:

(1) *Subjective factors*:
   (a) workers' qualitative capability—skills, technical ability, etc.
   (b) workers' quantitative capability—working effort, increase of working hours, etc.
(2) *Objective factors*—technological innovation such as factory automation, enhancing the work environments (quality of working life), etc.

1.6 Production Organisation

Organisation

Even in fully automated factories human work plays the fundamental role in operating those factories. A structure and framework of hierarchies and functions in which human work is effectively conducted is called an *organisation*.

Organising for Production

A *production organisation* consists of line and staff relationships, as depicted in Figure 1.12. The production management manager exercises his authority of allocating production resources, deciding an overall production plan and schedule, directing the schedule to individual line departments, and adjusting the performance such that the overall production objectives are met.

The staff organisation assists productivity in the work lines. One is concerned with planning activities for production through product, process, and production technologies, and the other deals with controlling for production progress, quality, cost and maintenance.

Notes

1. Chronological figures are mostly based upon *The Shorter Oxford English Dictionary of Historical Principles* (1973).
2. 'Nature is the mother of wealth' (W. Petty).

3. 'Material and immaterial production' and 'spiritual production' form the 'living production (*Lebensproduktion*)' (K. Marx).
4. The 'flow of materials' and the 'flow of information' were advocated by A.H. Church as early as in 1913, and H. Nicklisch coined the term, 'flow of value (*Fluß der Wert*)' in 1922.
5. Supply generates demand (Say's law).
6. This reads 'control' and 'controlling'. Square brackets for words/sentences play this role.
7. In modern economics 'production' is viewed as technical transformation of input vector **I** into output vector **O**, that is, $\mathbf{T}: \mathbf{I} \to \mathbf{O}$, where **T** is an operator.
8. A chain of production, distribution, exchange, and consumption, is the province of economics and was discussed by J. McVickar as early as 1825.
9. The symbol □ stands for the end of examples, algorithms, propositions, etc.
10. This term appeared in 1901 (J.G.K. Wicksell).
11. The importance of these three aspects was pointed out by A.H. Church in 1914.
12. Reference capable of affording the same product with a local currency.
13. This is discussed in Section 8.3.

References

DANØ, S. (1966) *Industrial Production Model* (Vienna: Springer), p. 5.

Encyclopedia Americana (1965) Vol. 22, pp. 632–632b.

HITOMI, K. (1972) *Decision Making for Production* (in Japanese) (Tokyo: Chuo-keizai-sha), p. 5.

HITOMI, K. (1975) *Theory of Planning for Production* (in Japanese) (Tokyo: Yuhikaku), pp. 3–5.

HITOMI, K. (1991) Strategic integrated manufacturing systems: the concept and structures, *International Journal of Production Economics*, **25** (1, 3).

HITOMI, K. (1994) Moving toward manufacturing excellence for future production perspectives, *Industrial Engineering*, **26** (6).

HITOMI, K. (1995) Production (in Japanese), in Okamoto, Y. (ed.), *Modern Management Dictionary* (2nd edn) (Tokyo: Dobunkan), Chap. X.

KOONTZ, H. and WEIHRICH, H. (1990) *Essentials of Management* (5th edn) (New York: McGraw-Hill), pp. 22–3.

OHARA, S. (1994) Strategic aspects of environmental management, *Proceedings, Tsukuba–Washington International Symposium on Management Issues and Challenges*, Tokyo, p. 32.

The Shorter Oxford English Dictionary of Historical Principles (1973) (Oxford: Clarendon Press).

TIMMS, H.L. and POHLEN, M.F. (1970) *The Production Function in Business—Decision Systems for Production and Operations Management* (3rd edn) (Homewood, IL: Irwin), pp. 5–7.

YOUNG, R.E. and MAYER, R. (1984) The information dilemma: to conceptualize manufacturing as information process, *Industrial Engineering*, **16** (9).

Supplementary reading

COHEN, S.S. and ZYSMAN, J. (1987) *Manufacturing Matters* (New York: Basic Books).

FLAIG, L.S. (1993) *Integrative Manufacturing* (Homewood, IL.: Business One Irwin).

KUROSAWA, K. (1991) *Productivity Measurement and Management* (Amsterdam: Elsevier).

CHAPTER TWO

Fundamentals of Systems

2.1 Basic Concepts of Systems and Chaos

Some Definitions of Systems

As described in the previous chapter, production/manufacturing is an input–output system; hence, a systems approach is effective for solving production/manufacturing issues.

What is a 'system'? This word appeared in 1619; its original meaning (Greek) is to 'set up' in 'sync'. It is defined as

> an organized or connected group of objects; a set or assemblage of things connected, associated, or interdependent, so as to form a complex unity; a whole composed of parts in orderly arrangement according to some scheme or plan; rarely applied to a simple or small assemblage of things; ...; a set of principles, ideas, or statements belonging to some department of knowledge or belief; a department of knowledge or belief considered as an organized whole; a comprehensive body of doctrines, conclusions, speculations, or theses; ...
>
> (*The Shorter Oxford English Dictionary*, 1973).

Synergy Effect of Systems

A system is an 'organised whole' of a plural number of units. The essential sense of this term captures its organic (or materialistic) characteristics, or the *synergy effect*; that is, the total optimisation is greater than the sum of the partial optimisations.

This effect was suggested by Laozi, a Chinese philosopher, about 2500 years ago; some time later it was noted independently by the Greek philosopher Aristotle. The German philosopher G.W.F. Hegel also mentioned this concept 200 years ago.

Chaos

A state which is not systematised is 'chaos'. This was first recognised and mentioned by the Chinese philosopher Zhuanzi about 2400 years ago. *Chaos* now means a mode which creates unforeseen irregular behaviour or pattern in spite of deterministic character following a certain specific rule/law.

Total System/System Integration

If a unit forming part of a system behaves with strong independence/autonomy, this unit is called a *module* or *holon*; a system consisting of autonomous modules is called a *total system*.

The total system often has a mode of *system integration* with the following three features (Matsuda, 1990):

(1) *syncretism*—integrating different fields whilst maintaining their own autonomy;
(2) *symbiosis*—obtaining symbiotic gain;
(3) *synergy*—synergistically obtaining amplification effects.

2.2 Definition of Systems

2.2.1 Attributes Characterising Systems

Characteristics of the System

Many characteristics are concerned with systems; some are size, complexity, totality, mission/functions/objectives, internal/external relationships, equilibrium/balance, hierarchy, dimensionality, dynamic behaviour, etc.

Four Basic Attributes of the System

From among many characteristics four[1] basic attributes which play basic roles to characterise the system are described in the following (Hitomi, 1971):

(1) *Assemblage*. A system consists of a plural number of distinguishable units (elements, components, factors, subsystems,[2] etc.), which may be either physical or conceptual, natural or artificial.

EXAMPLE 2.1

The solar system is composed of the Sun, the Earth, Mercury, Venus, Mars, Jupiter, Saturn, Uranus, Neptune, and Pluto. □

EXAMPLE 2.2

A 'manufacturing system' is composed of machine tools, jigs and fixtures, tools, operators, etc. □

---

(2) *Relationship*. Several units assembled together are merely a 'group' or a 'set.' For such a group to be admissible as a system, a *relationship* or an interaction must exist among the units.

EXAMPLE 2.3

'Physical relationship' is governed inherently by the laws of natural science, such as the law of gravitation in the solar system. ☐

EXAMPLE 2.4

'Logical relationship' is determined essentially by definitions and assumptions, such as the relation of production, inventory, and sales in a period:

$$\text{final inventory} = \text{initial inventory} + \text{production quantity} - \text{sales quantity}. \tag{2.1}$$

☐

EXAMPLE 2.5

'Institutional relationship' is specified by social institution, laws, and regulations, such as

$$\text{tax amount} = \text{profit (or income)} \times \text{tax rate}. \tag{2.2}$$

☐

(3) *Goal-seeking*. An actual system as a whole performs a certain function or aims at single or multiple *objectives*. Wherever these objectives are attained at their maximum/minimum levels, *system optimisation* is said to have been performed. For this purpose it is necessary to be able to measure, objectively or subjectively, the degree of attainment of the objectives. An objective that is measurable by any means is called a *goal/target*.

EXAMPLE 2.6

A 'manufacturing system' effectively converts resources of production into produced goods (products), attaining an objective that creates high utilities by adding values to the raw materials, resulting in superior quality, cost and delivery. ☐

EXAMPLE 2.7

A 'business management system' coordinates functional divisions—production, sales, personnel and finance, which constitute the system—and allocates limited resources available to those divisions. This system aims at organisational objectives such as profit maximisation, reasonable rate of return on capital, increase in market share, stable growth, public services (philanthropy), etc. ☐

(4) *Adaptability to environment*. A specific, factual system behaves so as to adapt to the change in its surroundings, or external environment. This *external environment* influences and is influenced by the system, in that matter and/or energy and/or information are received from and given to each other. A system that is capable of controlling itself in such a way as to be always optimal even under changes in the external environment, is called an *adaptive* (or *cybernetic*) *system*. If this system possesses dynamic adaptability, approaching a desired state with the least time lag by changing its internal structure and functions as the environment changes, it is a

self-organising system. In its structure a sort of 'fluctuation' exists; the system has a *dissipative structure*, evolving into a new state whenever fluctuation exceeds the critical limit.

EXAMPLE 2.8

'Human' is a complete adaptive system. □

EXAMPLE 2.9

A 'business system' is an adaptive system, in that it makes proper decisions so as to achieve its objectives under severe environmental situations (competitors, markets, industrial societies, economic and political conditions, international trends, etc.). The system often reacts to its environment to make its future behaviour more effective: e.g. it performs marketing activities, such as advertising and merchandising, to enhance potential demands in the market.

□

EXAMPLE 2.10

A 'business system' is a self-organising system, in that it generates a diversified variety of activities, resulting in economies of scope. □

2.2.2 Systems Defined

Four Definitions of Systems

On the basis of the foregoing considerations, the four essential definitions of systems can now be given as follows (Hitomi, 1975).

(1) *Abstract* (or *basic*) *definition*. On the basis of the first two attributes above, 'a *system* is a collection of recognisable units having relationships among the units'. Under this definition, *general system theory* has been developed, wherein things are deliberated theoretically, logically, and speculatively.

(2) *Structural* (or *static*) *definition*. On the basis of all four attributes, 'a *system* is a collection of recognisable units having relationships among the units, aiming at specified single or multiple objectives subject to its external environment'.

(3) *Transformational* (or *functional*) *definition*. From the last attribute, the effects of the environment upon the system are *inputs* (including unforeseen 'disturbances'), and, conversely, the effects in which the system influences the environment are *outputs*, as shown in Figure 2.1. From this consideration 'a *system* receives inputs from its environment, transforms them to outputs, and releases the outputs to the environment, whilst seeking to maximise the productivity of the transformation'.

(4) *Procedural* (or *dynamic*) *definition*. The process of transformation in the input–output system consists of a number of related stages, at each of which a specified operation is carried out. By performing a complete set of operations according to the precedence relationship on the stages, a function or task is

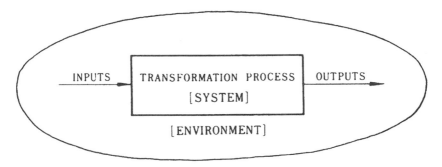

Figure 2.1 A system receives inputs from its environment, transforms them to outputs, and releases the outputs to the environment (transformational definition of the system).

accomplished. Thus, 'a *system* is a *procedure*—a series of chronological, logical steps by which all repetitive tasks are performed'.

Kinds of Systems

Systems are classified as:

- natural vs. artificial;
- real vs. conceptual (virtual);
- static vs. dynamic;
- deterministic vs. stochastic (probabilistic);
- control (cybernetic) vs. noncontrol;
- rigid vs. flexible, etc.

2.2.3 Why the Systems Approach?

Importance of Systems Approach

There are several reasons why the systems concept and approach have become more and more important in our life, as follows.

(1) Recently all organisations and societies such as manufacturing, management, economics, politics, international affairs, etc., have tended to become large and global; hence, it has become necessary to consider everything systematically in connection with its surroundings to achieve its functions and objectives. In manufacturing, an integrated system of materials procurement, production, inventory, distribution and sales, as shown in Figure 1.2, should be identified and analysed so as to attain the objectives. To coordinate individual functional divisions from the viewpoint of total systems optimisation, a logical systems approach is inevitable; otherwise, there is a danger of falling into the reverse—chaos.

(2) The recent progress of computers and information networks has enhanced the ability to gather, store, process, and transmit a large variety and quantity of data/information globally in less time than previously. This has contributed

significantly to the solution of complex problems. In the area of manufacturing, computer-integrated manufacturing (CIM) systems now play a role in automating the flow of information.
(3) Optimisation techniques, such as operations research, management science, systems engineering, and simulation techniques have been developed. With the aid of these soft sciences or technologies, systems thinking and optimisation, and hence rational and logical decision-making to provide optimum solutions for large-scale systems/problems, have become possible. To this end systems engineering has made a significant contribution.

Aim of Manufacturing Systems Engineering

Manufacturing systems engineering, which this book intends to describe, is a methodology associated with the optimum design, installation, and execution of large-scale manufacturing systems which are made economically feasible by utilising scientific laws and empirical rules which exist in manufacturing.

2.3 Basic Problems Concerning Systems

Input–Output System

As shown in Figure 2.1, the system (manufacturing system) can be thought of as a transformation (factory) **T** on inputs (resources of production, especially raw materials) **I**, which produces outputs (products) **O**; this input–output relationship is expressed symbolically by means of the following equation:

$$\mathbf{T}(\mathbf{I}) = \mathbf{O} \quad \text{or} \quad \mathbf{T}: \mathbf{I} \to \mathbf{O}, \tag{2.3}$$

where **T** is mathematically called an *operator*.

Five Basic Problems Concerning the System

Focusing on the above equation, questions concerning a system usually fall into one of the following three categories and five problems (Dorny, 1975):

(1) *System analysis*. Clarify contents of **T**, **I**, and **O** (investigate the factory, raw materials, and products).
(2) (a) *System operation*. Given **T** and **I**, find **O** (design the products to be made by given raw materials on an existing factory).
    (b) *System inversion*. Given **T** and **O**, find **I** (determine the raw materials to make a certain product on an existing factory).
    (c) *System synthesis* (or *identification*). Given **I** and **O**, determine a suitable **T** (given raw materials and a product to be produced, establish the suitable factory or determine an appropriate production process).
(3) *System optimisation*. Pick **I**, **O** or **T** so that a specified evaluation criterion is optimised (decide proper raw materials, products or production processes so as to minimise the total production time or cost or to maximise the total profit obtained).

2.4 Systems Design

2.4.1 Preliminary Considerations in Systems Design

What is Systems Design?

Systems design is to construct a new, useful system (static structure and operating procedure) under a specified evaluation criterion by the use of scientific disciplines and empirical laws concerning systems.

System Expression

A basic *system module* is depicted in Figure 2.2. 'Inputs' are converted into 'outputs' through 'transformation' processes run on a 'processor' to achieve a specified 'goal' under unexpected 'disturbances' or specified 'regulations'. A large-scale (multi-layer, multi-stage) system is constructed by connecting a suitable number of system modules in series or parallel by a set procedure such as structured analysis and design technique (SADT) or, for example, more specifically, IDEF (integrated computer-aided manufacturing (ICAM) definition). This is a top-down study to first consider the whole system and then conduct more detailed, structured functional analysis for individual subsystems from the hierarchical viewpoint, as represented in Figure 2.3.

Control System

The basic characteristic of systems design is 'operationality'. Components to be considered in this category are:

- *controllable variables* controlled as specified by the designer;
- *uncontrollable parameters* which cannot be controlled by the designer.

Systems with controllable variables are called *control* (or *cybernetic*) *systems*. By properly (optimally) setting values for controllable variables, the objective of the system is attained. In some cases this operationality is unlimited, whilst in others it is limited or restricted.

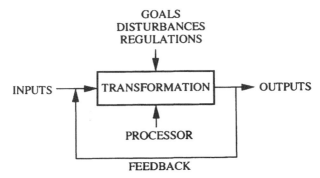

Figure 2.2 A basic system module: inputs are converted into outputs through transformation processes on a processor to achieve the specified goals under social regulations/unexpected disturbances.

FUNDAMENTALS OF SYSTEMS

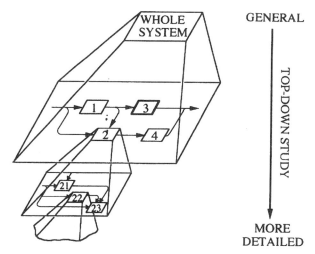

Figure 2.3 SADT is a top-down study to design a whole system in detail, layer by layer.

System Optimisation

Proper setting of values for controllable variables in a system is made so as to attain the highest measure of performance for the system's objectives. This is based upon the *optimising criterion* (or *principle*), and under this criterion 'systems optimisation' is achieved.

It is important to express the basic structure of a system. This is specified by the four attributes mentioned before and by the following two factors:

(1) *goals* of the system—attained by the function of the system made up of components;
(2) *constraints* on the system—both internal and external restrictions caused by the structure of the system itself and the relationship between the system and its external environment.

2.4.2 Model Building

What is a Model?

Goals and constraints can be described by models. A *model* is an abstract representation of a real situation or behaviour with a suitable language or expression. Since a model is an explicit representation of reality, it is generally less complex than reality, but it is important that it is sufficiently complete to approximate those aspects of reality to be investigated.

Representative Models

The following are major models that can be used in manufacturing systems design and analysis.

(1) *Physical models.* Those expressed stereoscopically are physical models. Scaled-down replicas, such as a small-sized wing in a wind tunnel or a small

ship in a water tank to which dimensional analysis can be applied, a mock-up of layout for a machine shop in plant engineering, etc. are examples.

(2) *Schematic* (or *graphical*) *models*. Those describing an actual situation in the form of diagrams, drawings, charts, graphs, etc. are called schematic or graphical models. Shapes, sizes, allocation, flow, loci, and others are demonstrated on figures, such as design drawings, flow charts, process charts, Gantt charts, PERT diagrams, break-even charts, decision trees, etc. These expressions aid decision-making in that they are often useful in sequentially deriving the near-optimal solution.

(3) *Mathematical* (or *analytical*) *models*. These are extremely precise, using the highest level of abstraction, and are the most effective method for performing systems optimisation analysis by means of functional expression.

(4) *Simulation models*. 'Simulation' is a mock-up of an actual situation, which in time-series manipulates the parameters, variables, constraints, and alternatives comprising the model expressed by computer-programming languages. It aids proper decision- and policy-making by efficiently and economically determining the system's structure and behaviour, operating procedures, and decision rules to meet the system's objectives. Physical models for this purpose are *simulators*; e.g. flight simulators for training pilots. Recently 'virtual (or artificial) reality' has played a role in simulating reality quickly; e.g. the *virtual factory*, which demonstrates virtual manufacturing operations on computer displays without utilising any actual production facilities.

Mathematical Model

A math[ematical] model typically expresses the two factors (1) and (2) mentioned in the previous subsection, maximising or minimising the goal (or objective) function subject to the constraints as follows:

$$\text{maximise } g(\mathbf{x}, \mathbf{y}) \tag{2.4}$$

$$\text{subject to } f(\mathbf{x}, \mathbf{y}) = 0 \tag{2.5}$$

$$\underline{\mathbf{x}} \leq \mathbf{x} \leq \bar{\mathbf{x}}, \tag{2.6}$$

where $\mathbf{x} = (x_1, x_2, \ldots, x_n)$ are controllable decision variables to be decided as a result of systems design, and $\mathbf{y} = (y_1, y_2, \ldots, y_m)$ are uncontrollable parameters which have been predetermined. $\underline{\mathbf{x}}$ and $\bar{\mathbf{x}}$ are lower and upper bounds allowable for $\mathbf{x}$, respectively. Constraint (2.4) is a measure of performance to be maximised or minimised—a technological measure such as maximum strength, economic measure such as minimum cost, maximum productivity, maximum profit, etc.

Constraint (2.5) is the functional constraint expressing the relationship between controllable variables and uncontrollable parameters, such as a structural relationship, laws of dynamics, etc. Constraint (2.6) is the regional constraint for controllable variables, such as size limitation, resource limitation, etc.

If $\mathbf{x}$ and $\mathbf{y}$ are not dependent on time, this model is *static*; otherwise, the model is *dynamic*. Examples are short-term production planning for a static case and long-run production planning for a dynamic case, as will be discussed in Chapter 13.

--- EXAMPLE 2.11 ---

(Model formulation) Formulate maximisation of a Cobb–Douglas type utility function under budget limitation C, if unit capital and unit labour costs are p_K and p_L, respectively.

Since amounts of capital and labour are K and L, we can build a model of *utility maximisation* as follows:

maximise $Y = aK^{\alpha}L^{\beta}$ (2.7)

subject to $p_K K + p_L L = C,$ (2.8)

where a, α and β are constants. In this formulation decision variables are: $\mathbf{x} = (K, L)$, and uncontrollable parameters are: $\mathbf{y} = (a, \alpha, \beta, C, p_K, p_L)$. This model is static and of constrained optimisation. □

2.4.3 System Optimisation

Types of System Optimisation

From an analytical standpoint, *system optimisation* is the determination of optimum values for decision variables $\mathbf{x}$, such that the goal function (2.4) is maximised or minimised subject to constraints (2.5) and (2.6) (*constrained optimisation*); it is often done without any constraint (*unconstrained optimisation*).

Types of Solutions

Values of decision variables which satisfy the constraints are called *feasible solutions*. An *optimal solution* is a feasible solution that maximises or minimises the goal function.

Optimisation

Unconstrained optimisation is easier than constrained optimisation, since in the case of maximising (2.4), we are only required to determine $\mathbf{x}^*$ (the optimal solution) such that:

$g(\mathbf{x}^*, \mathbf{y}) \geq g(\mathbf{x}, \mathbf{y}).$ (2.9)

In constrained optimisation, $\mathbf{x}^*$ should satisfy

$f(\mathbf{x}^*, \mathbf{y}) = 0, \quad \underline{\mathbf{x}} \leq \mathbf{x}^* \leq \bar{\mathbf{x}}$ (2.10)

as well. In this case, the optimal solution is generally given by:

$\mathbf{x}^* = \psi(\mathbf{y}),$ (2.11)

where ψ is a decision function indicating the *decision rule*, by which an *optimising algorithm* can be established so that the optimal solution $\mathbf{x}^*$ is obtained in a finite number of computational steps.

--- **EXAMPLE 2.12** ---

(Constrained optimisation) To obtain an optimal solution for the Cobb–Douglas type utility maximisation formulated in Example 2.11, first construct a Lagrangian function using (2.7) and (2.8) as follows (Dowling, 1992):

$$L(K, L, \lambda) = aK^{\alpha}L^{\beta} + \lambda(C - p_K K - p_L L) \qquad (2.12)$$

where λ is a Lagrange multiplier. Setting partial derivatives of the above function with respect to K, L and λ to 0, we obtain optimal solutions for capital K and labour L as follows:

$$K^* = \alpha C/p_K \quad \text{and} \quad L^* = \beta C/p_L. \qquad (2.13)$$

These optimal solutions are called *ordinary demand* (or *Marshallian*) *functions*, which are derived by means of utility maximisation subjected to a budgetary constraint (2.8). These are zeroth-order homogeneous functions[3] with respect to p_K, p_L, and C.

Then the maximum utility is:

$$Y^* = a(\alpha/p_K)^{\alpha}(\beta/p_L)^{\beta}C. \qquad (2.15)$$

This is called an *indirect utility function*. The expenditure function is expressed with utility as follows:

$$C = (p_K/\alpha)^{\alpha}(p_L/\beta)^{\beta}Y/a \qquad (2.16)$$

which is a first-order homogeneous function with respect to unit costs, p_K and p_L.

Additionally optimal values for K and L expressed with respect to Y are called *compensated demand functions*, which are:

$$K^* = (\alpha p_L/\beta p_K)^{\beta}Y/a \quad \text{and} \quad L^* = (\beta p_K/\alpha p_L)^{\alpha}Y/a. \qquad (2.17)$$

□

Optimisation Techniques

Typical optimisation techniques are as follows:

(1) *Extremum method*—most classical optimisation techniques, including differential calculus and Lagrange multipliers for conditional problems (employed in the above example), and calculus of variations for functionals, etc.
(2) *Mathematical programming*—optimisation techniques that play an important role in optimum decision-making, including linear programming (LP), which maximises or minimises a linear function subject to linear constraints; nonlinear programming which deals with nonlinear goal function and constraints; integer programming for integer-valued decision variables; fuzzy programming settling membership functions for vague decision variables; dynamic programming, which is a multistage decision process based upon the principle of optimality, and others.
(3) *Multiple-objective* (or *multicriterion*) *optimisation*—methods obtaining optimum solutions for a plural number of objectives. In general no *supremal solution* exists; an increase of a goal worsens another goal. This trade-off relation generates the *Pareto optimum* (*non-inferior*) *solutions* and a *preferred solution* is decided from among them, including the parametric (or combined) approach which generates a convex combination according to the relative

weights to the objectives, the lexicographic (or priority) approach which ranks the objectives in the order of importance, the satisficing approach which optimises the most important objective whilst holding the other objectives above their aspiration levels, goal programming in which all the objectives are driven as close as possible to their attainable goals, etc.

(4) *Network theory*—solution techniques by the use of networks or graphs, including project scheduling techniques such as PERT (*P*rogram *E*valuation and *R*eview *T*echnique), CPM (*C*ritical *P*ath *M*ethod) and GERT (*G*raphical *E*valuation and *R*eview *T*echnique), which are concerned with checking the feasibility and/or optimality of project completion time and/or cost; network flow analysis for the shortest-path and the maximal-flow problems; scheduling theory for flow and job shops; line balancing for constructing assembly lines, etc.

(5) *The maximum principle*—necessary condition for optimality by reducing the optimisation problem to the maximum problem of the Hamiltonian.

(6) *Functional analysis*—modern mathematical methodology for solving optimisation problems on the basis of concepts of function spaces and transformations in them and their applications.

(7) *Implicit enumeration*—iterative procedures for obtaining the optimal solution by means of efficiently reducing the computational steps, including branch-and-bound method, lexicographic search method, permutation search, etc. Genetic algorithms based on the concept of biological evolution are also usable.

Some of these optimisation techniques are employed and explained in this book.

2.4.4 Decision-making Criteria

Basic Criteria for Decision-making

System optimisation is made for well-structured problems, based upon the *optimising criterion* (or *principle*), but it is not always easy to make optimum decisions based upon this criterion.

Where optimisation is impossible for reasons such as theoretical infeasibility, huge time consumption, economical disadvantage, etc., the *satisficing criterion* (or *principle*) (Simon, 1977) is employed for ill-(or un)structured problems.

In cases where neither optimising nor satisficing criteria are utilised owing to the complexity of the problems, the *consistency criterion* (or *principle*) may be useful. Solutions are derived as logically and consistently as possible, and are required to be *robust*, in that 'robust' solutions can be employed suitably under the largely changing environment.

Feasibility Study/Heuristics

Two procedures are taken for decision-making under the satisficing criterion:

(1) *Feasibility study.* This derives feasible solutions. Solutions, which attain, but do not necessarily exceed, a certain level of aspiration that the decision-maker is willing to accept for each objective, are 'satisficing solutions'.

(2) *Heuristics* or *heuristic programming.* This is a method to reduce efforts of trial and error in the problem-solving process by sequentially diminishing a

difference between the present status of the high-rank problem and the levels of the lower-rank solvable problems. No unified procedure has been established for this process yet. A high-speed simulation of human thinking by neuro-computers may be effective in solving this sort of ill-structured problem, resulting in 'near-optimal solutions'.

2.4.5 Basic Approaches to Systems Design

Two Design Methods

Basically there are two approaches to systems design:

- *inductive design*—an analytical approach to derive a general solution for an actual system by identifying and investigating the cases of the existing system's reality;
- *deductive design*—an axiomatic approach to deduce a feasible or an optimal solution theoretically by first setting an ideal system based on universal disciplines and principles.

Work Design

Nadler (1970) proposed *work design* as one of the deductive-design methods for heuristically designing work systems from the viewpoint of the ideal system concept. Key points to the success of this design approach are how to determine the function, how to develop the theoretical ideal system, and how to reduce it to a technologically feasible ideal system which can be used in practice. In this approach seven system elements—function, inputs, outputs, sequence, environment, physical catalyst, and human agents—are expressed in a hopper-shaped diagram shown in Figure 2.4.

In work design 10 steps are processed following basic rules/check lists:

(1) determining the function;
(2) developing the ideal system;
(3) gathering information;
(4) suggesting alternatives;
(5) selecting a workable system;
(6) formulating the system;
(7) reviewing the system;
(8) testing the system;
(9) installing the system;
(10) measuring system performance.

2.4.6 Large-scale System Design

Two Basic Approaches to Large-scale System Design

Basic approaches to large-scale system problems are the following:

(1) *Modular method.* A large-scale system is first divided into several subsystems, each having as much independence as possible; this sort of system is often

FUNDAMENTALS OF SYSTEMS

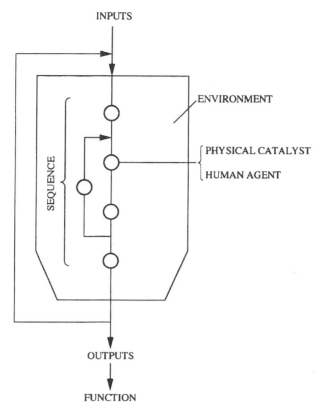

Figure 2.4 A work-design module: a deductive-design method by the ideal system concept.

called a *distributed system*. Then, suboptimisation or local optimisation can be carried out in each of the subsystems. Finally, all the subsystems are fully integrated by means of 'coordination'. The idea of the modular method is similar to that of the 'building-block system', which is often used in construction of mechanical systems such as transfer machines (see Section 25.1).

(2) *Hierarchical method*. As shown in Figure 2.5, subsystems with different functions are arranged vertically, such as a three-level system consisting of the execution system P for conducting actual implementation of the conversion process of inputs into outputs, the low-level management systems, $C_1, \ldots, C_N$, for performing tactical planning and control, and the high-level management system C_0 for dealing with strategic planning (Mesarovic *et al.*, 1970). This structure is called the *hierarchical system*. The higher-level subsystem has a priority on the system's behaviour and the decisions made here restrict the lower-level subsystems. Performances of the lower-level subsystems are reported to the higher-level subsystem, which again makes decisions for the next instruction.

Total Systems Approach

In constructing a large-scale management system, several modules, with distinct functions required for planning, implementation, and control of manufacturing, are

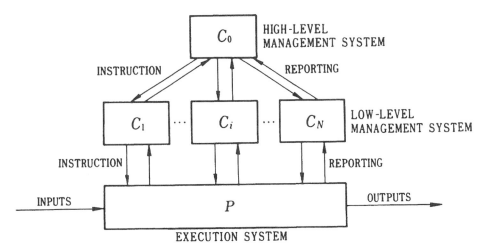

Figure 2.5 Basic structure of a hierarchical system: the execution system, P, converts inputs into outputs, the low-level management system, $C_1, C_2, \ldots C_N$, performs tactical (or operational) planning and control, and the high-level management system, C_0, deals with strategic planning. High-level subsystems make decisions based upon the performances reported from lower-level subsystems. The lower-level subsystems again behave according to instructions resulting from the high-level decisions.

integrated. This is the concept of the *total system*, which may be defined as

> an advanced approach in the design of management information systems for timely, optimum integration of administrative information, and this concept is closely associated with the use of electronic computers and data communications devices as 'systems tools' to process large quantities of data, converting them into useful, timely information for managerial decision-making in complex modern business organizations. These are conceived of as an integrated entity composed of interrelated systems and subsystems (Dickey and Senensieb, 1973).

2.5 Decision-making Procedures

2.5.1 What is Decision-making?

Rational Decision-making

Decision-making is the process of selecting one best plan from among several possible alternatives. Rational decision-making is structured under the following two premises as depicted in Figure 2.6 (Matsuda, 1969):

- *factual premise*—results which would be attained by implementing the possible means in the circumstances are predicted;
- *value premise*—values (measures of performance through implementation of the alternatives) are measured and ranking of the values can be made, either by ordinal numbers or by quantitative figures.

Then an alternative with the highest rank is chosen.

FUNDAMENTALS OF SYSTEMS

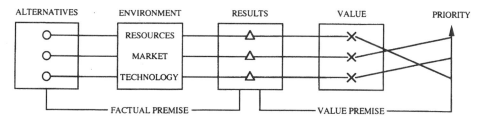

Figure 2.6 Rational decision-making process based upon the factual and value premises, through which a best alternative with top priority is chosen.

Two Modes of Decision-making

In the above rational decision-making the modes of decision are divided into two:

(1) *programmed decision-making*—repetitively routine decisions for well-structured problems;
(2) *non-programmed decision-making*—heuristically made decisions for ill-structured issues.

Types of Decision-making

Types of decision-making are also classified from the viewpoint of predictability of the future course of action as follows:

(1) *decision under certainty*—in the case that reliable information is available for the future: an alternative which generates the highest utility is chosen;
(2) *decision under risk*—in the case that probabilities of the occurrence of the future information are available: an alternative which generates the highest expected utility is chosen;
(3) *decision under uncertainty*—in the case that even probabilities of the occurrence of the future information are not known: criterion of pessimism (maximin or minimax or Wald), criterion of optimism (maximax), Hurwicz's criterion, Savage criterion, and Laplace criterion are applicable;
(4) *decision under conflict*—in the case that competitors take their actions against our plans: game theory is usable.

2.5.2 Optimum Decision-making Procedure

Three Phases/Ten Steps for the Decision-making Procedure

The following is a general procedure for optimum systems design by the use of a model, and involves three phases with ten steps, as represented in Table 2.1 (Hitomi, 1990). It is also effective in general problem-solving as well as optimum decision-making, where a best alternative is selected from several courses of action.

Table 2.1 A general procedure for systems design, which is the optimum design by model analysis (Hitomi, 1990). This procedure is also effective to the general problem-solving and decision-making.

Phase	Description	Step	Item
I	Problem analysis	1	Problem identification
		2	Factor analysis
		3	Information gathering
II	Problem-solving	4	Model building
		5	Testing the model
		6	Decision analysis
III	Evaluation	7	Prediction analysis
		8	Implementation
		9	Evaluating the performance
		10	Modification and redesign

Phase I   Problem Analysis

Step 1:   *Problem identification.* This identifies the range of the design issues, mainly the design specification, by the creative-thinking[4] ability, the judgement and intuition of the systems designer, thus clarifying the design specification.

Step 2:   *Factor analysis.* This identifies and lists the factors to be included in the design problem. It is important in this step to distinguish the controllable vs. uncontrollable factors and the qualitative vs. quantitative ones. Controllable and quantitative factors play fundamental roles in the optimum design of a system.

Step 3:   *Information gathering.* This is the activity of collecting data/information required for problem-solving in the next phase. It is important to gather objective and 'measurable' data/information concerned with as many events as possible identified by facts relating to the systems design.

Phase II   Problem-solving

Step 4:   *Model building.* In this step a suitable model is constructed to express the problem which has been clarified in the previous phase, thereby assisting in optimum design and optimum decision-making in the 'model world' (see Figure 2.7). The model contains only the important items (factors and their relationships) of the reality in the 'real world'. Good models are those by which optimum design/decision-making can be easily done so that, when the real system is installed and executed, its objectives are attained and its future behaviour can be suitably predicted.

Step 5:   *Testing the model.* The effectiveness of the model constructed in the previous step is tested and proved either deductively or inductively with the use of probability theory and statistics, simulation techniques, etc.

Step 6:   *Decision analysis.* A solution (the system structure and the operating procedure) is derived to meet the system's design objectives by analysing

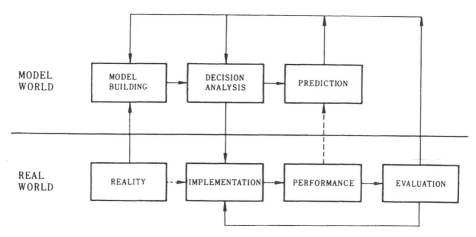

Figure 2.7 The model is utilised for the optimum system design and decision-making process, wherein through decision analysis in the 'model world' the designed system is installed and executed in the 'real world'. The advantage of the model analysis is the capability of predicting performance prior to the system's implementation.

the model. Where possible, optimality is sought according to the optimising criterion, thereby deriving an optimal solution. If an optimal solution is not obtained, a near-optimal solution is attempted according to the satisficing criterion. It is desired that solutions should be 'robust' to be flexible enough against changes in the system's environment.

Phase III   Evaluation

Step 7: *Prediction analysis.* One of the merits of applying a model in systems design is that it makes possible the prediction of system behaviour and performance at the implementation stage. In this prediction analysis, 'sensitivity analysis' is especially effective. If this result is not satisfactory, we return to step 4 or 6 for re-engineering (see Figure 2.7).

Step 8: *Implementation.* Based upon the best plan/schedule obtained as a result of decision analysis in step 6, a new system is installed and the procedure is executed in the real world.

Step 9: *Evaluating the performance.* Actual results of implementation in the above step are measured and evaluated by the following checking (or control) criteria:

- *reliability*—accurate performance of the specified functions and goals of the system when installed over a long period;
- *flexibility*—ability of the system to adapt to changes in the environment or disturbance;
- *stability*—ability of the system to maintain a stable state;
- *adaptability*—ability of the system to maintain optimality;
- *robustness*—ability of the system to operate even with substantial changes in the environment;
- *economical efficiency*—assurance of implementing the system economically.

Step 10: Modification and redesign. When the deviation between the actual performance and the standard established in the planning stage is in excess of a certain limit, modification is necessary. If simple modification is not sufficient, a re-design or re-engineering is performed by returning to the previous steps, as indicated in Figure 2.7. □

--- EXAMPLE 2.13 ---

Problem-solving for establishing a production plan. A 'production planning system' is designed as an example of following the above design methodology.

Step 1: For *problem identification* a short-term (static) production plan is to be established by determining the optimum amounts (x_1, x_2) of two products—P_1 and P_2, thereby maximising the total profit z.

Step 2: As *factor analysis* the following factors are taken into consideration:

- *observable factors*—kinds and amounts of production resources and revenues obtained through production of the products;
- *controllable factors*—acceptance/rejection of the two products and their production volumes;
- *uncontrollable factors*—available amounts of production resources and product demands;
- *qualitative factors*—workers' morale.

Step 3: Table 2.2 was obtained through *information gathering*.
Step 4: A mathematical *model* was built as follows:

$$\text{maximise } z = 5x_1 + 3x_2 \tag{2.18}$$

$$\begin{aligned}
\text{subject to} \quad & 4x_1 + x_2 \leq 2000 & \text{(a)} \\
& 7x_1 + 6x_2 \leq 4200 & \text{(b)} \\
& 13x_1 + 14x_2 \leq 9100 & \text{(c)} \\
& 16x_1 + 11x_2 \leq 8800 & \text{(d)}
\end{aligned} \tag{2.19}$$

$$x_1, x_2 \geq 0 \tag{2.20}$$

where x_1 and x_2 are production volumes for P_1 and P_2, respectively.

Step 5: For *testing the model* built in Step 4, linearity, reliability, and robustness were checked; thus the validity of the model was proven.
Step 6: *Decision analysis* was carried out—in this case through a graphical method (whose details will be explained in Section 13.1); an optimal plan was obtained as follows,

Table 2.2 Information gathered for production planning.

Resource per unit product	Product P_1	Product P_2	Resource limitation
Labour	4	1	2000
Energy	7	6	4200
Material	13	14	9100
Equipment	16	11	8800
Unit profit	5	3	

as depicted in Figure 2.8:

- optimum production volume for product P_1: $x_1^* = 471$
- optimum production volume for product P_2: $x_2^* = 114$ } total production = 585
- maximum profit: $z^* = 2700$.

Step 7: In *prediction analysis* the corporate top management considered that the social objective of responding to the market needs, that is, of maximising the total production amounts, was crucial. Hence, formulating the total production amounts as:

$$\text{maximise } y = x_1 + x_2 \qquad (2.21)$$

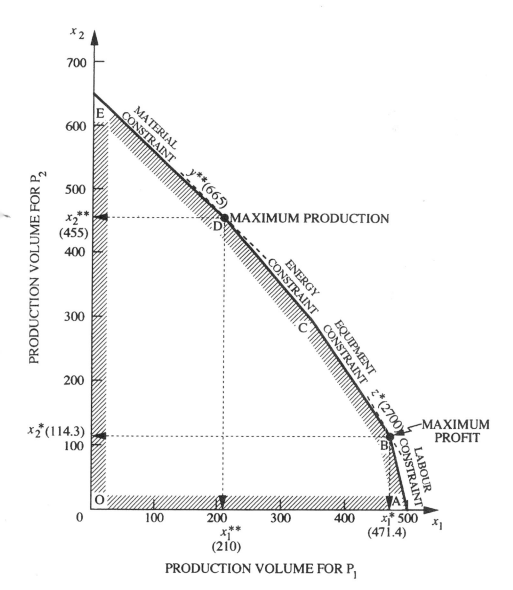

Figure 2.8 Graphical decision analysis for production planning by a linear program; optimal solutions for maximum profit and maximum production are indicated.

we obtained the (optimal) solution:

$$x_1^{**} = 210, \ x_2^{**} = 455, \text{ hence } y^{**} = 665 \ (z^{**} = 2415).$$

Step 8: In the *implementation* stage, the energy actually available was 4080, smaller than the anticipated amount in the information gathering. Hence the actual energy constraint was:

$$7x_1 + 6x_2 \leq 4080.$$

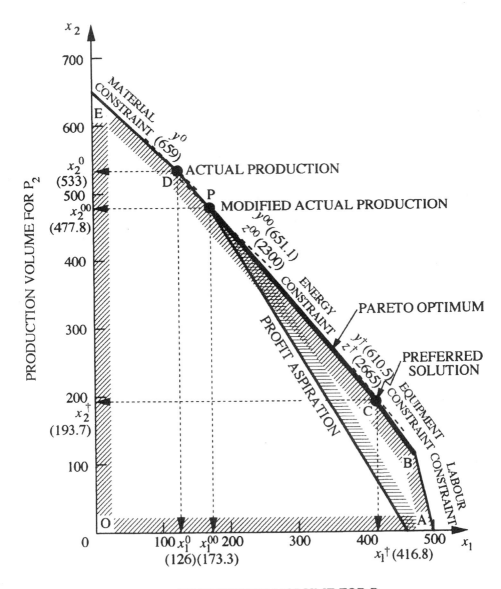

Figure 2.9 Obtaining the Pareto optimum and a preferred solution by a graph for production planning meeting the profit as well as social objectives.

Step 9: We then obtain the actual results:

- production of P_1: $x_1^0 = 126$
- production of P_2: $x_2^0 = 533$ } total production: $y^0 = 659$
- profit obtained: $z^0 = 2229$.

Hence, *evaluating the performance*, the profit obtained was much smaller than that anticipated, although the reduction of production amounts was small.

Step 10: Since the profit obtained was smaller than that anticipated, *modification and redesign* were necessary. Management decided that the total profit should be at least 2300: $z \geq 2300$. The modified production was $x_1^{00} = 173$, $x_2^{00} = 478$; hence the total production was $y^{00} = 651$, thereby resulting in ensuring the profit $z^{00} = 2300$.

For redesign with multiple (two) objectives (the profit objective and the social objective) an integrated goal function was formulated:

$$\text{maximise } w = \alpha y + \beta z \tag{2.22}$$

where α and β are the weights for the social and the profit objectives respectively: $0 < \alpha, \beta < 1, \alpha + \beta = 1$.

We obtain the 'Pareto optimum' as depicted in Figure 2.9. If we set the weights: $\alpha = \frac{2}{3}$ for the social objective and $\beta = \frac{1}{3}$ for the profit objective, we obtain $w = 7x_1 + 5x_2$. A socially appropriate production plan realising acceptable profit is

- production of P_1: $x_1^\dagger = 417$
- production of P_2: $x_2^\dagger = 194$
- total production: $y^\dagger = 611$
- total profit: $z^\dagger = 2665$. □

Notes

1. This figure is determined by the following three rules (Ichikawa, 1980): If a system S is decomposable into $S_1, S_2, \ldots, S_n$,
   (Rule 1: completeness) $S_1 \cup S_2 \cup \ldots \cup S_n = S$,
   (Rule 2: non-redundancy) $S_i \cap S_j = \phi$ (null set) $(i \neq j)$, and
   (Rule 3: simplicity) minimum n.
2. A part of a system; it is also made up of several units.
3. A function F is *kth order homogeneous* if

$$F(hx_1, hx_2, \ldots, hx_n) = h^k F(x_1, x_2, \ldots, x_n). \tag{2.14}$$

4. Break-through creative thinking includes: forced comparisons, attribute listing, morphological analysis, brain-storming, analytical approach, input–output method, synectic approach ('Make the strange familiar.'/'Make the familiar strange.') (ASTME, 1967), etc.

References

ASTME (1967) *Value Engineering in Manufacturing* (Englewood Cliffs, NJ: Prentice-Hall), pp. 20–23.
DICKEY, E.R. and SENENSIEB, N.L. (1973) Total systems concepts, in Heyet, C. (ed.), *The Encyclopaedia of Management* (2nd edn) (New York: Van Nostrand Reinhold), p. 1050.

DORNY, C.N. (1975) *A Vector Space Approach to Models and Optimization* (New York: Wiley), p. 3.

DOWLING, E.T. (1992) *Introduction to Mathematical Economics* (2nd edn) (New York: McGraw-Hill), p. 285.

HITOMI, K. (1971) The concept of system and its design (in Japanese), in Wakuta, H. (ed.) *Design of Accounting Information Systems* (Tokyo: Japan Management Association), Chap. 3.

HITOMI, K. (1975) *Theory of Planning for Production* (in Japanese) (Tokyo: Yuhikaku), Sec. 5.

HITOMI, K. (1990) *Principles of Manufacturing Systems* (in Japanese) (Tokyo: Dobunkan), Sec. 2.1.

ICHIKAWA, A. (ed.) (1980) *Multiple-objective Decision: Theory and Methods* (in Japanese) (Tokyo: Measurement and Automatic Control Society), pp. 85–86.

MATSUDA, T. (1969) *Planning and Information* (in Japanese) (Tokyo: Japan Broadcasting Association), pp. 36–38.

MATSUDA, T. (1990) From automation technology to system integration (in Japanese), *Automation Technology*, **22** (4).

MESAROVIĆ, M.D., MACKO, D. and TAKAHARA, Y. (1970) *Theory of Hierarchical, Multilevel, Systems* (New York: Academic Press), p. 56.

NADLER, G. (1970) *Work Design* (revised edn) (Homewood, IL: Irwin).

SIMON, H.A. (1977) *The New Science of Management Decision* (revised edn) (Englewood Cliffs, NJ: Prentice-Hall), p. 73.

The Shorter Oxford English Dictionary (3rd edn) (1973) (Oxford: Oxford University Press), p. 2227.

Supplementary reading

CHECKLAND, P.B. (1981) *Systems Thinking, Systems Practice* (New York: John Wiley).

GORE, C., MURRAY, K. and RICHARDSON, B. (1992) *Strategic Decision-Making* (London: Cassell).

CHAPTER THREE

Fundamentals of Manufacturing Systems

3.1 Meaning of the Term 'Manufacturing Systems'

Historical Background of Manufacturing Systems

Historically the phrase 'manufacturing system' was employed as early as 1815; a utopian socialist used this term meaning a 'factory system' (Owen, 1815). A manufacturing system also meant a series of inventions that were created during the Industrial Revolution in England about two hundred years ago (Going, 1911). In the early 20th century the system's view in management and manufacturing was emphasised as 'Scientific Management' by Taylor (1911).

Manufacturing Systems Today

Nowadays, the term manufacturing system signifies a broad systematic view of manufacturing. In 1961 the manufacturing system was conceived as the co-ordination of production engineering research with a series of activities of design, programming, control system, machine, and fabrication (Merchant, 1961).

It was also recognised as a production function that converts the raw materials into the finished products, and this function is controlled by the management system which performs planning and control (Hitomi, 1962).

It should be noted that from a wider viewpoint of manufacturing, *production systems* not only play a role inside each firm, but are part of the socially spatial interaction structure, settlement systems, and world systems as a whole (Nijkamp, 1977). Moreover, the *production structure* interacts with the security structure, the finance structure, and the knowledge structure of nations (Strange, 1988).

During the past few decades several books and technical journals on manufacturing systems have been published (see supplementary reading at the end of this chapter).

Three Aspects of Manufacturing Systems

On the basis of such concepts and views of the meanings of manufacturing and systems so far discussed, 'manufacturing (or production) systems' are now further defined through the following three aspects (Hitomi, 1975b):

- structural aspect—Section 3.2;
- transformational aspect—Section 3.3;
- procedural aspect—Section 3.4.

3.2 Structural Aspect of Manufacturing Systems

Plant Layout as a Manufacturing System

Based upon the structural (or static) definition of the system, the *manufacturing system* is a unified assemblage of hardware, which includes production facilities (including machine tools, jigs and fixtures), materials-handling equipment, workers, and other supplementary devices. This is supported by software, which is production information, namely the production method and technology.

This system performs on production objects (raw materials and components) to generate useful products having particular functions, thereby creating utilities so as to meet market demands.

Thus the structural aspect of the manufacturing system forms a static spatial structure (layout) of a plant. This influences the effectiveness of the transformation process in production; hence the optimum design of the plant layout, which will be discussed in Chapter 9, is the problem of the structural aspect of the manufacturing system.

—— EXAMPLE 3.1

A plant layout for parts-oriented production systems (see Chapter 27 for details) including the parts-machining system and the product-assembly system is represented in Figure 3.1.   □

Production System

The structural aspect of manufacturing systems can be also viewed as a 'production system'. This phrase appeared in 1907. Since 1943 it has also been used to mean the inference mechanism operated by knowledge-based systems in the field of artificial intelligence.[1]

3.3 Transformational Aspect of Manufacturing Systems

Production Process System as a Manufacturing System

Based upon a transformational (or functional) definition of the system, the *manufacturing system* is defined as the conversion process of the resources of

FUNDAMENTALS OF MANUFACTURING SYSTEMS

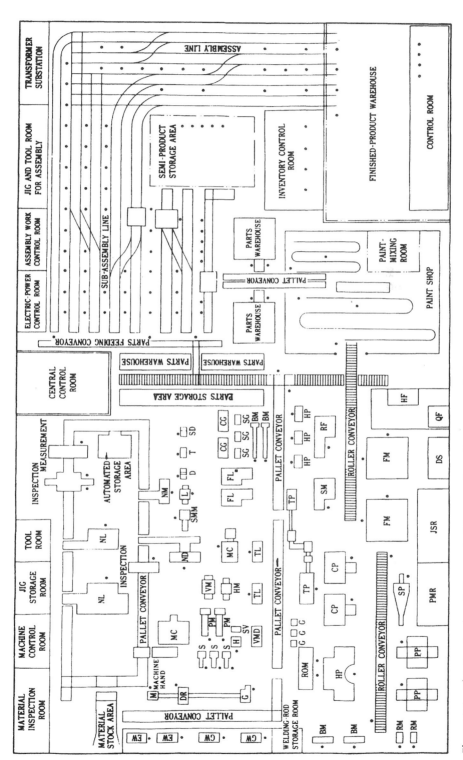

Figure 3.1 The structural or static aspect of the manufacturing system, which constitutes the layout structure of a plant.

production, particularly the raw materials, into the finished products, aiming at maximum productivity.

This system is concerned with the flow of materials and constitutes the 'production process system'.

— EXAMPLE 3.2

A car production system is schematically represented in Figure 3.2. ☐

The optimum decision-making for the transformational aspect of manufacturing systems mainly depends upon manufacturing technology, including production processes, machine tools, and industrial engineering techniques.

Logistic System Connecting Manufacturing and Marketing Firms

From the macro-economic and social viewpoints, final goods produced reach ultimate consumers through a long chain of organisations. The chain begins with the extractor of raw materials from nature. These raw materials are converted into the products, with the form desired by the ultimate user, successively through a series of organisations such as materials processor and fabricator. Then the products are brought to the ultimate customer where and when one wants them, through a series of marketing firms such as wholesaler and retailer, thus yielding the time and place utilities, as depicted in Figure 3.3.

This process system of production includes a chain of producers and distributors and is called a *logistic system* (a wide meaning of the material flow).

Logistic System of Acquisition–Production–Distribution

The scope of material flow limited to that for a manufacturing firm is depicted in Figure 3.4. The raw materials and components (inputs to a factory) are acquired from outside suppliers and stored as raw-materials inventory. At the conversion stage raw materials are processed by machining or other methods to fabricate parts and frequently stored as parts inventory. Finally the parts and bought-in components are assembled to produce finished products (outputs of the manufacturing system), resulting in the creation of form utility. Those products are commonly stored in the warehouse as finished product inventory and, according to the market needs, are shipped and delivered to the ultimate customers, thereby creating place and time utilities.

Logistic System as a Chain of Serial Functions

From a wide viewpoint the logistic flow constitutes a serial functional chain of procurement, production, distribution, inventory and sales.

3.4 Procedural Aspect of Manufacturing Systems

Management System as a Manufacturing System

Based upon a procedural definition of the system, the *manufacturing system* is considered as the operating procedure of production, which is the *management*

FUNDAMENTALS OF MANUFACTURING SYSTEMS 51

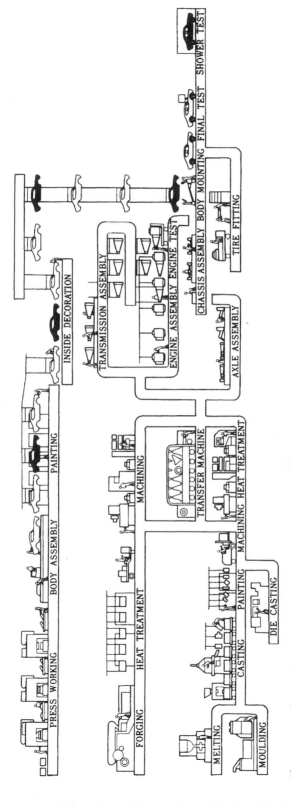

Figure 3.2 The production process system for manufacturing a car.

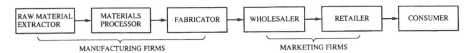

Figure 3.3 A logistic system: raw materials are first extracted from nature. They are processed and the products are fabricated and brought to the ultimate consumers through commercial activities.

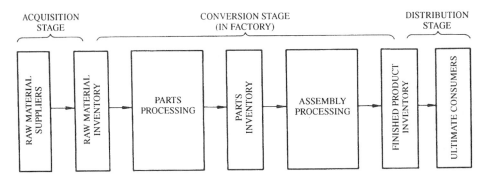

Figure 3.4 A manufacturing system as a material flow. Raw materials acquired are converted into finished products, which are distributed to ultimate customers.

system for manufacturing (or simply *manufacturing* (or *production*) *management*). This constitutes the so-called 'management cycle'—planning, implementation, and control[ling], as mentioned in Section 1.2.

This cycle was established in German management late in the 19th century and H. Fayol clarified the functions in 1916 (Fayol, 1916).

The manufacturing management system plans and implements the productive activities to convert raw materials into products to meet production objectives, and controls this process to reduce or eliminate the degree of deviation of actual performance from the plan.

Management

Management, in short, is an attempt to achieve organisational objectives through integrating resources, such as men, machines, materials, and money, which are not originally related to each other. The 'management process' is usually considered as a combination of several procedures, such as:

- establishing objectives;
- planning;
- organising;
- integrating;
- coordinating;
- allocating personnel;
- directing;
- supervising;
- motivating;

FUNDAMENTALS OF MANUFACTURING SYSTEMS

- measuring performances;
- controlling; etc.

Procedure of Production Management

The overall procedure of manufacturing management systems (production management) basically comprises two phases (see Figure 3.5) (Hitomi, 1975a).

(I) *Strategic production planning*: this deals with strategic production issues existing between the production system and the external environment and makes macroscopic decisions so as to adapt the production system adequately to the environment, usually in the long term.

This planning is relatively difficult since the solution procedure is generally unknown and left to the judgement and creative ability of the top management. The main problems to be solved in this field are as follows:

   (1) *Establishing production objectives*—the most basic of the decision-making activities in manufacturing. It is mainly concerned with the outputs to be produced; that is, it decides which commodities the production system should produce. Product planning, which is discussed in Chapter 7, plays an important role in this decision.

   (2) *Planning production resources*—production resources required to produce specified commodities, which are determined by the product planning function, include money, facilities and equipment, materials, and personnel. Determining requirements, planning for their acquisition from the external environment, and optimum allocation to the various divisions of the manufacturing system for economical production are appropriately made. This is called *resource planning*.

(II) *Operational* (or *tactical*) *production management*: this deals with operational production problems of the manufacturing system and makes optimal microscopic decisions, commonly in the short term, for effective production activities to be performed under the policy derived from the results of strategic production planning.

This operational production management procedure consists of the following five stages:

   (1) *Aggregate production planning*—determines kinds of product items and the quantities to be produced in specified time periods.

   (2) *Production process planning*—determines the production processes (or process routes) by which resources are effectively transformed into finished products, together with appropriate layout of these facilities for executing the determined sequence of production processes. This function will also interact with the product design function to ensure ease of manufacture of the product.

   (3) *Production scheduling*—determines an implementation plan for the time schedule for every job contained in the process route adopted; that is, when, with what machine, and who does what operation?

   (4) *Production implementation*—executes actual production operations according to the time schedule.

   (5) *Production control*—whenever actual production progress and performances deviate from the production standards (plans and

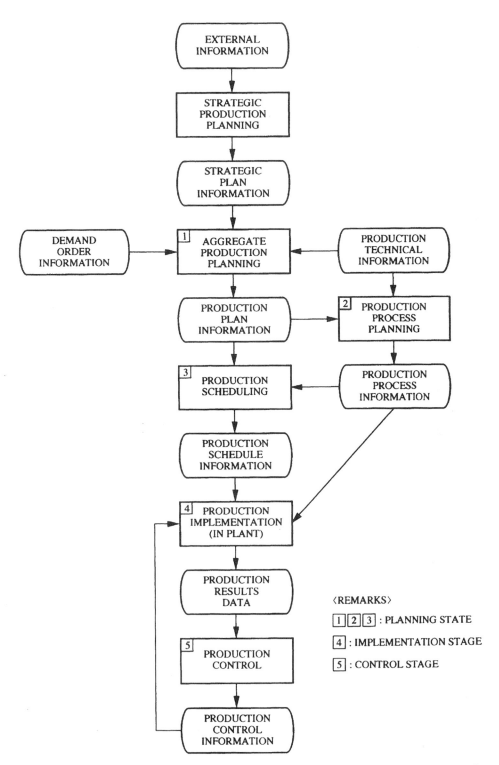

Figure 3.5 The procedural aspect of the manufacturing system, which constitutes the management function in production, that is, planning, implementation, and control stages.

schedules) set at the planning stages 1, 2 and 3, such deviations are measured and modifications are made.

The above is the general procedure of the production management system, and the connections of phases I and II and the five stages in II are represented in the block diagram of Figure 3.5. This flow is the 'flow of information' for effective and economical production.

Kinds of Information Flow

Production process planning stage (2) deals with 'intrinsic' production technology, hence the 'flow of technological information'; whilst phase I and stages 1, 3 and 5 of phase II are concerned with planning and control—management activities, hence the 'flow of managerial information'. Phase I especially treats the 'flow of strategic management information'.

The structure and operating procedures of the above information flow in an on-line, real-time basis with the use of computer facilities and information technology may be called *manufacturing information systems*.

3.5 Integrated Manufacturing Systems (IMS)

Microscopic Aspect of Mechanical Production

The detailed procedure of production—producing the assembled products with discrete parts—primarily consists of four stages as shown in Figure 3.6:

(1) *Product design stage.* This determines the technological specifications of products meeting the market needs through research and development activities, and completes the final design/drawing of parts/products. Industrial design features of aesthetic aspects and ergonomic ease of use are taken into consideration in addition to functional design for realising the function, quality, reliability and safety to be exhibited in the use of products.
(2) *Process planning stage.* This determines the procedures for processing and machining parts and assembling products, selects machine tools for implementing the operations and jigs and fixtures for holding workpieces, determines the sequence of operations to be carried out by production facilities, allocates the cutting tools and equipment required for every operation, and determines the machining or process conditions.
(3) *Production implementation stage.* This procures raw materials from outside suppliers, machines the raw materials into parts, assembles, adjusts and inspects the products, and delivers them to the market/customers (shipping).
(4) *Production management stage.* This prepares the aggregate production plan and the detailed schedule for the above production processes, and conducts the production control.

Integrated Manufacturing System (IMS) for Quick Response

The above-mentioned procedure of receiving the orders, producing the required products, and delivering them at the right time or as quickly as possible is referred to

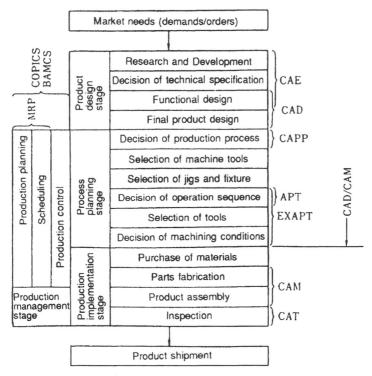

Figure 3.6 Functions of the integrated manufacturing system (IMS) consist of product design, process planning, implementation and management stages. Computer aids for manufacturing which will be explained in Part V, are also indicated.

as the *integrated manufacturing system* (IMS^2). Its original schema was presented by Martino (1972).

At present production aiming at the minimum time from the order receipt to the product delivery is called *quick-response manufacturing*. An integrated information system having computer aids for this purpose is CALS (Computer-aided Acquisition and Logistic Support/Commerce At Light Speed). This supports acquisition and manufacturing activities by computerising all the data concerning product life cycles through the product's development and design, its production, distribution, and maintenance. The Internet is commonly used for its implementation.

3.6 Manufacturing Systems Engineering—an Academic Discipline

Key Concerns of Manufacturing Systems Engineering

It is fundamentally important for the efficient and economical execution of production activities to unify completely material flow (production process system), information flow (production management system) and cost flow (production–economic system), as proposed in the Preface. This unified and integrated paradigm to production/manufacturing studies is called *manufacturing systems engineering* (*MSE*), hence the title of this book.

Six Aspects of Manufacturing Systems Engineering

The following six aspects/approaches are stressed in manufacturing systems engineering:

(1) to clarify the concept of manufacturing systems and their basic functions and structures—that is, the problem of designing the manufacturing systems, especially, the material flow (system engineering approach);
(2) to optimise manufacturing systems—that is, the problem of optimum decision-making for manufacturing (management science/operations research approach);
(3) to control manufacturing systems—that is, process control and the problem of automation of manufacturing—factory automation/computer-integrated manufacturing (control engineering approach);
(4) to process production information adequately for manufacturing systems within a strict time frame—that is, the problem of information flow for production management–management/manufacturing information systems (information technology approach);
(5) to clarify production/industrial economics for manufacturing—that is, the problem of cost flow in manufacturing/management (economics approach);
(6) to recognise the social aspects of manufacturing—that is, a problem of value flow as 'manufacturing excellence' for future production perspectives (social science approach).

Historical Review of Production/Manufacturing Studies

Both material and information flows were advocated by Church (1913), and the flow of value was pointed out by Nicklisch (1922). 'Manufacturing systems engineering' was advocated by Hitomi (1975b). Since then a few books on this subject have been published (see supplementary reading).

From the viewpoint of principles or concepts of technology, a study of manufacturing or production engineering is one of three major subjects in modern engineering, the others being fundamental engineering science and applied engineering (Honda and Suzuki, 1974). 'Fundamental engineering science' deals with the theories and scientific laws that govern structures and functions of machines, such as mechanics, dynamics, material sciences, fluid dynamics, thermodynamics, etc., and is principally connected with natural science, wherein 'true' or 'false' is the main concern. 'Applied engineering' describes the structure and function of machines, such as ships, aircraft, buildings, bridges, etc., and can be regarded as an art, wherein subjective judgements about what is 'good' or 'bad' are valuable. 'Production engineering', on the other hand, is concerned with technology and methods of production and includes such aspects as materials, tools, machine tools, production methods and organisation, etc. It is also connected to economics where 'cost-effectiveness' is of value. Design of products and production processes should be included in this field, and designing for manufacturability is an important consideration.

The term 'production (or manufacturing) engineering' appeared in 1910, and it often means intrinsic production technology; accordingly, 'manufacturing systems engineering' is a modern term unifying manufacturing technology and production management, as was indicated in the subtitle to the first edition of this book (Hitomi, 1979). Historically, attention has been paid to these two aspects for many years. A famous pioneer in the field of computers, C. Babbage (UK: 1792–1871) conducted research into the principles of metal-cutting tools as well as publishing a well-known book on the economy of manufacture (Babbage, 1832). F.W. Taylor (USA: 1856–1915), an industrial engineering pioneer, founded 'scientific management' as well as inventing the high-speed steel tool and proposing the tool-life equation. G. Schlesinger (Germany: 1874–1949) pioneered the development of machine tools as well as contributing to the promotion of industrial rationalisation, factory management, and psychotechnology. In Japan also, systems science on production and management was discussed in 1973 (Togino, 1973).

Notes

1 For this meaning it is an academic obligation to introduce a different terminology, because 'artificial intelligence' appeared much later than manufacturing/production [systems] historically.
2 Recently an IMS was also viewed as an 'intelligent manufacturing system' (this phrase appeared originally in 1978, hence not a new idea) as in the 'IMS international cooperative project' (budget: 150 billion yen (about 1.5 billion US dollars)) proposed by Japan's Ministry of International Trade and Industry (MITI). This project is intended to build a 'Factory of the Future' based on intelligent machines and expert systems, through joint research and development by Japan, the United States and the EU. Advanced manufacturing technologies thus obtained will be a worldwide common asset, and so may be transferred to other developing countries.

References

BABBAGE, C. (1832) *On the Economy of Machinery and Manufactures* (London: Charles Knight).
CHURCH, A.H. (1913) Practical principles of rational management, *Engineering Magazine*, **44** (6).
FAYOL, H. (1916) *Administration industrielle et générale* (in French) (Paris: Dunod).
GOING, C.B. (1911) *Principles of Industrial Engineering* (New York: McGraw-Hill), p. 11.
HITOMI, K. (1962) The concept of manufacturing management systems (in Japanese), *Business and Management*, **14**, 39–42.
HITOMI, K. (1975a) *Theory of Planning for Production* (in Japanese) (Tokyo: Yuhikaku), Sec. 9.
HITOMI, K. (1975b) *Manufacturing Systems Engineering* (in Japanese) (Tokyo: Kyoritsu Publishing), Chap. 1.
HITOMI, K. (1979) *Manufacturing Systems Engineering—A Unified Approach to Manufacturing Technology and Production Management* (London: Taylor & Francis).
HONDA, S. and SUZUKI, K. (1974) *Introduction to Technology* (in Japanese) (Tokyo: Asakura-Shoten), p. 6.
MARTINO, R.L. (1972) *Integrated Manufacturing Systems* (New York: McGraw-Hill).
MERCHANT, M.E. (1961) The manufacturing-system concept in production engineering research, *CIRP Annalen*, **10**, 77–83.
NICKLISCH, H. (1922) *Wirtschaftliche Betriebslehre* (in German) (Poeschel), p. 173.

NIJKAMP, P. (1977) *Theory and Application of Environmental Economics* (Amsterdam: North-Holland), p. 302.
OWEN, R. (1815) Observations on the effects of the manufacturing system, *Everyman's Library*, No. 799.
STRANGE, S. (1988) *States and Markets* (London: Pinter Publishers), p. 27.
TAYLOR, F.W. (1911) *The Principles of Scientific Management* (New York: Harper).
TOGINO, K. (1973) *Systems Science for Production Management* (in Japanese) (Tokyo: Kyoritsu Publishing).

Supplementary reading

ANJANAPPA, M. and ANAND, D.K. (eds) (1989) *Advances in Manufacturing Systems Engineering* (New York: American Society of Mechanical Engineers).
ASKIN, R.G. and STANDRIDGE, C.R. (1993) *Modeling and Analysis of Manufacturing Systems* (New York: Wiley).
BAK, J.H. (1978) *Manufacturing Systems Management* (in Korean) (Keimunsha).
BAUDIN, M. (1990) *Manufacturing Systems Analysis* (Yourdon Press).
BEBBETT, D. (1986) *Production Systems Design* (London: Butterworths).
BIGNELL, V. et al. (eds.) (1985) *Manufacturing Systems—Context, Applications and Techniques* (Oxford: Basil Blackwell).
BOER, H. (1991) *Organising Innovative Manufacturing Systems* (Aldershot: Avebury).
BROWNE, J., SACKETT, P. and WORTMANN, J. (1993) *The System of Manufacturing—A Prospective Study* (Galway, Ireland: University College Galway).
CARRIE, A. (1988) *Simulation of Manufacturing Systems* (New York: Wiley).
CHRYSSOLOURIS, G. (1991) *Manufacturing Systems—Theory and Practice* (Berlin: Springer).
COMPTON, W.D. (ed.) (1988) *Design and Analysis of Integrated Manufacturing Systems* (Washington, D.C.: National Academy Press).
Dictionary of Manufacturing Systems Terminology (1978) (in Japanese) (Tokyo: Yokendo).
ELMAGHRABY, S.E. (1966) *The Design of Production Systems* (New York: Reinhold Publishing).
GERSHWIN, S.B. (1994) *Manufacturing Systems Engineering* (Englewood Cliffs, NJ: Prentice-Hall).
HEIM, J.A. and COMPTON, W.D. (1992) *Manufacturing Systems—Foundations of World-class Practice* (Washington, D.C.: National Academy Press).
HITOMI, K. (1990) *Principles of Manufacturing Systems* (in Japanese) (Tokyo: Dobunkan).
HOFFMAN, T.R. (1967) *Production: Management and Manufacturing Systems* (Belmont, CA: Wadsworth Publishing).
International Journal of Manufacturing System Design (Singapore: World Scientific Publishing).
IWATA, K. et al. (1982) *Manufacturing Systems Studies* (in Japanese) (Tokyo: Corona-sha).
JAMSHIDI, M. and PARSAEI, H. (1995) *Design and Implementation of Intelligent Manufacturing Systems* (Englewood Cliffs, NJ: Prentice-Hall).
Journal of Manufacturing Systems (Dearborn, Michigan: Society of Manufacturing Engineers).
KUMAR, P.R. and VARAIYA, P. (eds) (1995) *Discrete Event Systems, Manufacturing Systems, and Communication Networks* (Berlin: Springer).
KUNISA, T. (1996) *Principles of Modern Manufacturing Systems* (in Japanese) (Tokyo: Senbundo).
KUSIAK, A. (1990) *Intelligent Manufacturing Systems* (Englewood Cliffs, NJ: Prentice-Hall).
LIU, F., YANG, D. and CHENG, J. (eds) (1995) *Manufacturing Systems Engineering* (in Chinese) (Beijing: Defence Industry Publishing).
Manufacturing Systems (Paris: CIRP).

MARE, R.F. de la (1982) *Manufacturing Systems Economics* (London: Holt, Rinehart and Winston).

MINATO, S. (1987) *Manufacturing Systems Engineering for Process Industries* (in Japanese) (Tokyo: Business & Technology Newspaper Company).

OSAKI, K. *et al.* (1981) *Manufacturing Systems Techniques* (in Japanese) (Tokyo: Kyoritsu Publishing).

O'SULLIVAN, D. (1994) *Manufacturing Systems Redesign—Creating the Integrated Manufacturing Environment* (Englewood Cliffs, NJ: Prentice-Hall).

RIGGS, J.L. (1976) *Production Systems* (New York: Wiley).

SATA, T. (ed.) (1973) *Manufacturing Systems* (in Japanese) (Tokyo: Business & Technology Newspaper Company).

Understanding Manufacturing Systems (Milwaukee, WI: Kearney & Trecker).

VERNADAT, F. (ed.) (1995) *Integrated Manufacturing Systems Engineering* (New York: Chapman & Hall).

VISWANADHAM, N. and NARAHARI, Y. (1992) *Performance Modeling of Automated Manufacturing Systems* (Englewood Cliffs, NJ: Prentice-Hall).

WATANABE, S. and AKIYAMA, Y. (eds) (1986) *Manufacturing Systems and Modern Automation Technology* (in Japanese) (Tokyo: Japan Industrial Press).

WILLIAMS, D.J. (1988) *Manufacturing Systems—An Introduction to the Technologies* (New York: Halsted Press/Open University Press).

WU, B. (1992) *Manufacturing Systems Design and Analysis* (New York: Chapman & Hall).

XIAO, C.Z. (1987) *Manufacturing Systems Engineering* (in Chinese) (Beijing: Mechanical Industry Press).

YOSHIYA, R. (ed.) (1967) *Handbook of Production System Design* (in Japanese) (Tokyo: Business & Technology Newspaper Company).

CHAPTER FOUR

Modes of Production

4.1 Types of Production

Production Mode

Designing the manufacturing system optimally and deciding its optimal operating procedures are subject to the unique characteristics of the reproduction mode of each firm, i.e. the type of production of the organisation under consideration. In analysing the various types of production, the following classifications are common (Timms and Pohlen, 1970).

Production to Order/for Stock Replenishment

Production to order is the production of items based upon customers' orders. *Production for stock [replenishment]* is undertaken in advance of receiving customers' orders, the products are then stored as inventory and shipped as orders are received.

Regardless of the volume of production, the distinct difference between these two types of production is low certainty of product specifications for production to order and high certainty for production for stock replenishment. In production for stock replenishment, product specifications are established in advance of order receipt with reasonable certainty as established by market research. Hence, production processes, kinds of operations, kinds and volumes of resources of production, and other conditions are known in advance, thereby resulting in easy production planning. On the other hand, the unique aspect of production to order is that exact product specifications are established only on the receipt of the customer's order, which results in difficulty in production planning and execution.

Jobbing/Intermittent/Continuous Production

This classification of production is related to the expected sales or production volume or quantity of the product demanded per period of time, say per month, per season or per year. If this volume is very low, production will be done on a jobbing or slow-moving basis (*jobbing production*); if the demand is expected to be very large, production will be on a continuous basis (*continuous* or *mass production*); between the two extremes, *intermittent* (or *batch*) *production* of lots or batches is appropriate.

Continuous and mass production may be applied where a variety of products are made. For example, in a car assembly line a mix of different models (size, colour, inside decoration, motor power, etc.) may be produced in a flow-type system. This is called *mixed production*.

Discrete-part/Process Production

This is a classification of production that is based upon the nature of the product. If the finished product is made up of a number of discrete parts or components, this is called *discrete-part production*. A feature of this type of production is that discrete-part products can be disassembled and reassembled. An example is car production. On the other hand, where components or ingredients cannot be readily identified in a finished product and the product cannot be disassembled, this is *process production*. Examples are steel, cloth, chemical products, etc.

4.2 Mass Production

4.2.1 A Short History of Mass Production

Early Age of Mass Production

Mass production was conducted as early as the 6th century BC by the Phoenicians in making bricks with a vast labour force. In Japan mass production of the wooden *Hyakuman*(million)-*tou*(tower) (a small tower with Buddhist scriptures inside) was implemented twelve hundred years ago; a million of these were made in five years.

American System of Manufacture

Mass production was advanced in the United States in the 19th century by employing the *principle of interchangeability* originated in France in the 18th century. With this principle standardised parts are produced within a specified tolerance; a product can be made up by assembling a set of parts selected randomly. In the middle of the 19th century the 'American system of manufacture' was established, and the United States became the world's top industrialised country in 1880.

Mass production and mass consumption formed a sort of production culture. Giant corporations quickened mass production; a prime example was Ford's model T car of which 15 million were produced during the period 1908–27. The establishment of conveyor systems in 1913 advanced productivity, reduced the cost/price of a car, and raised the wage rate. Incidentally conveyor systems were first created by O. Evans in 1787; he employed them for his automatic flour mill.

Recognising the Merit of Mass Production

The advantage and principle of mass production was recognised as early as the 17th century by A. Serra. In 1910, K. Bücher presented the principle that mass production is productive, as proved in the next section.

4.2.2 Principle of Mass Production

Economies of Scale

If the fixed cost is a and the unit variable cost is b, then the total cost producing the amount x is:

$$f(x) = a + bx \tag{4.1}$$

and the average unit production cost is:

$$u(x) = (a/x) + b, \tag{4.2}$$

which decreases with an increase of production volume x (see Figure 4.1), resulting in *economies of scale* or scale merit. This is an advantage of mass production.[1]

Six-tenths Factor

In process/chemical industries the fixed cost is not constant; it is usually given by ax^β, where α is a constant depending upon the production method. β is also a constant and takes a value of around 0.6 empirically; hence it is called the 'six-tenths factor' (Moore, 1959).

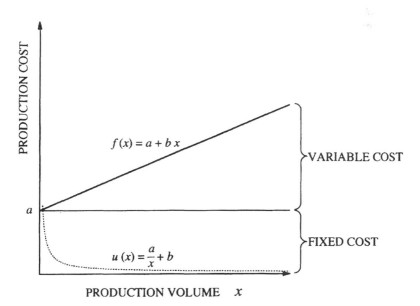

Figure 4.1 The total production cost consisting of the fixed and the variable costs increases with the production volume, while the unit production cost decreases—economies of scale.

Subadditivity Condition

Suppose that a total x of a single product is produced by N firms, each producing x_i ($i = 1, 2, \ldots, N$) and x in total, then $x_i < x$; hence from (4.2),

$$u(x) < u(x_i). \tag{4.3}$$

Multiplying this equation by x_i and summing with respect to i, we get:

$$f(x) < \sum_{i=1}^{N} f(x_i). \tag{4.4}$$

This expresses the *subadditivity condition*, which means the effect of production accumulation.

4.2.3 Optimum Production Scale

Optimum Production for the Silberston Curve

If the total production cost shows non-linearity, differing from (4.1), then, for example,

$$f(x) = a + bx^k \tag{4.5}$$

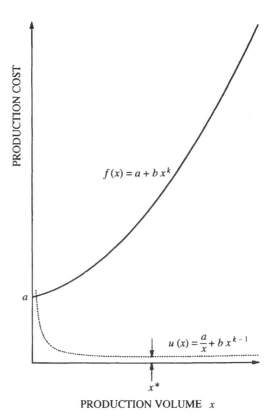

Figure 4.2 The total production cost with increasing returns holds the optimum production scale where the unit production cost is minimum—the Silberston curve.

where $k > 1$ (increasing returns). Then the unit production cost is:

$$u(x) = (a/x) + bx^{k-1}. \tag{4.6}$$

With an increase of production volume x this unit cost decreases at first; it is minimum when

$$x^* = (a/(k-1)b)^{1/k}. \tag{4.7}$$

The unit cost increases thereafter.

The production volume which minimises the unit cost is the *optimum production scale*. This phenomenon was first discovered in car production. The diagram which indicates this mode is called the 'Silberston curve' (Maxcy and Silberston, 1959).

—— EXAMPLE 4.1 ——

Figure 4.2 demonstrates the Silberston curve. □

4.3 Multi-product, Small-batch Production

4.3.1 Definition and Features of Multi-product, Small-batch Production

Why Multi-product, Small-batch Production?

As economies grow and societies become affluent, the desire of the individual to possess special goods different from those of other people increases, and so does the demand for specially ordered products, whilst the life cycle of products decreases.

This tendency results in a wide variety of low-volume (small-batch) production. In addition, lead time from the receipt of the order to the shipment of the product is expected to be as small as possible (quick response) in order to win in the competitive edge.

—— EXAMPLE 4.2 ——

Whilst the Ford Motor Company consistently persisted in producing model T cars, General Motors established a policy of 'full line' to manufacture a variety of cars, which raised their market share. □

—— EXAMPLE 4.3 ——

Japanese auto makers produced 85 models in 1969; the number of models increased to 420 in 1986. Meanwhile the production volume per model decreased from 32 000 to 18 000. Each model makes up a market—*market segmentation*. □

Hence, nowadays, jobbing and intermittent (small-batch/non-mass) production is more important than single-product, mass production. Statistics indicates that mass production accounts for only 15% of production, the other 85% being production volumes of less than 50.

Multi-product, Small-batch Production Defined

Multi-product (or *item*), *small-volume* (or *batch*) *production* is non-mass, variety production—a type of production in which a great variety of products/parts are manufactured in a specified short time period, and the production volume for each product/part is very low.

A factory or machine shop that deals with this type of production is called a *job shop*.

Most of the multi-product, small-batch production is production to order, producing items based upon customers' orders. However, these days multi-product, small-batch production may also be directed to production for stock replenishment.

Characteristics of Multi-product, Small-batch Production

Characteristics of non-mass, multi-product, small-batch production are as follows (Hitomi, 1982).

(1) *Variety of product items.* Various product items are manufactured, and their production volumes and due-dates are also diversified.
(2) *Variety of production processes.* The conversion process of raw materials into products ('material flow') is also varied, and often complicated.
(3) *Complexity of productive capacity.* Due to the dynamic nature of demand for a variety of products, the productive capacity in the job shop is often in excess or in deficiency, resulting in over- or under-working time. On some occasions Japanese-style 'human-power tactics' performed by a number of unskilled workers are used when there is a large deficiency in the productive capacity.
(4) *Uncertainty of outside conditions.* There are frequent changes in product specifications, volumes and due-dates for ordered products, generation of special jobs, lateness of the arrival of purchased raw materials and parts, etc.
(5) *Difficulty of production planning and scheduling.* Imprecise information due to the dynamic situation of orders, a variety of production processes, frequent changes in ordered products, etc., make optimum production planning and scheduling, as well as cost estimation for jobs, difficult.
(6) *Dynamic situation of implementation and control of production.* Owing to the uncertainty of an optimal plan and schedule in variety production, executing production activities in the job shop is complex. Machine breakdown, absence of workers, manufacture of inferior products, etc. occur on many occasions. Consequently, production activities must rely upon past experience and intuition rather than upon an exact plan/schedule based upon optimum decision-making. Production control is also complex and liable to change.

To overcome the difficulties involved in multi-product, small-batch production in the job shop, several effective techniques have been developed. These are briefly mentioned in the following (Hitomi, 1989).

4.3.2 Techniques of Multi-product, Small-batch Production

Industrial Engineering (IE)

This is a traditional methodology of production organisation and management based upon the pioneering works in the field of 'scientific management' by F.W. Taylor (1856–1915) and others at the end of the 19th century and the beginning of the 20th century. Among the several definitions of IE is the generally accepted one developed in 1955 by the [American] Institute of Industrial Engineers:

> Industrial engineering is concerned with the design, improvement, and installation of integrated systems of people, materials, equipment, energy, and information. It draws upon specialized knowledge and skill in the mathematical, physical, and social sciences, together with the principles and methods of engineering analysis and design, to specify, predict, and evaluate the results to be obtained from such systems.

This definition is broad enough to cover the diverse activities of the practitioners in the field of production/manufacturing. IE techniques are also employed in service industries. IE emphasises the following:

- *standardisation* standardises parts, products, production processes, etc.;
- *simplification* simplifies parts, products, production processes, etc. as much as possible;
- *specialisation* specialises parts, products, production processes and skills.

'Standardisation' is the most basic in IE; it makes task/work/operation proceed according to the 'standard' settled as a measure of performance.

One effective application of IE to multi-product, small-batch production is the *learning effect*; that is, the unit operation time decreases as the operation is repeated. This is due to increase in the operator's level of skill, improvement of the production methods such as modification of operation contents and sequences, use of effective jigs and fixtures, resetting of machining conditions, improved management practices, etc. The effect is large where the cumulative production quantity is small, and decreases as this quantity increases. (Refer to Section 8.3.)

Group Technology (GT)

This is a technique to increase a production lot (or batch) size by grouping various parts and products with similar shape, dimension and/or process route. This technology was first proposed and developed in Russia by S.P. Mitrofanov, who aimed at similarity for a number of different parts and classified them into several group cells.

The fundamental requirement of GT is the *classification and coding system*. It describes the characteristics of a part by its geometrical shape and/or process route (setup, sequence of operations such as metal fabrication, machining and assembly, inspection and measurement) for producing a part with a code number (this is discussed in Section 7.2). By collecting parts with the same code number and grouping them into a group cell as a 'parts family', design, process planning, production and cost estimation are made more systematic.

A most significant advantage of applying GT to production activity is that setup time and setup cost for each job are reduced, because several jobs are grouped and processed in sequence, and the same jigs and tools may be used.

With GT the job flow is expected to be a flow-shop pattern; hence, the times and costs for material handling and waiting between stages are eliminated, and process planning and scheduling are simplified. This scheduling is called *group scheduling*, which will be discussed in Section 28.1.

Since the flow-shop pattern is expected with GT, the machine tools can be laid out as a group (cell); it is called a *GT* (or *cellular*) *layout*. This considerably simplifies the material flow and handling. (Refer to Section 9.1.)

Parts-oriented Production [System]

Demands for a variety of products with a short lead time can be overcome if all such products are produced and stored as finished-product inventory. However, this increases inventory costs and stored products go unsold when there is little or no demand.

A method to cope with this problem is 'parts-oriented production'. Since various products often contain the same components, such components are produced according to the results of demand or sales forecasts to be held as stock in advance of receiving orders and are positively stocked as parts inventory. Various products are assembled on the receipt of orders by suitably combining these standard parts and other special parts quickly produced to order. In this way, production lead time from the order receipt to the shipment of finished product can be little more than the time required for assembly.

In the firm's activities, in addition to the production function, the marketing (or sales) function takes on an important role. A parts-oriented production system can be regarded as an 'order-entry system' from the marketing viewpoint. In this system, the customer can effectively decide the product style and components required at the time of ordering. The manufacturer then checks the existence of products and/or parts in store. Parts explosion, materials planning, production orders for unavailable parts, assembly scheduling, due-date control, inquiry into the progress of production and other managerial and operational works are carried out accurately and at the right time typically by a host computer connected with remote terminals in each business office on an on-line, real-time basis. Car manufacturers and home-appliance makers employ such systems. Information systems for this purpose have been developed and used in practice. (Refer to Chapter 27.)

MRP

This is an acronym for 'material requirements planning'; the wider development is manufacturing resource planning' (MRPII). MRP is widely used computer-based software providing a systematic procedure for production planning applied to large job-shop production in which multiple products are manufactured in lots through several processing steps.

First, MRP establishes a 'master production schedule (MPS)', which is an aggregate plan showing required amounts versus planning periods for multiple end items to be produced. Then, with the use of a bill of materials (BOM) and an inventory file, production information for dispatching multiple jobs is generated on the hierarchical multiple-stage manufacturing systems by considering the common parts and the substitution of many parts included in multiple products. Thus, MRP schedules and controls a total flow of materials from the raw materials to the finished products on a time basis (usually, weekly).

The possibilities of assembling products and of machining parts are also examined in relation to the capacity of production facilities. This function is called *capacity requirements planning (CRP)*.

The general procedure of MRP is:

(1) calculate the requirements for products/parts;
(2) make up the lot;
(3) subtract lead times;
(4) run CRP.

(An example of MRP and CRP is mentioned in Section 13.6.)

Lot Scheduling

In intermittent (or batch) production the demand for a commodity is small compared with the production capacity. The product is then manufactured periodically in a quantity which will meet the demand for some time period until the next production run. In the interval between two production runs the production facilities may be utilised for other works in a similar fashion.

The quantity to be manufactured periodically is called a *lot* (or *batch*) *size*, as analysed in Section 13.4. Optimisation analysis of lot size has been the subject of much research since F.E. Raymond's first analysis in 1931.

Modular Production

Many different resources of production that enter the production process are combined to produce a catalogue of parts. The essence of the modular concept is to design, develop and produce those parts which can be combined in the maximum possible number of ways. The term 'modular production' was used to indicate a concept of producing a variety of products to meet the market needs by optimally combining standard parts (Star, 1965), which is the same idea as 'parts-oriented production' mentioned previously. However, here *modular production* is meant as a procedure to optimally sequence multiple items on an assembly line as well as optimally allocating work elements (modules) on each of the workstations on the line; this is called *mixed-model line balancing*.

A method of deciding an assembly sequence of multiple items is (Muramatsu, 1976):

(1) find the greatest common divisor q of production volumes Q_j (j = item index) of the multiple items,
(2) calculate $q_j = q/Q_j$, and
(3) heuristically decide a sequence by repeating selection of an item having the smallest q_j value and adding this value to q_j each time the item is selected. (If there are ties, select an item with smaller j so as not to select the same item continually.)

--- **EXAMPLE 4.4** ---

Amounts to be produced for three product items, 1, 2, and 3, are 400, 200, and 100. In this example (1) $q = 100$; (2) $q_1 = q/Q_1 = \frac{100}{400} = \frac{1}{4}$; (3) $q_2 = 1/2$, and (4) $q_3 = 1$. The above procedure (3) generates a heuristic sequence, as indicated in Table 4.1. The ratio of

Table 4.1 Iteration process for determining a heuristic sequence for multiple-item assembly.

Iteration No.	Product item			Sequence
	1	2	3	
1	$\frac{1}{4}$*	$\frac{1}{2}$	1	1
2	$\frac{1}{2}$	$\frac{1}{2}$*	1	1 2
3	$\frac{1}{2}$*	1	1	1 2 1
4	$\frac{3}{4}$*	1	1	1 2 1 1
5	1	1*	1	1 2 1 1 2
6	1*	$1\frac{1}{2}$	1	1 2 1 1 2 1
7	$1\frac{1}{4}$	$1\frac{1}{2}$	1*	1 2 1 1 2 1 3

production volumes for three product items at iteration no. 7 is 4 : 2 : 1, which is the same as that for the total production volumes—400 : 200 : 100. A sequence determined is 1211213 (repeated 100 times). □

Flexible Automation

'Automation' is a term created in 1936 by D.S. Harder; it has been very effective in producing large quantities of a single item automatically. However, automation is now required to cope with multi-product, small-batch production. This new type of automation is called 'flexible automation'.

To achieve flexible automation, as will be explained in detail in Section 25.4, four activities—machining, loading and unloading, transfer and storage of workparts—need to be performed automatically by employing hardware devices with high flexibility.

An automatic production facility can be constructed by combining both these high-flexibility machine tools or process equipment and an automated materials-handling system together to a control computer. This is called the *flexible manufacturing system (FMS)* or in earlier times the 'computerised manufacturing system (CMS)'.

Flexible Manufacture

The wide variety of product specifications demanded often makes for considerable uncertainty as to resources of production (production facilities, human labour and raw materials). Hence, one strives for the highest degree of flexibility in these inputs (Timms, 1966). Since machines are less flexible than human labour, relatively fewer machines may be used and more human labour. The human labour must be highly skilled in the sense that it must be able to perform various types of operative work effectively and efficiently. Human skill of the narrow specialist type is not appropriate. In addition, managerial labour for process planning, scheduling and controlling operative work must be narrowly specialised and very flexible for quick decision-making adaptable to dynamic operations in the job shop, because product specifications are not known before the receipt of the order, and the conversion from a raw material into the product must be performed to meet the competitive delivery date. A reserve of 'utility men' from the unskilled labour force rather than the skilled labour force can often be used in various operations as flexible inputs to perform 'human-power tactics'.

When product specification permits, raw materials used should be of high machinability to minimise machining time, since the production lead time is often short. Owing to the uncertainty of product specification in advance of the receipt of the order, the status of the raw-material inventory is not definite; hence, great flexibility in materials procurement enabling quick deliveries of raw materials must be provided.

For technological reasons, production facilities for this type of production must be of a most flexible character; for example, general-purpose machines are used whenever possible. In addition to the basic level of facilities to meet the average demand, an appropriate supplementary level of facilities is required to meet changes in market demand and to reduce production lead times in order to compete effectively on a due-date basis. This level of production facilities is *production* (or *physical*) *capacity*. Since a high capital investment is needed, production capacity should be held to a competitive minimum, and 'lease (rent) versus buy' decisions become an important factor in planning facilities. Subcontracting or outsourcing certain special production operations, for which the firm has none of the required facilities, is one way of maintaining flexibility with no investment. Small special-purpose tools or automatic tools are often effective to use.

The above-mentioned flexibility for resources of production is an effective way to cope with non-mass jobbing production. This type of production is called 'flexible manufacture' or 'human-centred production'. It emphasises human skills with the use of high technology, which is more humanised than full automation. Technology-centred full automation, in spite of high capital investments, is often less able to provide innovative flexibility. It is vulnerable to machine failure, and, in eliminating skilled workers, removes the pride and pleasure in workmanship which can contribute to the culture of a manufacturing environment.

On-line Production Management

An approach to dynamic, multi-product, small-batch production is to control individual operations in the job shop under a dynamic environmental situation by installing computer terminals at work centres, which communicate with a control computer in an on-line mode.

Thus, as various production data are generated, they are collected at the work centres and transmitted into the control department, where rapid data processing by the computer makes it possible to generate a new schedule relating to future production activities. This is then transmitted back to the work centres. This system is called an *on-line production management* (or *production control*) *system*, and with this effective management for non-mass jobbing production is possible. (Refer to Chapter 29 for more details.)

Just-in-time (JIT) Production

This system is based upon two criteria (Ohno, 1978). One is *just-in-time* (*JIT*) which refers to the production and supply of the required number of parts when needed.[2] Hence, this methodology is often called the *JIT production [system]*. Since work-in-process inventories including parts and products are expected to decrease (even to zero), this system is also called the *stockless* (or *zero-inventory*) *production [system]*.

The other criterion is *Jidoka* (self-actuation). This implies that when unusual events happen in a production line, the worker in charge stops the line and removes the cause of trouble.

This technology has been developed over the past thirty years through the efforts of Toyota Automobile Industries; hence it is often called the *Toyota production system*, which is now known worldwide (Monden, 1983).

In general, production scheduling and dispatching may be achieved in two ways. Traditionally, production control focuses on the beginning of the production line, the schedule pushing work through operations from beginning to end ('push system'). The JIT production system employs the pull system, which focuses on the end of the line and pulls work through from preceding operations. Workers perform the required operation on the material/workpart drawn from the preceding workstation at the necessary time with the use of a *kanban* (instruction card). Therefore, work-in-process inventory is minimised, and over-production can be eliminated. For the purpose of providing production information, two kinds of kanban are used—'withdrawal' and 'production ordering' information (refer to Section 16.3).

Thus, stabilisation of a process for multi-product, small-batch production is achieved by production smoothing. Improvements in setup operations have to be made by keeping the setup time to within ten minutes.[3]

4.4 Product Diversification

Economies of Scope with Diversification

Diversification is a philosophy that seeks to multiply the number of new and different kinds of products and/or services to be produced by a firm and is now commonly a top-level corporate strategy.

By producing and selling a number of products rather than a single product, the total production cost becomes less. This phenomenon is called *economies of scope*. Under this principle the following relation holds for cost function $f(x_1, x_2, ..., x_N)$, which expresses the total production cost for producing quantities, $x_1, x_2, ..., x_N$, for N kinds of products:

$$f(x_1, x_2, ..., x_N) < f(x_1, 0, ..., 0) + f(0, x_2, ..., 0) + \cdots + f(0, 0, ..., x_N). \quad (4.8)$$

EXAMPLE 4.5

Canon Corporation originally manufactured only cameras, but of total sales in 1991 (about US$13.8billion), camera sales were only 14%; other products were electronic office equipment at 40%, duplicators at 39%, and optical equipment at 7%. □

Japan's History of Diversification

Japan has a long history of diversification. Early in the 20th century, large conglomerate institutions with the name *Zaibatsu* became giant financial, industrial, marketing entities that included banking, commercial trade, and manufacturing—iron and steel, mechanical, chemical, and electrics/electronics, etc. Japanese diversification exceeded that in the United States and the United Kingdom at that time.

Notes

1 This was expressed in *Cheaper by the Dozen* by Frank B. and Lillian M. Gilbreth. The book of this title was published in 1948 and a film followed in 1950.
2 This principle, with the inclusion of the words 'at the lowest cost', was advocated as early as 1928 by L.P. Alford.
3 This is called a 'single' setup, analogous to setting the tee in preparation to drive a golf ball.

References

HITOMI, K. (1982) *Multi-product, Small-batch Production—A Text* (in Japanese) (Tokyo: Business & Technology Newspaper Company).
HITOMI, K. (1989) Non-mass, multi-product, small-sized production: the state of the art, *Technovation*, **9** (5).
MAXCY, G. and SILBERSTON, A. (1959) *The Motor Industry* (London: Allen & Unwin).
MONDEN, Y. (1983) *Toyota Production System* (Norcross, GA: Industrial Engineering and Management Press).
MOORE, F.T. (1959) Economies of scale: some statistical evidence, *Quarterly Journal of Economics*, **73** (2).
MURAMATSU, R. (1976) *Production Management* (in Japanese) (Tokyo: Asakura Shoten), pp. 97–104.
OHNO, T. (1978) *Toyota Production Systems* (in Japanese) (Tokyo: Diamond-sha).
STAR, M.K. (1965) Modular production: a new concept, *Harvard Business Review*, **43** (6).
TIMMS, H.L. (1966) *The Production Function in Business* (revised edn) (Homewood, IL: Irwin), pp. 25–30.
TIMMS, H.L. and POHLEN, M.F. (1970) *The Production Function in Business—Decision Systems for Production and Operations Management* (3rd edn) (Homewood, IL: Irwin), pp. 21–28.

Supplementary reading

HOUNSHELL, D.A. (1984) *From the American System to Mass Production, 1800–1932* (Baltimore: Johns Hopkins University Press).

CHAPTER FIVE

Integrated Manufacturing and Management Systems

5.1 Basic Functions and Structures of Management Systems

Management System

It is clear that manufacturing systems are deeply concerned with management, which achieves organisational objectives by utilising production resources. In order to utilise those resources efficiently, a systems approach is considered to be effective. This is the *management system.*

Since various functions and operative works are done in a management system, it is appropriate to divide the structure of this complex management system into several subsystems from two aspects—hierarchical and functional, as shown in Figure 5.1.

Hierarchical Structure of the Management System

This structure is based upon the level of management. Basically the three levels are as follows (Hitomi, 1972):

(1) *Strategic* (or *administrative*) *planning level*—a non-routine aggregate planning relating the system and the environment. It includes: establishing the firm's philosophy; recognising and evaluating the system's environment; establishing management policies and objectives; clarifying management strategies; organising personnel; determining new products; long- and medium-term profit planning and budgeting; facilities planning and investing; planning new plant installation; wide-range programming for attaining the management goals; allocating overall resources; evaluating production performances, etc.

(2) *Management level*—routine tactical planning and control based upon the strategic decisions made at the strategic planning level. It includes: establishing functional control objectives for attaining the management goals; planning and selecting the course of action; acquiring resources and allocating them into divisions and departments; coordinating interdivisional interests; preparing

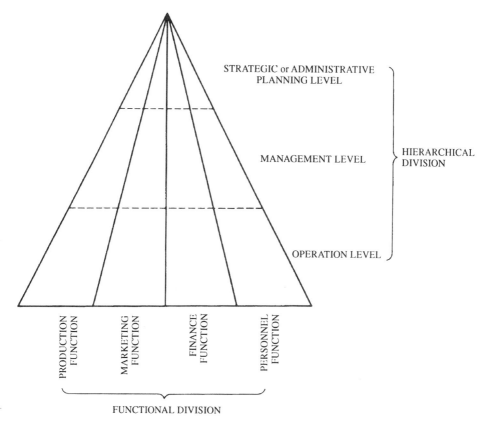

Figure 5.1 A management system can be viewed from two aspects: hierarchical and functional structures.

detailed functional work programmes; evaluating performance of operative work and determining modification plans for activities deviating from the established goals and standards, etc.

(3) *Operation level*—routine execution, namely operative work based upon the tactical decisions made at the management level. It includes: implementation of operative work according to the indicated programme; routine processing of transactions; modifying small errors of actions; measuring performances; preparing performance reports, in particular, about exceptional or unusual performances, etc.

Functional Structure of the Management System

This structure is viewed from basic functions constituting management activities; basically, the following four functions play fundamental roles (Timms and Pohlen, 1970).

(1) *Production function*. This provides form, time and place utilities, directly, by converting raw materials into finished products, thereby supplying outputs (goods and services) to the market.

(2) *Marketing* (or *sales*) *function*. This directly provides possession utility by selling the products in the market through advertising and merchandising activities.
(3) *Personnel function*. This acquires operative and managerial human skills and aids in their use.
(4) *Finance function*. This acquires capital funds for business activities and aids in their use.

5.2 Basic Framework of Integrated Manufacturing Management Systems

Manufacturing Management System

Structures and functions of a manufacturing management system differ, depending upon the types of economic goods that the manufacturing firm produces as outputs (products) and on types of production.

In this section a basic, general pattern of the integrated manufacturing management system (IMMS) is explained, following a framework depicted in Figure 5.2 (Hitomi, 1978).

Logistic System

The most important function in the manufacturing firm is the production function which is concerned with the material flow; that is, resources of production, especially raw materials, are acquired from the external environment, and converted into finished products that are brought to the market as commodities. This is the 'logistic system', which is a chain of procurement, production and distribution, and sales and, in addition, inventory as follows (also see Figure 1.2).

(1) The *procurement* (or *purchasing*) *subsystem* acquires production resources from the external environment, especially raw materials and parts, required for manufacture in specified time periods, taking raw-materials inventory into account.
(2) The *production subsystem* produces finished products (commodities) required by the sales division, in relation to finished-product inventory.
(3) The *sales subsystem* delivers required quantities of commodities, with reasonable prices within the due dates or at appropriate delivery periods, according to consumers' orders or market demands for customer satisfaction.
(4) The *inventory subsystem* plays the role of buffer for a smooth material flow, in the form of raw-materials inventory, work-in-process inventory, and finished-product inventory. These are concerned with storage between material procurement and production subsystems, between stages of the production process, and between production and sales subsystems, respectively.

Operation Planning System

Effective production implementation conducted in the logistic system is designed through the operation planning system. The major functions in this system are

INTEGRATED MANUFACTURING AND MANAGEMENT SYSTEMS

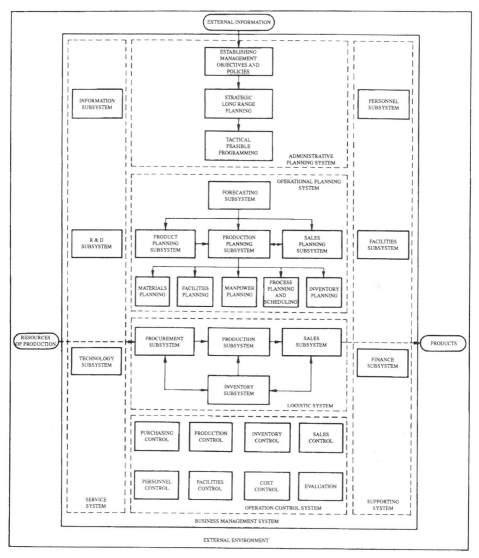

Figure 5.2 A basic framework of the integrated manufacturing management system (IMMS), which consists of a basic production function–logistic system to deal with the material flow, operation planning and control systems to manage the operative production work, and an administrative planning system concerned with strategic decision-making. Service and supporting systems help perform effective operative and managerial work.

production planning, sales planning, and product planning, which are undertaken according to the results of forecasting.

(1) The *forecasting subsystem* forecasts the future demands/orders.
(2) The *product planning subsystem* designs competitive products.
(3) The *sales planning subsystem* plans the future marketing activities and sales volume in connection with production planning.
(4) The *production planning subsystem* establishes detailed feasible plans/ schedules for operating the logistic system in order to produce the quantity of

specified kinds of products established as sales goals. Materials planning, facilities planning, manpower planning, process planning and scheduling, and inventory planning are carried out.

Thus, procurement of resources of production, especially raw materials, plant layout, man–job assignment, conversion process of raw materials into finished products, etc. are designed in this system.

Operation Control System

Whilst the logistic system is actually implemented according to the operation plan established in the operation planning system, unforeseen situations often occur, such as the breakdown of production facilities, absence of workers, delivery delay of raw materials, etc. As a result, the logistic system cannot be executed as designed; the operation control system conducts follow-up or control. An audit of the progress and modification of the deviation of actual performances from the planned standards are performed:

(1) controlling the flow from raw materials to products—purchasing control, production control, sales control, and inventory control;
(2) controlling production resources—personnel control, facilities control, and cost control.

Administrative (or Strategic) Planning System

This top management role supervises the stages of planning, implementation, and control done in the logistic system and the operation planning and control systems. It issues appropriate directions and orders to such lower levels of the system. In this system proper management objectives and policies have to be established, so that the firm can grow under dynamically changing external and internal environments, social conditions, political circumstances, economic trends, and market situations.

(1) *Strategic long-range planning* executes a long-range management decision with cost-benefit analysis.
(2) *Tactical feasible programming* calculates a proper profit by realising the strategic plan. The programs constructed are directed to the lower-level systems, where the feasible programs are put into practice.

Service System

This executes, smoothly and effectively, the various functions of the administrative planning system, operation planning and control systems, and the logistic system.

(1) The *information subsystem* provides accurate information for various decision-making points in the system at the right times.
(2) The *technology subsystem* handles professional knowledge of production technology, management technique, industrial engineering, quality engineering, value analysis, etc. as a staff activity.
(3) The *research and development (R & D) subsystem* deals with developing new products.

Supporting System

This backs up the above-mentioned systems by acquiring necessary resources (men, machines, and money), and aiding in their effective use.

(1) The *personnel subsystem* acquires and allocates the operative and managerial human skills.
(2) The *facilities subsystem* procures production facilities, and particularly constructs new plant.
(3) The *finance subsystem* acquires and employs capital funds necessary for business activities.

Integrated Manufacturing Management Systems

The integrated manufacturing management system of a manufacturing firm is operated by the unification and coordination of the various interrelated functions and activities described above, the whole adapting itself to changes in the external environment (markets, competitors, etc.).

5.3 Framework of an Integrated Manufacturing System

5.3.1 Functions Concerning Manufacturing

Essential Functions in Manufacturing

'Manufacturing' has its most basic meaning when the products manufactured are delivered as commodities to the market and are bought by consumers. Prior to 'production' of the products, 'design' of the products and their components (parts) is required. In addition, 'management' (often including 'strategic planning' functions) is the function of identifying the needs of the market and transmitting them to the design and production divisions. The relation of these three activities in manufacturing is depicted in Figure 5.3.

Four Major Functions in an Integrated Manufacturing System

The following four main functions are to be executed in an integrated manufacturing system (Hitomi, 1995):

(1) The *production function* converts the resources of production (especially raw materials) into products.
(2) The *design function* plans and draws the products/parts.
(3) The *management function* plans and controls design and production activities.
(4) The *strategic planning function* consists of decision-making of strategic issues.

5.3.2 Outline of Integrated Manufacturing System Activities

Basic Activities in an Integrated Manufacturing System

An outline of the above four fundamental functions is explained in the following, using Figure 5.4. (Terms in parentheses refer to the boxes in this diagram.)

(1) The 'production function' is the key activity in the manufacturing industries. This procedure (*production*) consists of fabrication of parts and assembly of

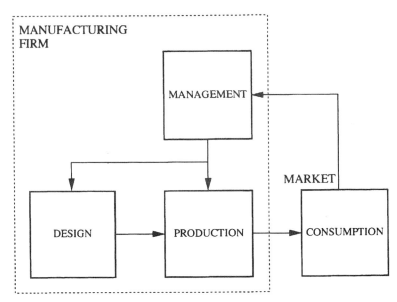

Figure 5.3 Relation of three basic functions in a manufacturing firm—design, production, and management—to meet the market/consumers' needs.

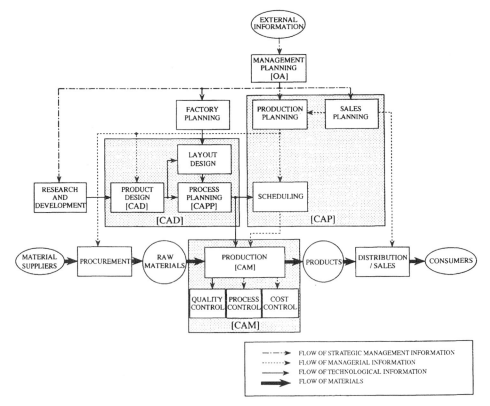

Figure 5.4 Framework of relation of the basic functions involved in manufacturing—flow of materials, technological information, managerial information, and strategic management information. Computer aids—CAD, CAM, CAP, and OA are also indicated.

products in the mechanical industries or a series of unit operations in the process industries. The raw materials supplied (*procurement*) are changed in shape, pass through the production stages, are eventually stored in the warehouse and become the final products (in some cases the materials for subsequent manufacturing firms). This is called the 'flow of materials'.

(2) The 'design function' designs and draws the product and parts (*product design*) and decides the procedure by which to produce them (*process planning*). On some occasions required production facilities (machine tools, jigs and tools) are laid out (*layout design*). This function relates to intrinsic production technology. Hence it is called the 'flow of technological information'.

(3) The 'management function' establishes a master plan to determine the kinds of products and amounts to be manufactured (*production planning*) according to the long-term forecast and marketing (*sales planning*). Based upon this production plan, it is decided which machine performs what operation (*scheduling*). Following this schedule the actual operation is performed. However, this activity might be disturbed for many reasons including machine breakdown, absenteeism, late delivery of raw materials, tool failure and malfunction of the machines. Modifying the difference between these actual results and the plan and schedule are control activities such as:
  (a) *process control*, concerning control of due dates and production amounts;
  (b) *quality control*, concerning the quality of the product and the control of processes;
  (c) *cost control*, concerning the total production cost.

The above-mentioned functions are related to management, so a series of these functions is called the 'flow of managerial information'.

(4) The 'strategic planning (or management) function' establishes a strategic plan in both short and long terms, based upon the external information. *Management planning* in its wider meaning decides the long-term business plan, profit planning, product planning, personnel planning and factory planning. It serves as the basis of production and sales planning. *Sales planning* is closely linked to production planning and activates *distribution and sales* of the products manufactured through the production division. The strategic management function also makes decisions concerning factory location and establishes production processes (*factory planning*). This relates directly to layout design in the design function. On the other side, *research and development* (R & D) concerning creation of new products is promoted. This links to product design in the design function.

The above activities construct the 'flow of strategic management information'.

Four Important Flows in Manufacturing and CIM

As mentioned above, manufacturing is executed through the flow of materials and the flow of information; the latter is further divided into the flow of technological information, the flow of managerial information, and the flow of strategic management information as was mentioned in Section 3.4.

These four flows are combined as shown in Figure 5.4 and integrated into a whole, generally via computer networks with a common data/knowledge base, such that the organisation can be operated efficiently and economically. This is a structure and procedure for computer-integrated manufacturing (CIM) and information systems, which will be described in Chapter 23.

References

HITOMI, K. (1972) *Decision Making for Production* (in Japanese) (Tokyo: Chuo-keizai-sha), pp. 32–52.

HITOMI, K. (1978) *Production Management Engineering* (in Japanese) (Tokyo: Corona), Sec. 6.1.

HITOMI, K. (1995) Computer-integrated manufacturing: its concept, structures, and for future production perspectives, in Sakuma, A. and Oniki, H. (eds), *Impacts of Information Technology on Management and Socioeconomics* (Tokyo: Maruzen), Chap. 2.

TIMMS, H.L. and POHLEN, M.F. (1970) *The Production Function in Business—Decision Systems for Production and Operations Management* (3rd edn) (Homewood, IL: Irwin), pp. 6–7

Supplementary reading

HITOMI, K. (1975) *Theory of Planning for Production* (in Japanese) (Tokyo: Yuhikaku).

HITOMI, K. (1990) *Principles of Manufacturing Systems* (in Japanese) (Tokyo: Dobunkan).

PART TWO

Process Systems for Manufacturing

This part describes basic principles of production process technology—'material flow' and 'technological information flow'. The 'flow of materials' is the key activity of manufacturing. Its essentials, together with the 'flow of technological information', are introduced in Chapter 6.

Fundamentals of a chain of functions in technological information flow—product planning and design for new product development (Chapter 7), process planning and design dealing with the effective conversion of raw materials into finished products (Chapter 8), and layout planning and design concerning spatial allocation of production facilities (Chapter 9)—are mentioned.

Logistic planning and design (Chapter 10) introduces how to solve the transportation problem by linear programming. The travelling salesman problem is also explained.

Manufacturing optimisation (Chapter 11) decides optimum machining conditions for single-stage manufacturing and for multistage manufacturing systems, based upon three fundamental criteria—the minimum time, the minimum cost, and the maximum profit rate.

CHAPTER SIX

Material and Technological Information Flows in Manufacturing Systems

6.1 Logistic Systems

Structure of Logistic Systems

The activity of manufacturing is regarded as the 'flow of materials'. From the macroeconomic and social viewpoints this is the *logistic system*, consisting of the following three subsystems as shown in Figure 6.1.

(1) The *materials-supply system* deals with transportation of raw materials and parts from the raw-material suppliers to the manufacturer.
(2) The *materials-handling system* deals with handling and transfer of workpieces inside the factory.
(3) The *physical-distribution system* transports and distributes produced goods to ultimate customers.

6.2 Material Flow

Basic Activities Inside a Factory

Concerning the material flow inside a factory, the manufacturing system conducts the following three basic activities, shown as the 'manufacturing system' in Figure 6.1:

(1) *conversion*—function of converting the form of materials or workpieces;
(2) *transportation* (or *transfer*)—function of moving workpieces between workstations;
(3) *storage*—function of elapsed time (no change of form and place of the workpiece).

These activities generate the form, place, and time utilities mentioned in Section 1.4.

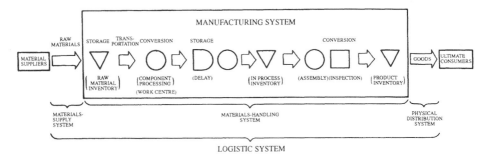

Figure 6.1 A total system of material flow in manufacturing. This is made up of the material-supply system between the raw-material suppliers and the manufacturing system, the materials-handling system between production stages inside the factory, where three basic activities (conversion, transfer, and storage) are performed, and the physical-distribution system between the manufacturing system and the ultimate customers.

Conversion

This is done through activities called *operations*. An operation is done at a specific workstation or work centre, where machine tools, jigs and fixtures, tools, operators, etc. are suitably arranged. A product or job is normally completed through a series or a set of operations, which constitutes a multiple-stage production process.

Types of Operations

Typical form-conversion operations included in the field of manufacturing are as follows:

(1) *Transformational operations* produce raw materials for manufacturing, such as iron and steel, plastics, etc., by changing their nature or extracting them from ores.
(2) *Metamorphic operations* produce parts/products by an accompanying change of form, shape, or structure without quantitative mass change of the original materials; for example, casting, forging, moulding, extrusion, powder metallurgy, metal-forming, press-working, high-energy forming, etc.
(3) *Joining operations* produce parts/products by attaching another material to the original workpiece; for example, welding, adhesive bonding, plating and coating, painting, etc.
(4) *Removal operations* produce parts/products by taking off a portion from the original workpiece as a chip, thus reducing the volume and weight, as in metal cutting, grinding, stamping, precision machining (lapping, superfinishing, electro-discharge machining, supersonic machining, electron-beam machining, laser-machining, etc.), cutting-off by gas, arc, or plasma, filing by hand, etc.
(5) *Treatment operations* harden, quench or clean the surfaces of the work material by heat treatment, surface treatment, etc.
(6) *Assembly operations* produce products by putting together the discrete parts or components of the product, so as to facilitate disassembly.
(7) *Supplementary operations* consist of additional operations such as quality tests, inspection, packaging, etc.

Raw materials are converted into products, changing their form and structure by the various conversion processes mentioned above on passing through the manufacturing system, thereby successively increasing form utility and value.

Transportation

Transportation or transfer is often called *materials handling* (*MH*). Usually all the required operations converting the raw material into the product can not be performed at one workstation; hence, materials handling is necessary. Although this activity generates place utility, it is not a direct operation; it adds cost but not value. Hence reduced materials handling will increase productive efficiency. The amount of materials-handling activity in production processes is fairly large at present.

--- **EXAMPLE 6.1** ---

The MH cost typically comprises 30–75% of the total cost. Effective materials handling can reduce this cost to 15–30% (Sule, 1988). □

--- **EXAMPLE 6.2** ---

The transfer and waiting times can be 95% of the total production time; only 5% is on the machine. Of that 5% of the total production time setup, inspecting, and idle times share 70%. Hence actual machining is only 1.5% of the total production time (Carter, 1971). □

--- **EXAMPLE 6.3** ---

The ratio of the physical distribution cost (procurement 12%, transfer inside the factory 60%, sales 28%) to sales was about 10% in Japan (1975) and 20% in the US (1963) (Nishizawa, 1977). □

Reduction of Transportation

From the above discussion it is necessary to eliminate transportation for productivity increase. This can be achieved in the following ways:

(1) Reduce the transfer distance by establishing a smooth material flow and an optimal plant layout, or by performing several conversion operations on a single stage, utilising multiple functional machine tools, such as a machining centre.
(2) Decrease the number of transfer activities; handling materials as a group with standard shape, volume, weight, etc.; using the 'unit-load' method to establish a maximum transportation unit.
(3) Determine an optimal transfer route and speed by constructing a straight continuous route with minimal delays, crossings, and counterflow and by setting a balanced or synchronised transfer activity, generating no in-process inventory. This can be more easily achieved by mechanised or automated transfer than normal means.

Storage/Inventory

Storage is stock, for example, at the warehouse, temporary stay between workstations (delay or waiting), etc. In general, it is associated with imbalance between the conversion and transportation functions. Stagnation of material flow can occur between the time of raw-material supply and the start of its use for manufacture, between two successive stages of the production processes, or the time to finish the product and to ship it to the market. These are called raw-materials inventory, [work-]in-process (WIP) inventory, and finished-product inventory, respectively. Thus *inventories* play essential roles as buffer stocks for a smooth, flexible material flow, and to generate time utility.

Role of Inventory

Finished-product inventory absorbs the difference between the individual activities of production and marketing. It enables shipping of products on time according to the customers' orders and wants, while, on the other hand, setting the manufacturing system for its stable and optimum utilisation.

Another utilisation of the storage function is in the 'parts-oriented production system' introduced in Section 4.3. This system actively uses the parts inventory, which plays a more important role than just storage of parts as in other manufacturing systems.

On the other hand, the just-in-time (JIT) system, which was also described in Section 4.3, attempts to lessen or eliminate any stock in the logistic system. However *keiretsu* (obedient subcontractor)-type companies are forced to hold warehouses near giant corporations so as to be able to deliver even a single part at any time requested. (Notice that this requirement of JIT for very frequent deliveries can introduce transportation problems and possibly environmental issues.)

—— **EXAMPLE 6.4** ——

(*Keiretsu*) Subcontractors to Toyota Automobile Industries are: 231 first level (large and medium-sized), about 4000 second level (small) and over 30 000 third level (very small). □

6.3 Technological Information Flow

Functions Constituting the Technological Information Flow

For smooth material flow the proper establishment of the 'technological information flow' is required prior to actual production. This flow constitutes a chain of functions: 'product planning and design', 'product process design' and 'plant layout design', as depicted in Figure 5.4. These three functions are explained in detail in the following three chapters.

References

CARTER, C.F. (1971) Trends in machine tool development and application, *Second International Conference on Product Development and Manufacturing Technology*, Paper No. 4.

NISHIZAWA, O. (1977) *Cost Accounting for Physical Distribution* (in Japanese) (Tokyo: Chuo-keizai-sha), Chap. 7.

SULE, D.R. (1988) *Manufacturing Facilities* (Boston, MA: PWS-KENT Publishing), p. 189.

Supplementary reading

KALPAKJIAN, S. (1989) *Manufacturing Engineering and Technology* (Reading, Mass.: Addison-Wesley).

NIEBEL, B.W., DRAPER, A.B. and WYSK, R.A. (1989) *Modern Manufacturing Process Engineering* (New York: McGraw-Hill).

CHAPTER SEVEN

Product Planning and Design

7.1 Product Planning

What is Product Planning?

End products to be produced by the manufacturing system are decided as a result of strategic planning in phase I of the production management procedure mentioned in Section 3.4 (also see Figure 3.5). The function for this decision-making, conducting research and developing new material goods that meet the market needs, is *product planning*. It is a continuous function in the dynamic environment of a competitive situation, since any single product generally possesses a finite life cycle.

Product Life Cycle

The *product life cycle*, the period between conception of a product and the time when it ceases to be profitable, tends to be decreasing owing to a high tendency for multi-product, small-batch production and diversification, as mentioned in Chapter 4.

--- **EXAMPLE 7.1** ---

Ford's model T car lasted 20 years (1908–27) and 15 million were produced and sold. Product life cycles for today are: two years for sewing machines, one year for air conditioners, six months for telephones, TV sets, and only three months for Japanese word processors. Life cycles are longer for capital goods; e.g. 2–3 years for machine tools (*Nikkei Mechanical*, 30 April 1990). □

As shown in Figure 7.1 the sales volume and revenue for a certain new product tend to rise in the initial period after its introduction. There follows a period of growth as the customers' recognition and acceptance have increased with a rapid increase of purchasing power. In the period of maturity the sales volume still increases; however, as additional competitive products appear in the market, the profit tends to decrease due to reduction of the selling price. In the saturation and

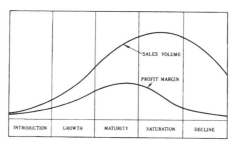

Figure 7.1 A life cycle for a product. Sales volume and revenue rise in several periods after introduction and then decline; hence, modification of the product or development of a new product is required as a function of product planning.

decline periods, freshness of the product may be lost and the volume of sales and hence the profit decrease. In such periods, efforts should be directed toward cost reduction with utilisation of effective methods such as value analysis, and the sales volume and profit should be maintained by sales promotion and by performing minor changes or modification of the present product with new ideas.

Planning Systems for a New Product

In the planning system for a new product, as depicted in Figure 7.2, the market needs predominate. With information based upon market research, or data from the marketing and sales departments of the firm, the 'general specifications' of a new product are established. When ideas for the product are proposed for the market commodity, the 'technical specifications' can be determined by the creative abilities of product research and development (R & D), thus leading to the final design. At this stage, some change of the product's general specifications previously determined may be allowed. The reason is that the design ideas for a new product are not always directly related to reality. Therefore, the decision made to bring a new product to actual production depends upon whether the compromise made for practical manufacturing receives market acceptance.

Design tools such as quality function deployment (this is explained in Section 17.2) can be used to trade off between market requirements and the realities of achievable design and manufacturing characteristics. Following research and development, *product design*, which determines the detailed shape of a product, is completed at the final stage of product planning so that the product can perform a specified function based upon the technical specifications for the product.

The product design is converted into products through production. The manufactured goods are distributed to the market, then to the ultimate consumers.

Product Research and Development (R & D)

New products are created through a chain of functions: product research, product development, and product design.

(1) *Product research* includes basic and applied researches.
    (a) Basic research is the discovery of new scientific laws/knowledge; it often does not lead to a concrete product.

(b) Applied research is done to create a new product within time and cost restrictions.
(2) *Product development* aims to build prototypes/models of the new product generated in product research. This is followed by product design which is described in the next section.

Industrial Property and Product Liability

When a new product is invented a patent should be taken out. The patent for the newly invented product is one of these *industrial properties*. Others are utility models, designs and trademarks. *Intellectual property* is also subject to copyright protection. On the other hand, 'know-how' lies in the realm of industrial and technological secrets, and is not covered.

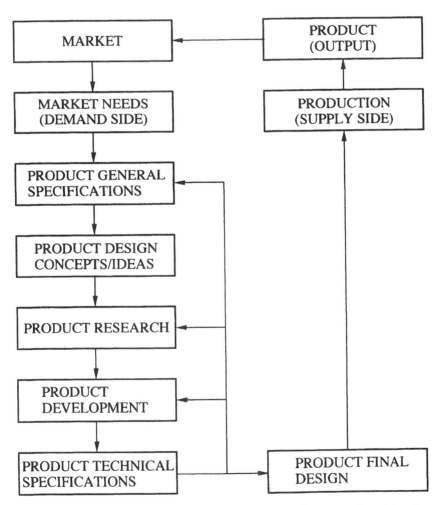

Figure 7.2 Planning new products: market needs are reflected in research and development, which result in product design. Production is run based upon the design, and the produced goods are delivered to the market.

If a product sold at the market has caused personal/physical damages during its use, product designers/companies have to assume legal responsibility—*product liability*.

Software and Services Aspects of Products

The inclusion of software aspects is increasingly required for present-day products (hardware); e.g. computers (hardware) do not operate without useful programs (software).

—— EXAMPLE 7.2 ——————————————————————————————

The production of software compared with hardware is 1.71 times in the United States and 1.44 times in the EU, and only 0.64 times in Japan (Input Company: *Japan Economic Newspaper*, 24 October 1992, p. 22). ☐

As economies progress, the industrial structure of a nation changes from primary industry to secondary industry (goods production) and further to tertiary industry (service production). Greater profit margins are obtained in service production (commerce) than in goods production (agriculture and manufacturing), as was described in China as early as in 68 AD and by W. Petty in 1691. This is now called the *Petty–Clark law*.

—— EXAMPLE 7.3 ——————————————————————————————

In 1991 the American Telephone and Telegraph Company (AT&T) earned more from communications services (profit margin: 13.8% of the revenue) than from products and systems (only 3.0%) *(Business Week*, 20 January 1992). ☐

7.2 Product Design

7.2.1 Fundamentals of Product Design

What is Product Design?

Product design is the function of creating drawings or other graphical representations of products/parts which performs necessary functions based on the 'technical specification', which has been established in the product planning stage.

Three Aspects of Product Design

Product design contains the following three aspects/phases:

(1) *Functional design.* This is the design expressing the shape and defining the function which the product to be manufactured must possess. *Function* means functional characteristics which the product should possess while it is in operation; hence the use value is generated and the possession utility is provided to the ultimate users. Function is divided into:
    (a) basic function, without which there is no need for the product, and
    (b) secondary function, which supports the basic function and often brings about product differentiation.

Product differentiation is the production of a variety of products and it creates new markets, called *market segmentation*. Additionally, the product may often involve an 'unnecessary' function, which has no functional characteristics. This function is often built into the product in an attempt to raise attractiveness to the market, but should be eliminated as much as possible for cost reduction and resource saving.

(2) *Production design* (or *design for manufacture[ability]*). This aspect of design is concerned with easy and economical production relative to the conversion of raw material into the finished product, from the viewpoints of both manufacturing technology and production process planning. The effectiveness of this design phase is closely related to productivity increase and production cost cutting. *Design for disassembly*, which aims at easy disassembly of the products at the end of their useful life to promote materials recycling, is becoming more and more important from the environment-preserving and the recycling viewpoints.

(3) *Industrial* (or *marketing*) *design*. This is the design of form, shape or styling, which is used to influence people to buy the product because of its eye-appeal. It also includes design for ease of use of the product by the ultimate customers, based upon the human–engineering aspect (*ergo design*). In this design stage, care should be taken to ensure that aspects of functional and production design are not overlooked or unacceptably compromised. Industrial design often results in increased cost and should provide secondary and 'unnecessary' functions which will increase market appeal and sales revenue.

Important Features in Product Design

The following features are emphasised in product design:

- effective function;
- quality, reliability, and maintainability;
- economical production;
- easy handling by users (human–engineering aspect);
- aesthetic appearance;
- resource-saving aspects;
- minimum disposal.

7.2.2 Product Quality

Important Factors Involved in Finished Products

The most important factors to be considered in the finished product are:

- *quality* to perform the functions and have those features which the customer requires and for which the product is designed;
- *reliability* to maintain such quality over the product life time;
- *safety* to avoid danger (hazard) from the use of the product.

Costs tend to rise with increasing quality and reliability. Therefore, as indicated in Figure 7.3, a suitable level of quality and reliability should be decided in connection

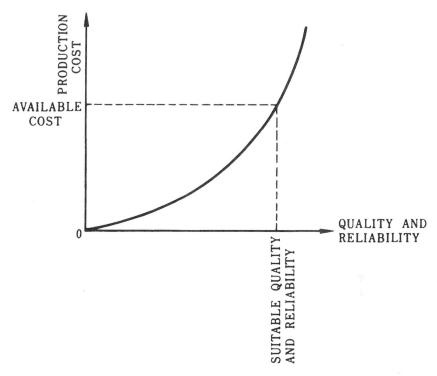

Figure 7.3 Relationship between quality and reliability of a product and production cost required. Production cost increases excessively as quality and reliability increase.

with the cost that is acceptable. However, in a highly competitive market environment, design must seek to improve both quality and cost.

What is Quality?

This is difficult to define, but in a word, *quality* is the 'goodness' of a product, how well it satisfies customer requirements, and includes the dominant characteristics which distinguish one product from another. Two kinds of quality are:

(1) measurable quality—characteristics which can be expressed quantitatively and are required to be greater than a certain satisfactory level;
(2) quality that is not physically quantified—perceived to some extent by sensory tests.

Quality Design

The product quality level is a matter of product design, and also of process design. The former is the design quality level, and the latter, the production quality level.

(1) *Design quality level* is directly related to those market segments which top management wishes to serve; quality as high as possible relative to the price is desirable, or a large product value. Product designers are prone to err by designing products of excessive quality. Since this causes excess cost and impaired marketability, it is desirable to prevent this error by setting clear general specifications.

(2) *Production quality level* is concerned with the number of rejected or inferior goods produced during the production process. Quality engineering, which will be mentioned in Chapter 17, is a powerful tool in this process.

7.2.3 Product Reliability

What is Product Reliability?

Product reliability is concerned with failure-free operation and the ability of the product to keep its quality and satisfactorily perform its function throughout a specified period of time under specified working conditions. It is important for this purpose to prevent failure (high reliability) and to design for ease of repair when failure has occurred (maintainability). Thus the 'product life', which is the duration of satisfactory operation given the necessary planned technological servicing, including various types of repairs, is increased.

Mathematical Expression of Product Reliability

The reliability of a product that is composed of N individual components each with reliability R_j ($j = 1, 2, \ldots, N$) is expressed as follows:

- *Lusser's reliability formula* for the *serial system*—N components are in series (Figure 7.4(a)):

$$R_s = \prod_{j=1}^{N} R_j \tag{7.1}$$

- reliability of the *parallel system*—N components are redundantly connected (Figure 7.4(b)):

$$R_p = 1 - \prod_{j=1}^{N}(1 - R_j). \tag{7.2}$$

Reliability Design

There are several design methods to increase the product reliability as follows.

(1) *Redundant systems.* It is clear from (7.2) that the reliability of a redundant (parallel) system increases as the number of components increases; hence, to raise redundancy is one method for establishing a high system reliability. But it should be noted that the redundancy causes increase in the weight of the product and the total production cost, together with an increase in the occurrence of a human error due to a greater complexity of the product.
(2) *Modular methods.* To enhance *maintainability*, which deals with easy and quick repair of a failed part/product, it is important to design the product so that repair or renewal or replacement of worn parts of the product can be made easily. *Modular* (or *building-block*) *methods*, in which independent/autonomous components/units are employed to construct a product are one way of achieving this aim.

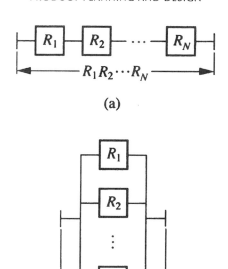

Figure 7.4 System reliability for (a) a serial system and (b) a parallel system.

── EXAMPLE 7.4 ──

Common parts/subassemblies used to construct a car are modules; they are easily replaced in case of failure. ☐

(3) *Fail-safe design.* For increasing reliability and for safety design, *fail-safe design* is a design method wherein, for example, if the load that a product incurs is supported by several parts, even if a part fails, the other parts will be rearranged to continue to support the load, thus catastrophic failure of the product is prevented. Even when failure occurs, the product function will remain safe. This design method is useful in cases where failure would result in a large monetary loss or human danger.

── EXAMPLE 7.5 ──

A red signal should appear every time a control loop of the railway signal fails. ☐

(4) *Foolproof design.* Such design incorporates a mechanism by which human error is prevented.

── EXAMPLE 7.6 ──

Most cameras now have devices which prevent the user making double exposures. ☐

7.2.4 Eco-design

Concept of Ecological Design and Life-cycle Assessment

From now on *eco-design* or *green design* is crucial in order to lessen the use of natural resources and prevent public nuisance (disutility). This is achieved by increasing recycling of parts after use and by reducing dumping of wastes. It is done by taking the life cycle of a product: raw-material procurement–production–consumption–recycling/dumping into account in the early stage of design—design for the product life cycle/environmental elegance (Overby, 1990).

The evaluation of the total burden to the environment through the product life cycle is called *life-cycle assessment* (*LCA*). The product design is required to minimise the total cost throughout its life. (Also see Section 36.4.)

7.2.5 Design Cost Reduction

Design Costs

The cost required in the design itself is typically only 5% of the total cost; however, bad design can influence the total cost by 70% (Boothroyd, 1988). Hence good design cost reduction is important. Effective techniques for reducing costs in the design stage are mentioned in the following.

Simplification Design

Standardisation and simplification, which are two key factors in 'industrial engineering (IE)', as mentioned in Section 4.3, are useful to economical design and production. Interchangeability, product reliability and maintainability are increased by standardising products, parts, raw materials, process routes, etc. As was shown by equation (7.1), reliability decreases as the number of serially connected components increases; hence, the number of components for a product should be reduced and the mechanism and structure of the product should be simplified as much as possible. This is called *simplification design*. This design method will also improve the manufacturability of a design.

Value Analysis (VA)

Usually there are several alternatives for part structure, raw materials, work flow, and other factors in the manufacturing process of the product to produce a specified function. Raw materials should be selected by taking both material cost and manufacturing cost into account. (See Example 8.6 in Section 8.2.)

Value analysis/engineering (*VA/VE*) is an effective tool for this purpose; that is, a complete system for identifying and dealing with the factors that add cost or effort in making products but do not contribute to the product's value (Miles, 1972). This is a useful tool in resource-saving design and production, in that this technique generates the highest value for a part or product having the necessary function at the least cost. Here 'value' is defined:

$$\text{value} = \frac{\text{quality}}{\text{cost}}. \tag{7.3}$$

PRODUCT PLANNING AND DESIGN 99

Accordingly, the higher the quality of goods produced at the specified cost, or the less the cost of goods produced with the specified quality, the higher the value of those goods.

Five steps are taken in the process of value analysis:

(1) what is the product (selection of the product)?
(2) what is the function of the product (investigation of the function and removal of unnecessary functions)?
(3) how much does the product cost (test of the cost)?
(4) is there any alternative (search of the alternative)?
(5) how much does the alternative cost (comparison of the costs)?

Then a product/part or a process route with less cost but providing the required function is chosen.

7.2.6 Group Technology (GT)

Group Technology Defined

Group technology (GT) is a technique and philosophy to increase production efficiency by grouping a variety of parts having similarities of shape, dimension, and/or process route, as mentioned in Section 4.3. This contributes to efficient multi-product, small-batch production as well as reducing design costs. In addition, GT is now applicable to production management problems, such as optimum machine loading and production scheduling (Ham *et al.*, 1985).

Group technology was first proposed and developed in Russia by Mitrofanov (1966), who noticed the similarity of a number of different parts and classified them into several group cells (parts families); e.g. as shown in Figure 7.5. The principles and practice of GT have been applied throughout the industrialised world with significant benefit to manufacturing and design.

Advantages of GT

There are several advantages of using group technology as follows:

(1) *Mass-production effect*. By grouping various different types of parts, with the use of an effective parts-classification coding system, the type of production is changed from jobbing production to lot or mass production.
(2) *Possibility of flow-shop pattern*. With GT the work flow is expected to be a flow-shop pattern; hence, the times and costs for materials handling and waiting between stages are reduced, and process planning and scheduling are simplified. This scheduling is called [*the*] *group scheduling* [*technique*] (Hitomi and Ham, 1977), which will be explained in Chapter 28.
(3) *Reduction of setup time and cost*. A most significant advantage of applying GT to production activity is that setup time and setup cost for each job are reduced, because several jobs are grouped and processed in sequence, and the same jigs and tools may be employed. In practice, special tooling and fixtures can be effectively employed to reduce setup time.

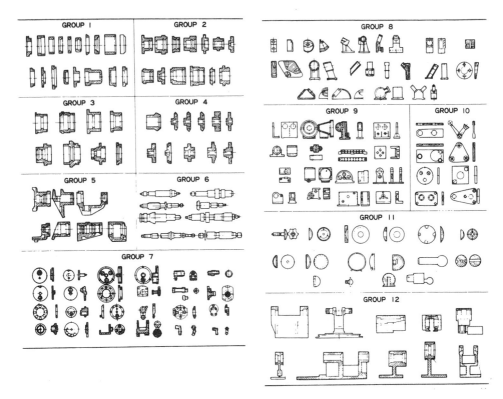

Figure 7.5 Grouping many different parts into several group cells by similarity of shape (after Mitrofanov, 1966). This is a case for a turning operation.

(4) *GT (or cellular) layout.* Since the flow-shop pattern is expected when GT is applied, it is assumed that the machine tools will be laid out as a group (see Figure 9.2(c)). This considerably simplifies the material flow and handling.
(5) *Economy.* GT significantly saves production cost. Since in GT the parts with similar shape, dimension, and/or process route are grouped and fabricated in the same lot, it sometimes occurs that a part is made far ahead of its due date, or perhaps late, incurring inventory cost or some amount of penalty. This drawback can be avoided by an appropriate GT program.

Classification and Coding Systems

The fundamental requirement of GT is the *classification and coding system*. It describes through a code number the characteristics of the part by its geometrical shape and/or process route (setup, sequence of operations such as metal fabrication, machining, and assembly, inspection and measurement) for producing the part. By gathering parts with the same code number and grouping them into a group cell (*parts family*), design, process planning, manufacturing, and cost estimation are made more systematically, often with the use of computers.

In group technology application, the system for classifying and coding parts plays an essential role in its success. However, no one coding system seems to be suitable

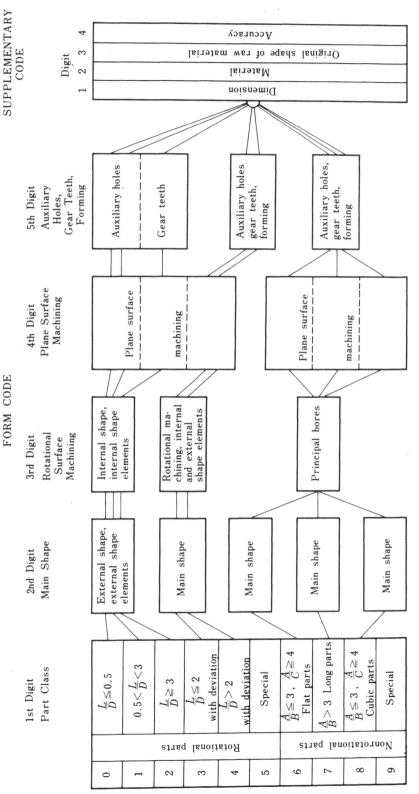

Figure 7.6 A representative classification and coding system—the Aachen system (after Opitz (*Werkstücksystematik und Teilefamilienfertigung*, 1965) reproduced by permission of Verlag W. Girardet). This consists of a form code of five digits and a supplementary code of four digits.

PROCESS SYSTEMS FOR MANUFACTURING

Form code: 13232
4 × on circumference

	1st Digit	2nd Digit	3rd Digit	4th Digit	5th Digit
	Part class	External shape, external shape elements	Internal shape, internal shape elements	Plane surface machining	Auxiliary holes and gear teeth
0	$L/D \leq 0.5$	No machining	Without bore, without through hole	No plane surface machining	No auxiliary bore
1	$0.5 < L/D < 3$	Smooth, no shape elements	With through hole	External plane surface and or surface curved in one direction	Axial holes without indexing
2		No shape elements	No shape elements	External plane surfaces related to one another with a pitch	Axial holes with indexing
3		Screwthread (stepped to one end, smooth or multiple increases)	Screwthread (stepped to one end, smooth or multiple increases)	External groove and or slot	Axial and or radial holes and or in other directions
4		Functional groove and or functional taper (and screw thread)	Functional taper (radial groove) (and screw thread)	External spline (polygon)	Axial and or radial holes with indexing and or in other directions
5		No shape elements (stepped to both ends, multiple increases)	No shape elements (stepped to both ends, multiple increases)	External spline, slot and or groove	Spur gear teeth without auxiliary holes
6		Screwthread	Screwthread	Internal plane surface and or groove	Spur gear teeth with auxiliary holes
7		Functional groove and or functional taper (and screw thread)	Functional taper (radial groove) (and screw thread)	Internal spline (polygon)	Bevel gear teeth
8		Operating thread	Operating thread	Internal spline, external groove and/or slot	Other gear teeth
9		Others	Others	Others	Others

(Rotational parts: 0, 1)
(No gear teeth: 0–4; With gear teeth: 5–9)

Figure 7.7 Details of the form code of the Aachen classification and coding system for the case that general classification of parts (1st digit) is 0 and 1 (rotational) (after Opitz (*Werkstücksystematik und Teilefamilienfertigung*, 1965), reproduced by permission of Verlag W. Girardet). An example of coding—part description by group technology—is also shown.

for general-purpose use. As a typical example the Aachen system (*Werkstücksystematik und Teilefamilienfertigung*, 1965) is shown in Figure 7.6. The basic structure of this system is the form code, which expresses a general classification of parts (1st digit), the relation of detailed shape of parts and processing by machines (2nd–5th digit), the supplementary code expressing dimensions (6th digit), material (7th digit), original shape of raw materials (8th digit), and accuracy (9th digit).

--- **EXAMPLE 7.7** ---

For rotational parts having 0 and 1 for the 1st digit in the Aachen system of coding and classification the detailed coding system for the 2nd–5th digits is shown in Figure 7.7 along with an example of coding for a part. □

7.3 Product Structure and Explosion

Hierarchical Structure of a Product

Most mechanical products are assembled from several manufactured or purchased parts. Some parts are complex parts (sub-assemblies) composed of several discrete piece parts (units) which cannot be dismantled further. Hence, a product is a final assembly of several sub-assemblies and single parts, which may be either manufactured from raw materials or purchased from the outside, as indicated with a hierarchical structure in Figure 7.8.

Bill of Materials

To describe a mechanical product fully, data on items (sub-assemblies, single parts and raw materials used to construct the product) and data on product structure, by

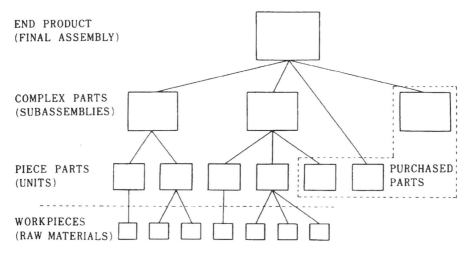

Figure 7.8 An illustration of a hierarchical product structure. A product is made up of several complex parts made up of parts that are manufactured from raw materials or workpieces. Some parts are purchased from the outside.

which the product is made up, are required. This may be shown diagrammatically, as indicated in Figure 7.9, by using a tree, an incidence matrix, and a [directed] graph.[1]

The detailed make-up of the end product and where-used position of the parts constitutes a *bill of material[s]* (BOM) as follows:

- *Summarised bill of material* (or *product-summary*) *list*. This shows the total usage of each item collected into a single list for the product, as represented in Table 7.1.
- *Single-level bill of material* (or *product-structure*) *list*. This indicates the relationship between the product, complex and piece parts as well as their structure and quantity at one level, as displayed in Table 7.2.

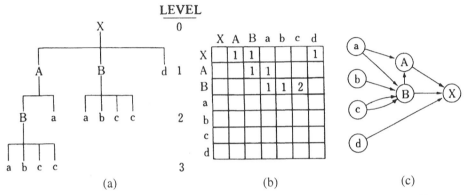

Figure 7.9 Diagrammatical expression of the product structure or the bill of material by a tree (a), an incidence matrix (b) and a graph (c).

Table 7.1 The summarised bill of material or product-summary list. The quantity of parts and materials constructing an item of product is expressed in this list, neglecting complexity and construction of parts.

Product X

Item	Quantity
A	1
B	2
a	3
b	2
c	4
d	1

Table 7.2 The single-level bill of material or product-structure list. It indicates the relationship between the product and parts as well as their structure and quantity at one level.

Product X		Part A		Part B	
Item	Quantity	Item	Quantity	Item	Quantity
A	1	B	1	a	1
B	1	a	1	b	1
d	1			c	2

Table 7.3 Calculating net requirements of complex parts (A and B) and piece parts or raw materials (a, b, c, and d) for constructing an end product X when ten pieces of the product are required.

				Product/Part structure					
Item	Requirement	Inventory	Net requirement	Item	Quantity	Item	Quantity	Item	Quantity
X	10	2	8	A	1	B	1	d	1
A	8	3	5	B	1	a	1		
B	13	1	12	a	1	b	1	c	2
a	17	0	17						
b	12	5	7						
c	24	4	20						
d	8	0	8						

Based upon BOM configurations, *product data management (PDM)* systems play an important role as a common database in running CAE/CAD/CAM, which will be described in Chapter 24.

Parts Explosion

After the products to be manufactured have been decided upon as a result of production planning, the required kinds and quantities of complex/piece parts and raw materials are determined. This is done by the [*parts*] *explosion*, which generates a parts list by 'exploding' the bill of materials. Then, together with the inventory lists, information about net requirements for parts and raw materials is obtained. This procedure is utilised in material requirements planning (MRP), which will be mentioned in Chapter 13.

Calculating Net Required Parts/Materials

The quantity of parts and raw materials required during a specified production period can be thus obtained by multiplying the quantity of the end products decided upon by the production planning function by the quantity of parts/materials for those products indicated in the bill of materials. By subtracting the inventory amount from that quantity, net requirements of parts/materials are calculated.

── **EXAMPLE 7.8** ──────────────────────────────────

Table 7.3 illustrates the procedure of calculating net requirements of parts/materials when ten pieces of product X are required as a result of production planning. □

Notes

1 A *graph* (or more precisely a *linear graph*) is an ordered pair: $G = (V, E)$, where $V = N_1$, $N_2, \ldots$ is a set of vertices and $E = E_1, E_2, \ldots$ is a set of edges. An edge E_j is connected (incident) to the two neighbouring vertices, N_i and N_k. Where edges are directed, or every edge is mapped onto some ordered pair of vertices and indicated with an arrow, such a graph is called a *directed graph* (or *digraph*).

References

BOOTHROYD, G. (1988) Estimate costs at an early stage, *American Machinist*, **132** (8), 54–57.
HAM, I., HITOMI, K. and YOSHIDA, T. (1985) *Group Technology—Applications to Production Management* (Boston: Kluwer-Nijhoff).
HITOMI, K. and HAM, I. (1977) Group scheduling technique for multiproduct, multistage manufacturing systems, *ASME Journal of Engineering for Industry Trans.*, Series B, **99** (3), 759–765.
MILES, L.D. (1972) *Techniques of Value Analysis and Engineering* (2nd edn) (New York: McGraw-Hill).
MITROFANOV, S.P. (1966) *Scientific Principles of Group Technology*, Parts 1–3 (English translation from the Russian by E. Harris) (Boston Spa, Yorks: National Lending Library for Science and Technology).
OVERBY, C.M. (1990) Design for the entire life cycle—a new paradigm? *ASEE Annual Conference Proceedings*, pp. 552–562.
Werkstücksystematik und Teilefamilienfertigung (2nd Conference) (1965) (Essen: Verlag W. Girardet).

Supplementary reading

ANDERSON, D.M. (1990) *Design for Manufacturability* (Lafayette, CA: CIM Press).
BOOTHROYD, G. (1994) *Product Design for Manufacture and Assembly* (New York: Dekker).

CHAPTER EIGHT

Process Planning and Design

8.1 Scope and Problems of Process Planning

What is Process Planning?

Production processes or process routes—a sequence of production operations through which raw materials are effectively converted into the planned products—must be determined after aggregate production planning and completion of product design. This decision-making is called *process planning*. It depends upon the kinds and quantities of products to be finished, kinds of raw materials and parts, production facilities and technology on hand, etc.

Basic Functions of Process Planning

The task of process planning includes the following two basic stages (Timms and Pohlen, 1970):

(1) *Process design* is macroscopic decision-making of an overall process route for converting the raw material into a product.
(2) *Operation design* is microscopic decision-making of individual operations contained in the process route.

Decisions concerning process design are often fed back to the product planning system, calling for a modification of product design and further product development. Results of operation design lead to production implementation, namely actual works/operations done in a factory.

8.2 Process Design

8.2.1 Key Problems of Process Design

Function of Process Design

The main decision problems in process design are:

(1) to analyse the work flow for converting raw material into a finished product—*flow(-line) analysis*;

(2) to select the workstation for each operation included in the work flow.

These two aspects of process design are interrelated, and so must be decided simultaneously. In an existing plant the work flow is to be decided, based upon operations of which the existing production facilities are capable. On the other hand, in the case of establishing an entirely new plant, an effective work flow is first designed, and then workstations are determined in connection with capital investment.

8.2.2 Representation and Analysis of Work Flow

Work Flow

Work flow is a set and sequence of operations which constitute the conversion process of raw material into the product. This is determined by production and manufacturing technologies, and forms the basis of operation design and layout planning.

Precedence Relationship

There are two possible relationships between any two operations.

(1) A partial order exists between two operations; this order is called a *precedence relationship* and is represented by Figure 8.1.

--- **EXAMPLE 8.1** ---

Reaming must be performed after drilling. □

(2) No precedence relationship exists between two operations and they may be performed concurrently and in parallel.

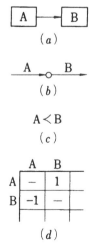

Figure 8.1 The precedence relationship between two successive operations or activities (operation A precedes operation B or operation B follows operation A) is represented by (a) flow diagram, (b) arrow diagram, (c) ordering symbol, or (d) precedence matrix.

PROCESS PLANNING AND DESIGN

The work flow is a sequence of operations among which specified precedence relationships exist.

Work-flow Patterns

We have the following three intrinsic patterns for work flow, as indicated in Figure 8.2 (Harrington, 1973).

(1) *Sequential* (or *tandem*) *process pattern*—linear series of individual operations from the receipt of one major input material until the finished product is completed; e.g. fabrication of metal parts, cotton processing, etc.

--- **EXAMPLE 8.2** ---

A sequential process pattern of gear production is depicted in Figure 8.3. ☐

(2) *Disjunctive* (or *decomposing*) *process pattern*—a pattern of progressive disassembly into several products and by-products from a single raw material, often called *joint production*; e.g. oil refinery, ore-processing plant, etc.

--- **EXAMPLE 8.3** ---

A disjunctive process pattern of coal refinery is represented in Figure 8.4. ☐

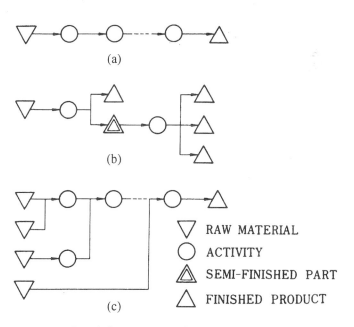

Figure 8.2 Basic types of work-flow pattern (after Harrington, 1973). (a) The sequential or tandem process pattern is of flow-type manufacturing; (b) the disjunctive or decomposing process pattern is seen in chemical industries; and (c) the combinative or synthesising process pattern is represented by an assembly shop.

Figure 8.3 Gear production—sequential process pattern.

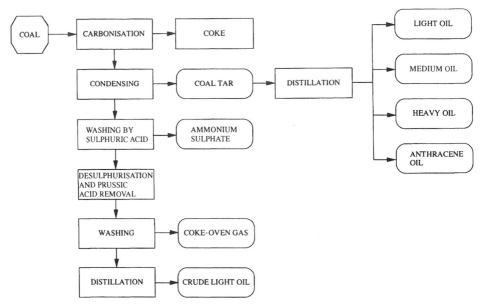

Figure 8.4 Coal refinery—decomposing process pattern.

(3) *Combinative* (or *synthesising*) *process pattern*—a pattern of producing a major product by combining (assembling) several workpieces and/or parts manufactured by several parallel operations; e.g. assembling a motorcar, a TV set, etc.

—— EXAMPLE 8.4 ——————————————————————

A combinative process pattern of car assembly is shown in Figure 3.2. □

Process Symbols

Activities involved in the work flow are classified under five headings—operations, transportations, inspections, delays, and storages—in accordance with the Standard for Operation and Flow Process Charts adopted by the American Society of Mechanical Engineers in 1947. Figure 8.5 shows the symbols for these activities.

Process Charts

Using the above symbols a chart indicating the sequence of operations comprising the work flow of multistage manufacturing is drawn; it is called a *process chart*. This

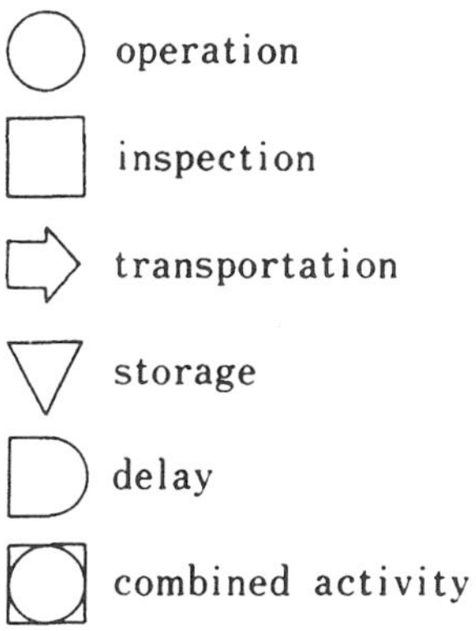

○ operation

☐ inspection

⇨ transportation

▽ storage

D delay

⊙ combined activity

Figure 8.5 Symbols employed for process chart activities.

may be an operation process chart, flow process chart, multiple activity (man and machine process) chart, workplace (right- and left-hand) chart, assembly process chart, or simultaneous motion cycle (simo) chart.

---- EXAMPLE 8.5 ----

A flow process chart is illustrated in Figure 8.6. □

Efficient Work Flow

There are, in general, several alternatives for work flow in converting raw material and components into a finished product, depending upon production quantity (demand volume, economic lot size), existing production capacity (available production technology, degree of automation), product quality (accuracy, tolerance), raw material (property of material, machinability), etc. The best workflow (including selection of raw material and optimum process route) is selected by evaluating each of the alternatives based upon criteria for minimising the total production time or cost, which are expressed:

$$\text{total production time} = \Sigma[\text{transfer time between stages} \\ + \text{waiting time} + \text{setup time} + \text{operation time} + \text{inspection time}]; \quad (8.1)$$

$$\text{total production cost} = \text{material cost} + \Sigma[\text{cost of transfer between stages} \\ + \text{setup cost} + \text{operation cost} + \text{tool and jig cost} + \text{inspection cost} \\ + \text{work-in-process inventory cost}], \quad (8.2)$$

where Σ is taken over all the stages for production.

---- EXAMPLE 8.6 ----

A result of investigating the total costs for two kinds of materials, A and B, is indicated in Table 8.1. The minimum-cost principle suggests selection of material B having better machinability in spite of higher material cost. □

FLOW PROCESS CHART

Item: Crank Shaft Quantity: 100 Chart No.: 201
Drawing No.: B-123 Part No.: C-479 Analyst: K. Hitomi
Chart Begins: Raw materials storage Date: 16. Feb. 1978
Chart Ends: Parts warehouse Sheet No.: 1 of 2

DISTANCE (m)	UNIT TIME (min)	SYMBOL	DESCRIPTION
		▽ 1	Storage of forged bars.
5	0.30	○ 1	Bars loaded on truck upon receipt of requisition.
50	1.50	⇒ 1	Bars on truck to crank-shaft lathe.
1	0.05	○ 2	Bars removed from truck and stored on rack.
	7	D 1	Waiting for operation to begin.
1	3.25	⊙ 1	Bars machined by lathe and inspected.
2	0.20	⇒ 2	To press by operator.
	1.20	□ 3	Inspect completely.
35		⇒ 5	To parts warehouse.
		▽ 2	Stored till requisitioned.

SUMMARY

EVENT	NUMBER	TIME (min)	DISTANCE (m)
Operation	8	5.00	
Inspection	1	1.20	
Combined Activity	1	3.25	
Transportation	5		110
Storage	2		
Delay	3	30	

Figure 8.6 A flow process chart. This chart indicates the process for manufacturing the crankshaft.

Table 8.1 Total cost analysis—material B should be selected based upon the minimum-cost principle.

Cost element	Material A	Material B
(1) Raw-material cost ($/pc*)	100	120
(2) Chip cost ($/pc)	5	8
(3) Machining time (min/pc)	10	7
(4) Labour cost and overhead ($/min)	8	8
(5) Machining cost ($/pc) (3) × (4)	80	56
Total cost ($/pc) (1) − (2) + (5)	175	168

*pc denotes 'piece'.

8.2.3 Selection of Workstation

Economical Selection of Workstation

Workstations for the required work flow are selected economically by choosing, from the alternatives available, the least-cost combination of production facilities or machine tools together with operative manpower capacity.

Optimum Selection of Production Facility by the Break-even Analysis

Since the volume of output influences the total production cost, an efficient production facility must be selected according to production quantity. Figure 8.7

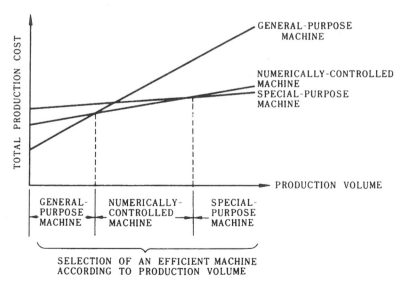

Figure 8.7 Break-even chart for selecting an efficient production facility according to production volume. It is economical to select the general-purpose machine tool, the numerically controlled machine tool, or the special-purpose machine tool, respectively, according to small, medium, or large volume of production.

shows how to select an optimum machine tool according to the expected production or sales volume by means of the break-even chart and analysis. As seen in this figure, the fixed cost associated with performing an operation increases as the degree of automation increases, whilst the variable cost per unit of output, including power charge and labour cost, decreases. The cost-efficient workstation will be the general-purpose machine tool for a small volume of production, the numerically-controlled (NC) machine tool for a medium volume of production, and the special-purpose machine tool for a large volume of production.

8.3 Operation Design

8.3.1 Problems of Operation Design

Function of Operation Design

The next step after process design is *operation design*. This is concerned with the detailed decisions of production implementation, that is, the types of operations to be performed in the production process (the content of each operation and the method of performing it). The content of each operation is determined in connection with process design and may be broken down into several steps, such as loading the workpiece into the chuck of a machine tool, starting the machine, and unloading the workpiece from the chuck and placing it on a conveyor.

Operation Analysis

The method of operation can be analysed from the viewpoints of a combination of machine elements and human elements (*man–machine system*), operative workers (who may have large differences in skill), and work simplification.

8.3.2 Analysis of Man–Machine system

Aims of Man–Machine System Analysis

Man–machine system analysis is performed to identify and reduce or eliminate idle times for either the worker or the machine constituting the selected workstation. The man–machine combination is essentially chosen to minimise the overall operation cycle time with least idle times.

Man–Machine Chart

The primary tool of analysis of the man–machine system is a *man–machine* chart. This is a graphical representation of the operation content and method for simultaneous activities of the worker and the machine. The chart indicates the periods of cooperative work, independent work, and idle times along a time scale, as shown in an example in Figure 8.8. Each of the idle times would be examined for the possibility of reducing or eliminating it, thus resulting in a revised man–machine chart with a consequent method improvement.

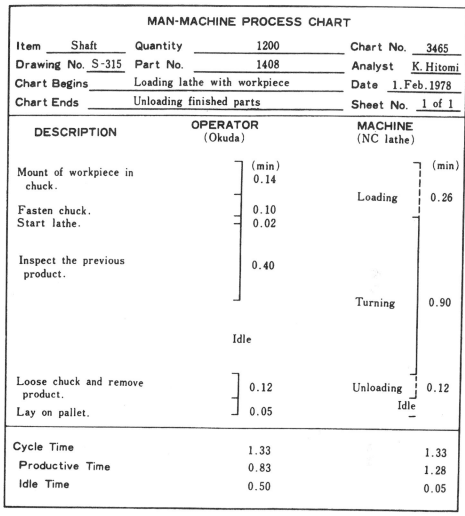

Figure 8.8 A man–machine chart. Attempts should be made to reduce inefficient times for both the worker and the machine in a general effort towards method improvement.

With automated manufacturing the roles of the operative workers are becoming less important except for the activities of raw-material supply, tool and jig setting and replacing, productive maintenance, product inspection etc.

8.3.3 Analysis of Human Factor

Tools for Human-factor Analysis

Where the role of an operative worker is essential to productivity, such as in assembly operations, human-factor analysis is made to determine efficient motions and proper times for the work. The following two techniques are effective for this purpose:

(1) *Motion* (or *method*) *study* is a systematic procedure of dividing work into the most basic elements possible, studying these elements separately and in

relation to one another, and synthesising the most efficient method of performing the work.
(2) *Time study* is a systematic procedure for measuring the length of time required and establishing an allowed time standard to perform work, based upon measurement of work content of the prescribed method by an operator of average skill, working with average effort, under normal conditions.

Motion study and time study were pioneered by Frank B. and Lillian Gilbreth and Frederick W. Taylor, respectively, in the late 19th century and early 20th century. These are basic techniques of traditional industrial engineering and closely linked. Generic terms for these techniques for the study of human work and the investigation of all the factors which affect the efficiency of human work are *time and motion study*, *work study* or *methods engineering*.

One technique for the evaluation of an operative worker's task is right- and left-hand analysis, from which a *right- and left-hand chart* (or *workplace chart* or *operator process chart*) is prepared to indicate the worker's simultaneous body motions in detail for the purpose of eliminating ineffective motions, as illustrated in Figure 8.9.

Motion Economy and Therblig

After obtaining information of the operation of a task, through the right- and left-hand analysis, every work element is critically evaluated to eliminate wasteful or fatiguing motions. To assist in the evaluation the *principles of motion economy*, which were first set forth and developed by the Gilbreths and improved by others, are useful and are applied especially to physical movements. The basic principles of motion economy include 22 statements arranged in three major subdivisions:

(1) the use of the human body;
(2) the arrangement and conditions of the workplace;
(3) the design of the tools and equipment.

as listed in Table 8.2 (Niebel, 1982).

When it is necessary to obtain very detailed information concerning a particular operation, *micromotion study*, based upon the analysis of motion pictures, is employed. This microscopic analysis of human work was also developed by Gilbreth, and he concluded that any and all movements are made up of 17 basic motion elements which he called *therbligs* (his own name spelled backwards). Symbols and colour designations of therbligs are shown in Table 8.3. A reasonable and efficient human work pattern is created by identifying and analysing the individual work elements with these therbligs, eliminating ineffective therbligs, and synthesising them into an integrated work plan.

Methods of Time Study/Work Measurement

Time study or *work measurement* is used to specify a time required to do each operation and to determine an operation standard, thus establishing 'a fair day's work' (proposed originally by C. Babbage in the first half of the 19th century and later by F.W. Taylor in the latter half of the century). The principal methods of

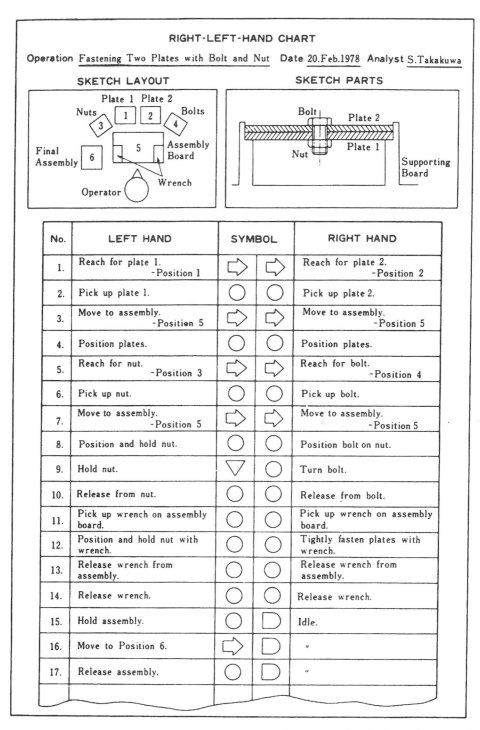

Figure 8.9 A right- and left-hand or operator process chart. This chart indicates the worker's simultaneous body motions for fastening two plates with a bolt and nut.

Table 8.2 The principles of motion economy, based upon Gilbreth's pioneering work. Motion economy eliminates wasteful movements.

(i) *Use of the human body*
1. The two hands should begin as well as complete their motions at the same time.
2. The two hands should not be idle at the same time except during rest periods.
3. Motions of the arms should be made in opposite and symmetrical directions and should be made simultaneously.
4. Hand motions should be confined to the lowest classification with which it is possible to perform the work satisfactorily. General classifications of hand motions are as follows:
    (a) finger motions;
    (b) motions involving fingers and wrist;
    (c) motions involving fingers, wrist, and forearm;
    (d) motions involving fingers, wrist, forearm, and upper arm;
    (e) motions involving fingers, wrist, forearm, upper arm, and shoulder.
5. Momentum should be employed to assist the worker wherever possible, and it should be reduced to a minimum if it must be overcome by muscular effort.
6. Smooth continuous motions of the hands are preferable to zigzag motions or straight-line motions involving sudden and sharp changes in direction.
7. Ballistic movements are faster, easier, and more accurate than restricted or controlled movements.
8. Rhythm is essential to the smooth and automatic performance of an operation, and the work should be arranged to permit easy and natural rhythm wherever possible.

(ii) *Arrangement of the workplace*
9. There should be a definite and fixed place for all tools and materials.
10. Tools, materials, and controls should be located close to and directly in front of the operator.
11. Gravity feed bins and containers should be used to deliver material close to the point of use.
12. Drop deliveries should be used wherever possible.
13. Materials and tools should be located to permit the best sequence of motions.
14. Provisions should be made for adequate conditions for seeing. Good illumination is the first requirement for satisfactory visual perception.
15. A chair of the type and height to permit good posture should preferably be arranged so that alternate sitting and standing at work are easily possible.
16. A chair of the type and height to permit good posture should be provided for every worker.

(iii) *Design of tools and equipment*
17. The hands should be relieved of all work that can be done more advantageously by a jig, a fixture, or a foot-operated device.
18. Two or more tools should be combined wherever possible.
19. Tools and materials should be pre-positioned wherever possible.
20. Where each finger performs some specific movement, such as typewriting, the load should be distributed in accordance with the inherent capacities of the fingers.
21. Handles such as those used on cranks and large screwdrivers should be designed to permit as much of the surface of the hand to come in contact with the handle as possible. This is particularly true when considerable force is exerted in using the handle. For light assembly work the screwdriver handle should be so shaped that it is smaller at the bottom than at the top.
22. Levers, crossbars, and handwheels should be located in such positions that the operator can manipulate them with the least change in body position and with the greatest mechanical advantage.

Taylor & Francis
c/o Returns Dept
7625 Empire Dr.
Florence, KY 41042-2919
United States

Ship To
TEXTBOOK
COMP COPIES
BOCA, FL 33431   USA

	Phone #	Customer PO	Or
		CAS0001688128	02
Quantity	SKU	Title/Author	
1	0748403248	Manufacturing Systems Engineering A L	

In the unlikely event that you must return product, please adhere to th
*All special promotion products are non-returnable, unless otherwise st
*You are responsible for shipping charges.
*Return items prepaid and insured. Taylor & Francis will not accept COD
*Be sure package is secure
*All returned items MUST BE IN SALEABLE CONDITION.
*Shipping and handling charges are non-refundable.

Forward Returns To:
Taylor & Francis
c/o Returns Department
7625 Empire Drive
Florence, KY 41042-2919

umber	Acct #: 429532787		
6-0001	ShipVia: FedEx Ground		
	Price	Discount	Amount
Approa	0.00	0.00%	$0.00
structions and policies.		**Total:**	**$0.00**

ns.

Table 8.3 Symbols and colour designations for therbligs made up of 17 basic motion elements, by which a human work is constructed.

Therblig name	Therblig symbol	Colour designation	Symbol	Description
Search	S	Black	◯	Eye search to locate article, tool, etc.
Select	SE	Grey, light	→	Hand is ready to pick the required object.
Grasp	G	Lake red	∩	Object is grasped by open hand.
Reach	RE	Olive green	⌣	Empty hand is moved towards object, tool, etc.
Move	M	Green	⌢	Hand moves object from place to place.
Hold	H	Gold ochre	⌒	Object is held firmly.
Release	RL	Carmine red	⌐	Object is dropped from hand.
Position	P	Blue	9	Movement ensures that object can be placed in its proper location.
Pre-position	PP	Sky blue	ð	Placing a tool or object ready for further use so that 'position' will subsequently not be needed.
Inspect	I	Burnt-ochre	0	Examination of results or performance.
Assemble	A	Violet, heavy	#	Putting together two or more components.
Disassemble	DA	Violet, light	#	Separating the combined components.
Use	U	Purple	U	Hand performing the desired operation.
Unavoidable delay	UD	Yellow ochre	⌐	Idleness beyond operator's control.
Avoidable delay	AD	Lemon yellow	⌐	Idleness which the operator can eliminate without altering the sequence of events.
Plan	PL	Brown	β	Delay owing to operator thinking.
Rest	R	Orange	℞	To overcome fatigue.

analysis are described below.

(1) *Stopwatch time study*. The work elements are timed directly with the aid of a stopwatch. From this analysis a time standard is assigned to each work element. Videotape recorders may be used for this purpose.

(2) *Synthetic basic motion times* or *predetermined time standards* (*PTS*). These are a collection of valid time standards which are assigned to the fundamental motions of which a task consists. They are used, in advance of beginning the task, to estimate rational work elements and their rates quickly and often accurately without using a stopwatch or other time and motion recording devices.

Two principal procedures for establishing predetermined time standards are:

(1) *Work-factor (WF)*. This system recognises the following major variables that influence operation time, such as
   (a) the body member making the motion,
   (b) the distance moved,
   (c) the weight carried, and
   (d) the manual control required.
   Work-factor motion-time values measured in units of 0.0001 minutes are determined as an incentive pace and enable the effect of these major variables to be determined.

(2) *Methods-time measurement (MTM)*. This is a procedure which analyses any manual operation into the basic motions of reach, move, make a crank motion, turn, apply pressure, grasp, position, release, disengage, make a body motion, and eye motion. This system assigns to each motion at normal pace a predetermined time standard measured in time units of 0.000 01 hours which is determined by the nature of the motion and the conditions under which it is made. The steps of MTM are: first, all motions required to perform the job are summarised properly for both the left and the right hands, and then the standard time for each motion is determined from the methods-time data tables.

8.3.4 Productive Operation

Operation Sheets

By determining for a combination of machine and worker, the task to be performed by the operative worker, and the time standard for the assigned task or job, a sequence of operations required to perform the job is identified. This is listed in a document called an *operation sheet*.

--- **EXAMPLE 8.7** ---

Figure 8.10 depicts an operation sheet for making shafts—the steps of work elements constituting the content of the job specified in the proper sequence, along with such detailed information as machines, jigs and tools used, machining conditions, and standard times for setup, machining, and one cycle of the operation. □

Together, the process sheet and the operation sheet provide all the information needed to specify the total manufacturing details of production processes.

Standard Time

For each individual operation a *standard time* may be specified. This is the total time in which an operation should be completed at standard performance and consists of the time elements as indicated in Table 8.4. The observed time values of the operation, recorded under time study, are modified to the time required to perform the operation at the 'standard', or 'normal' pace by a technique called *performance rating*. This time is the 'basic time', or 'normal time'. To determine the standard

PROCESS PLANNING AND DESIGN 121

Table 8.4 Structure of standard time.

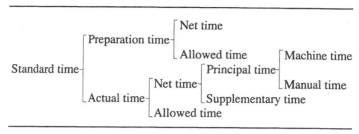

OPERATION SHEET

Item	Shaft	Quantity	1500		Chart No.	346
Drawing No.	C-103	Part No.	116700		Analyst	K. Okuda
Chart Begins	Lathe Dept.				Date	15. Sept. 1977
Chart Ends	Lathe Dept.				Sheet No.	1 of 1
Machine No.	#791	Machine Type	Lathe		Material	Steel
Normal Cycle Time	3.283	Pieces per hour	18.3		Coolant	Oil
Setup Time	.9					

No.	Element Description	Speed	Feed	Jigs, Tools
1	Centering	75	hand	Centre drill
2	Turning	400	0.015	Right-hand cutting tool
3	Turning	400	0.015	Left-hand cutting tool
4	Turning	85	0.0025	Forming tool
5	Screw-cutting	240	pitch 0.138	Threading tool
6	Cutting-off	85	0.002	Cutting-off tool

Machine Maintenance
  Oil   once/day
  Grease once/week

Remarks   No special.

Figure 8.10 An operation sheet. It shows the steps of work elements which constitute the work content in the proper sequence with information as to machines, jigs, and tools used, as well as machining conditions and standard times for setup, machining, and cycle times.

time it is necessary to add time allowances for personal needs, unavoidable fatigue, and extra delays. Further allowances may be made for other reasons, for example as part of a payment scheme, and are added to the standard time to establish the 'allowed time'.

8.3.5 Division of Labour and Job Design

Background to Division of Labour

A recent trend in thinking that the value of labour has been neglected in our society has resulted in significant changes in the role of manual work. The importance and effectiveness of [technological] 'division of labour' was already pointed out by Xenophōn as early as 394 BC and in the modern age it was recognised by Adam Smith in 1876. The concept was put into practice by H. Ford in 1913 using the conveyor system to achieve mass production for model T cars, and this new high-productive method was named the 'Ford system'. Such repetitive inhumane work was criticised in the famous film, *Modern Times,* in 1936 by Charlie Chaplin and is now being reconsidered in the light of providing greater human respect and dignity at work. In 1972 Volvo Automobile Company in Sweden abolished the conveyor system, and used autonomous groups of workers to assemble cars.

From this trend have developed two industrial methods:

- establishing automated manufacturing as much as possible to eliminate low-skill work;
- designing valuable jobs for the worker and labour flexibility.

Job Design for Better Quality of Working Life

With the latter method above, the work environment is improved by *job design* such as:

- *job rotation*, which alters the worker's job periodically;
- *job enlargement*, which enlarges the worker's job contents;
- *job enrichment*, which improves the worker's job content.

Job design efforts are made to improve and enhance the quality of manual work and establish a better 'quality of working life (QWL)', such that an individual worker or a group of workers carry out their work tasks under a system of self-control and with motivation. Examples are cellular one-operator or group assembly systems, which perform multiple-product assembly efficiently and economically.

8.3.6 The Learning Effect

Meaning and Background of Learning Effect

It is generally observed that, even for the simplest of operations, the unit operation time decreases as the operation is repeated. This is due to an increase in the operator's level of skill, improvement of the production methods such as modification of operation contents and sequences, use of effective jigs and fixtures, resetting of machining conditions, improved management practices, etc. This phenomenon is called the *learning effect*. The effect is large where the cumulative production quantity is small, and decreases as the cumulative production quantity increases. Accordingly, this characteristic is especially important in jobbing and low-volume lot production.

---- **EXAMPLE 8.8** ----

Production time for a ship is reduced by about 40% for the tenth vessel compared with the initial production. Around 30% of cost reduction is expected by doubling the production quantity for semiconductors/integrated circuits. □

Learning Curves

The learning effect is represented by a chart called the *learning* (or *progress* or *manufacturing-time forecasting*) *curve*. It is illustrated in Figure 8.11—a graphical representation showing the relationship between the cumulative average unit production time and the cumulative production volume.

Once the flat section of the learning curve is reached, the problem of performance rating is simplified; however, it is not always possible to wait this long in developing a time standard, especially in jobbing production. In this case the analyst must have acute observation ability and mature judgement. Alternatively, standard data based upon MTM may be used.

Production-progress Function

It is often recognised that linearity exists between the cumulative production quantity (independent variable) M, and the cumulative average unit production time (dependent variable) t, on bilogarithmic paper. This phenomenon was discovered by Wright in 1936.

This property results in the following mathematical expression for the learning effect, often called the *production-progress function*:

$$t = t_1/M^k, \qquad (8.3)$$

where t_1 is the unit production time required to produce the first unit and k is the learning coefficient showing the slope of the learning curve on bilogarithmic paper, since the above equation is converted to the following expression by taking logarithms of both sides:

$$\ln t = \ln t_1 - k \ln M. \qquad (8.4)$$

---- **EXAMPLE 8.9** ----

The learning curve illustrated in Figure 8.11 is expressed as:

$$t = 600/M^{0.074}. \qquad (8.3')$$

This curve is converted to a straight line on a bilogarithmic paper as represented in Figure 8.12.

□

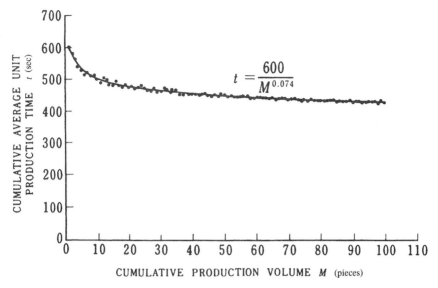

Figure 8.11 A learning curve. It shows that the cumulative average unit production time decreases with increase in the cumulative production volume. This learning effect is such that a linearity exists between these variables on a bilogarithmic paper. (See Figure 8.12.)

The total production time required to produce a predetermined number of units M can then be estimated from (8.3) as follows:

$$T_M = tM = t_1 M^{1-k}. \tag{8.5}$$

It follows that the unit production time required to produce the Mth unit is:

$$t_M = T_M - T_{M-1} \doteq t_1(1-k)/M^k. \tag{8.6}$$

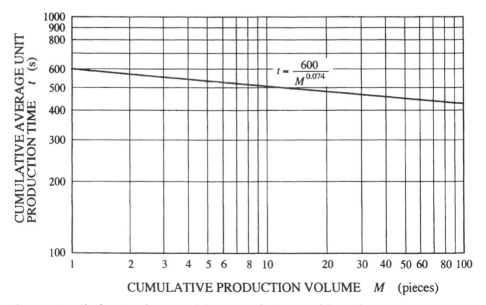

Figure 8.12 The learning diagram exhibits a straight line on a bilogarithmic paper.

EXAMPLE 8.10

From (8.3') the unit production time for the 100th unit is:

$t_{100} = 600(1 - 0.074)/100^{0.074} = 395$ s $= 6$ min 35 s.

This shows a great deal of reduction of production time. □

The ratio of cumulative unit production time when a certain quantity is produced against that for doubled production quantity is called the *learning rate*. It is expressed as:

$$r_1 = \frac{t_1/(2M)^k}{t_1/M^k} = 2^{-k}. \tag{8.7}$$

If a reduction of $100x\%$ in labour hours per unit production occurs between doubled quantities, it is represented by a $100(1-x)\%$ learning curve. This is expressed as:

$$1 - x = 2^{-k}. \tag{8.8}$$

EXAMPLE 8.11

For a 90% learning curve, the above equation becomes $0.9 = 2^{-k}$. Hence, k equals 0.136. □

8.4 Optimum Routing Analysis

8.4.1 Problem Setting

Meaning of Optimum Routing Analysis

It is usual that there are several alternative process routes or patterns for converting raw material into the finished mechanical product. To select one best process route from among the alternatives is *optimum routing analysis* (or *optimum process planning*).

EXAMPLE 8.12

Alternative routings made up of four stages—turning operation at the first stage, planing operation at the second stage, drilling operation at the third stage, and finish operation at the fourth (final) stage—can be represented in the form of a network as shown in Figure 8.13. Notice that arrows represent operations with definite work contents and operation times (or costs). Further, a set of successive arrows from the start point 1 (source), which signifies the raw material, to the end point 9 (sink), which stands for the finished product, indicates a production process route for producing a finished product from a raw material. Selecting a route with least time (or cost) is the optimum routing analysis. The selected route is the *optimal production process* (or *route*). □

Procedures for Optimum Routing Analysis

An optimum routing problem has a finite number of combinations, and a solution for this type of combinatorial problem can be always obtained by calculating the total

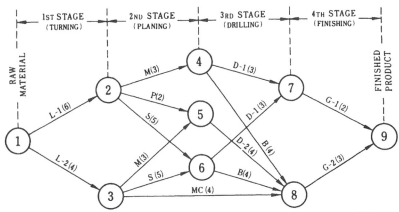

⟨NOTATION⟩ L: LATHE, B: BORING MACHINE, D: DRILLING MACHINE, G: GRINDER,
M: MILLING MACHINE, MC: MACHINING CENTRE, P: PLANER, S: SHAPER

FIGURES IN PARENTHESES INDICATE PROCESSING TIMES

Figure 8.13 Network representation of alternative production process routes for producing a product from a raw material. This conversion process is made up of four stages. To select one 'best' plan is the optimum routing analysis.

production times (or costs) for all possible routes by trial and error and, then, selecting a route with a least time (or cost). This complete-enumeration procedure requires a substantial computational effort when the problem includes a large number of arrows. Dynamic programming and the network technique are two methods for solving this type of problem with less computational effort.

8.4.2   Solving by Dynamic Programming

Background of Dynamic Programming

Dynamic programming (*DP*) is a multistage decision theory, as discussed in Section 2.4.3. It is an optimisation technique developed by R. Bellman in 1957. This technique first solves a small part of the problem, then based upon that optimal solution, enlarged problems are successively generated and solved, finally the optimal solution for the total problem is obtained. The optimal solution is calculated from a recursive functional equation with max or min deduced by the 'principle of optimality'.

Dynamic Programming Technique

Basic procedures for dynamic programming are explained below.

(1) The state of the system at stage i is expressed by 'state variable' $x(i)$.
(2) At each stage a 'decision' is made to change the state of the system. Only the present value of the state variable is associated with this decision. For a deterministic, discrete system, the state of the system at stage $i+1$ is expressed by stage i, the state $x(i)$, and decision $u(i)$ at that stage as follows:

$$x(i+1) = T[x(i), u(i), i] \tag{8.9}$$

where T is transformation.

(3) A measure of performance to be maximised or minimised is established for the multistage decision process.

(4) A possible sequence of decisions, e.g. [$u(1), u(2), \ldots, u(N)$], where N is the number of stages, for a deterministic, discrete case, is the 'policy'. A policy such as maximising or minimising the measure of performance is the 'optimal policy', which results in an optimal solution for the multistage decision problem.

(5) The *principle of optimality* is a basic principle for obtaining an optimal solution by the dynamic programming approach. It states that 'an optimal policy has the property that whatever the initial state and initial decision are, the remaining decisions must constitute an optimal policy with regard to the state resulting from the first decision' (Bellman, 1957).

(6) Based upon the principle of optimality, an optimal sequence of decisions for determining the optimal value for a measure of performance, which is to be obtained as a function of the initial state and the number of stages, is successively determined by the recursive functional equation with max or min.

Applying the Principle of Optimality for Constructing the Recursive Functional Equation

For the routing analysis illustrated in Figure 8.13, the minimum time required to reach the sink 9 from node i is denoted by $f(i)$. Let us consider a case where node j is selected just next to node i. By denoting the processing time for moving from node i to node j by τ_{ij}, the time for moving from node i to node j, and further to the sink 9, is $\tau_{ij} + f(j)$, where $f(j)$ is the minimum time for moving from node j to the sink, as depicted in Figure 8.14. Hence, to find a minimum time route from node i to the sink is to determine j such as to minimise $\tau_{ij} + f(j)$; following the principle of optimality, we get:

$$f(i) = \min_j [\tau_{ij} + f(j)]. \tag{8.10}$$

Using this recursive relation, an optimal solution—minimum time $f(1)$ and optimum route—is obtained by the forward or backward calculation method.

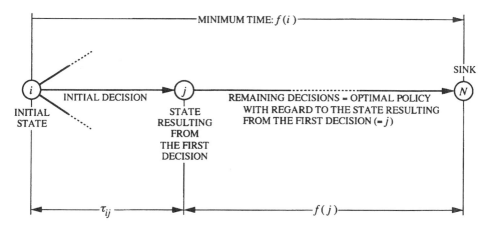

Figure 8.14 Application of the 'principle of optimality' constructs the recursive functional equation (8.10) for determining the optimum process route.

EXAMPLE 8.13

To solve the problem of Example 8.12 the backward calculation method is employed as follows:

To begin with, $f(9) = 0$.
Then, by using (8.10)

$$f(8) = \min_{j=9}[\tau_{8j} + f(j)] = \tau_{89} + f(9) = 3\,(8 \to 9)$$

where $i \to j$ signifies that the optimal route is from node i to node j. Similarly, $f(7) = 2\,(7 \to 9)$.
Next,

$$f(6) = \min_{j=7,8}[\tau_{6j} + f(j)] = \min[\tau_{67} + f(7), \tau_{68} + f(8)] = [3 + 2, 4 + 3]$$
$$= 5\,(6 \to 7).$$

Similarly, $f(5) = 7\,(5 \to 8)$ and $f(4) = 5\,(4 \to 7)$.
Further,

$$f(3) = \min_{j=5,6,8}[\tau_{3j} + f(j)] = 7\,(3 \to 8)$$

and

$$f(2) = \min_{j=4,5,6}[\tau_{2j} + f(j)] = 8\,(2 \to 4).$$

Finally, we get:

$$f(1) = \min_{j=2,3}[\tau_{ij} + f(j)] = 11\,(1 \to 3).$$

Thus, this optimum routing analysis results in the optimal solution, which shows that an optimal route for minimising the total production time is $1 \to 3 \to 8 \to 9$ (4 hours for turning operation $\to$ 4 hours for machining centre work $\to$ 3 hours for grinding operation) and that the required minimum time is 11 hours. □

8.4.3 Solving by Network Technique

Network Technique

As illustrated in Figure 8.13, a combination of a set of *nodes* (or vertices) represented by circles, and a set of *arcs* (or edges or branches or activities) represented by line segments or arrows connecting two neighbouring nodes, is called a *network*. An illustration of this figure is a case of a network with directed arcs or arrows.

Where there exists an arc E_{ij} which connects two nodes N_i and N_j, these two nodes are connected. If there exists a sequence of nodes and arcs such as $N_1, E_{12}, N_2, E_{23}, \ldots, N_{k-1}, E_{k-1,k}, N_k$, this is called a *chain* or *path* from node N_1 to node N_k. When $N_1 = N_k$, it constructs a *directed cycle*. An *acyclic directed network* is one having directed paths and containing no cycles or loops, as in Figure 8.13.

The above definition for a network is almost similar to one for a graph; however, a positive value d_{ij}, called a *capacity*, is defined on arc E_{ij} for the network. In Figure 8.13 this capacity is processing time, which is indicated in parentheses on each arrow. In addition, the network often contains two special nodes called a 'source' (the start point of the network) and a 'sink' (the end point of the network), which are indicated by nodes 1 and 9 in the case of Figure 8.13.

Shortest-path [Algorithm]

The problem we are currently considering is how to find a path from node 1 to node 9 which will have a minimum sum of capacities along the path. In this sense, this is called the problem of the *shortest path* (or *chain*).

For an acyclic directed path we number the nodes as shown in Figure 8.13 such that all activities lead from smaller-numbered nodes to higher-numbered nodes (see also PERT which will be mentioned in Section 14.3). The present problem can then be solved by the following algorithm of finding the shortest path from node h to node k (Moore, 1962).

The shortest-path algorithm
Step 1: Set as the evaluation for node h: $e_h = 0$.
Step 2: Continue to label the remaining nodes (chosen in ascending order) according to the following formula:

$$e_j = \min_{i = h, h+1, \ldots, j-1} (e_i + d_{ij}). \tag{8.11}$$

Step 3: When node k has been labelled, e_k is the value of the shortest path from node h to node k. The path is determined by tracing backwards from node k to all nodes such that

$$e_i + d_{ij} = e_j, \quad j = k, k-1, \ldots, h+1, h. \tag{8.12}$$

— **EXAMPLE 8.14** —

Let us apply the above algorithm to the routing analysis illustrated in Figure 8.13, which has already been solved by the dynamic programming approach in the previous example.
To begin with, from step 1, $e_1 = 0$.
Then, proceeding to step 2,

$$e_2 = \min_{i=1}(e_i + d_{i2}) = e_1 + d_{12} = 6(1 \rightarrow 2).$$

Similarly, $e_3 = 4(1 \rightarrow 3)$ and $e_4 = 9(2 \rightarrow 4)$.
Next,

$$e_5 = \min_{i=2,3}(e_i + d_{i5}) = \min(e_2 + d_{25}, e_3 + d_{35}) = \min(6+2, 4+3)$$
$$= 7(3 \rightarrow 5).$$

Similarly, $e_6 = 9(3 \rightarrow 6)$ and $e_7 = 12(4 \rightarrow 7$ or $6 \rightarrow 7)$.
Further,

$$e_8 = \min_{i=3,4,5,6}(e_i + d_{i8}^*) = 8(3 \rightarrow 8).$$

Finally, we get:

$$e_9 = \min(e_7 + d_{79}, e_8 + d_{89}) = 11(8 \rightarrow 9).$$

Consequently, the minimum total production time is 11 hours, and the corresponding optimal route is: $1 \rightarrow 3 \rightarrow 8 \rightarrow 9$ by step 3. This result coincides with the one obtained in Example 8.13. □

8.5 Line Balancing

8.5.1 Scope of Line Balancing

Meaning of Line Balancing

In assembly lines production stages are tightly connected e.g. by conveyor lines; each production stage is dependent on prior stages such that the production time must be equalised for all stages, to assure smooth production flow. This type of continuous production is called a *line-production system*. Line balancing is concerned with this system and aims at optimum decision-making in regard to:

- cycle time;
- number of workstations or production stages;
- grouping of work elements by assigning them on a same workstation such that their precedence order is assured.

Line balancing for the case of continuous mass production, where a single product item is assembled on a single assembly line, is the most basic problem—single-item (or model) line balancing. Recently, however, it is required to make up several product items on a single assembly line, even for cases of mass production such as in automobile assembly, as was mentioned in Section 4.1—'mixed production'. For such multiple- or mixed-item (or model) line balancing the optimum line balance, as well as the optimal sequence for assembling those product items, has to be decided—a sort of scheduling which will be explained in Chapter 14. (A method of deciding an assembly sequence of multiple items was mentioned as 'modular production' in Section 4.3.)

Problems of Line Balancing

In order to determine a proper line balance, the following basic information is required:

- product items and their production quantities;
- operations or work elements, their times and their sequence for completing each product item;
- structure of assembly line (number of workstations) and its technological abilities, etc.

Then the following two approaches to the assembly-line-balancing problem may be adopted (Moore, 1962):

(1) Find the optimal number of workstations under a fixed cycle time.
(2) Minimise the cycle time, hence, the total delay or idle time given a fixed number of workstations.

Criteria for Line Balancing

The above-mentioned *total delay time* is expressed as:

$$D = \sum_{k=1}^{N}\left(\tau - \sum_{i \in I_k} t_i\right) = N\tau - T \tag{8.13}$$

where τ is the cycle time; N is the number of workstations; t_i is the time required to complete the ith operation (or work element) ($i = 1, 2, \ldots, I$); I is the number of work elements; I_k is a set of work elements assigned on the kth workstation ($k = 1, 2, \ldots, N$), and $T = \sum_{i=1}^{I} t_i$ is the total sum of operation times.

Then the *balancing index* is:

$$E = T/N\tau = 1 - D/N\tau. \qquad (8.14)$$

The line balance is regarded good if this index is above 95%.

8.5.2 Optimal Number of Workstations

Basic Calculation for the Number of Workstations

To begin with, it is necessary to obtain a solution to the line-balancing problem such that the operation time for each work element is equal to or less than the cycle time:

$$t_i \leq \tau \qquad (i = 1, 2, \ldots, I). \qquad (8.15)$$

The minimum possible number of workstations for an assembly line is given as

$$N_{\min} = \min\left(\text{integer } N \,\bigg|\, N \geq \sum_{i=1}^{I} t_i/\tau\right) = \lfloor T/\tau \rfloor \qquad (8.16)$$

where $\lfloor A \rfloor$ stands for a least integer equal to or larger than A.

If the above equation is met, the assembly line is in a minimal balance, which is the most optimal solution; in practical cases, however, this balance is seldom achieved.

It is necessary that the number of workstations is never less than the number of work elements having operation time greater than half of the cycle time; hence, the minimum feasible number of workstations is:

$$N_{\text{feas}} = (\text{number of } i \,|\, t_i > \tau/2). \qquad (8.17)$$

On the other hand, the maximum possible number of workstations for an assembly line is the number of work elements when their operation times are almost equal to the cycle time:

$$N_{\max} = I. \qquad (8.18)$$

As a result, we must settle the optimal number of workstations N^* in the following range:

$$\max(N_{\min}, N_{\text{feas}}) \leq N^* \leq N_{\max}. \qquad (8.19)$$

--- **EXAMPLE 8.15** ---

Find the optimal number of workstations for assembling a product with an assembly sequence and operation times as shown in Figure 8.15, where the cycle time is given as 10 min.

In this case, $I = 10$, $T = 47$, $N_{\min} = 5$, and $N_{\text{feas}} = 4$. Hence, $5 \leq N^* \leq 10$.

The optimal number of workstations is settled in a minimal balance such that $N^* = 5$. The grouping is achieved by constructing a chain of collection of work elements assignable to workstations step by step, as represented in Figure 8.16 (Jackson's step-by-step enumeration (Jackson, 1956)). The result is indicated by broken lines in Figure 8.15. □

8.5.3 Minimum Cycle Time

Basic Calculation for Cycle Time

The cycle time is given by:

$$\tau = \begin{cases} H/Q & \text{(a)} \\ \max_{1 \leq k \leq I} \sum_{i \in I_k} t_i & \text{(b)} \end{cases} \quad (8.20)$$

where H is the preassigned production period and Q is the planned production quantity. (b) shows the maximum of the sum of processing times for the work elements assigned to I workstations.

Work elements are grouped and assigned to each workstation, satisfying the technological precedence relationship, such that the sum of idle times—$\tau - \sum_{i \in I_k} t_i$—which occur at the workstations, or the total delay time D, is minimised subject to the cycle time obtained above.

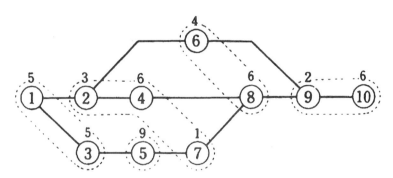

Figure 8.15 Precedence graph for a line-balancing problem. Numbers in circles designate the indices of work elements or operations constituting the assembly work, and those above circles are operation times of work elements in minutes. Grouping of work elements for the minimum number of workstations, in this case, 5, is indicated by broken-line enclosures.

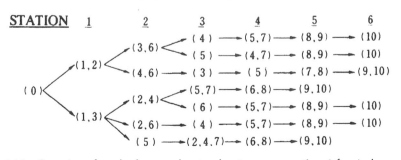

Figure 8.16 Grouping of work elements by step-by-step enumeration (after Jackson, 1956). In each of the parentheses a set of work elements assigned to each workstation is represented. The minimum number of workstations is found to be 5.

EXAMPLE 8.16

Find the minimum cycle time for assembling a product with an assembly sequence and operation times as shown in Figure 8.17, such that a weekly demand of 650 pieces can be met with a 40-hour week.

The theoretically maximum permissible cycle time is:

$\tau_{theo} = 40 \times 60/650 = 3.69$ (min).

Since $T = 13.7$ (min), the corresponding minimum number of workstations is, from (8.16):

$N_{theo} = \lfloor 13.7/3.7 \rfloor = 4$.

A solution for this is illustrated in Table 8.5; the grouping of work elements on the four workstations is indicated by broken lines in Figure 8.17. Thus a minimum of four workstations results in the minimum cycle time of 3.6 minutes, producing $40 \times 60/3.6 = 666$ pieces, which suffices for the demand from the market.

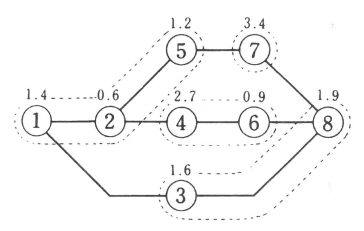

Figure 8.17 Precedence graph for a line-balancing problem. Numbers in circles designate the indices of work elements or operations constituting the assembly work, and those above circles are operation times of work elements in minutes. Grouping of work elements for the minimum cycle time, in this case, 3.6 min, is indicated by broken-line enclosures.

Table 8.5 Assignment of work elements to four workstations and calculation of the minimum cycle time as well as idle time.

Workstation number	1	2	3	4	
Work element	1 2 5	4 6	7	3 8	
Total work time (min)	3.2	3.6	3.4	3.5	
Cycle time (min)		3.6 3.6			
Idle time (min)	0.4	0	0.2	0.1	Total 0.7

In this case, $D = 1.9$. Then the balancing index is, from (8.14):

$$E = \left(1 - \frac{0.7}{4 \times 3.6}\right) = 95\%.$$

Hence this line balance is considered good.

(Note that, if the number of workstations is set at 5, the cycle time reduces further, but the total delay time increases: Question 8.21.) □

References

BELLMAN, R. (1957) *Dynamic Programming* (Princeton, NJ: Princeton University Press), p. 83.
HARRINGTON, J., Jr. (1973) *Computer Integrated Manufacturing* (New York: Industrial Press), pp. 36–38.
JACKSON, J.R. (1956) A computing procedure for a line balancing problem, *Management Science*, **2** (3).
MOORE, J.M. (1962) *Plant Layout Design* (New York: Macmillan), Chap. 19.
NIEBEL, B.W. (1982) *Motion and Time Study* (7th edn) (Homewood, IL: Irwin), pp. 174–176.
TIMMS, H.L. and POHLEN, M.F. (1970) *The Production Function in Business—Decision Systems for Production and Operations Management* (3rd edn) (Homewood, IL: Irwin), p. 301.

Supplementary reading

CHOW, W.-M. (1990) *Assembly Line Design* (New York: Marcel Dekker).
HALEVI, G. and WEILL, R.D. (1995) *Principles of Process Planning* (London: Chapman & Hall).

CHAPTER NINE

Layout Planning and Design

9.1 Scope and Problems of Layout Planning

Meaning of Layout Planning

After the optimum work flow has been determined to change the form from raw material and bought-in components to the finished product through the conversion process, as established by process planning, the next problem is to determine a spatial location for a collection of physical production facilities associated with that specified work flow, in connection with the operators, the plant location and the site. This decision system is called *layout planning/design*. In particular, when referring to the design of layouts for production plants, the term 'plant layout' is used.

Aims of Layout Planning

(1) *Efficiency of production*. Products can be produced in such amounts and at such times required by production release based upon market demands.
(2) *Stability of utilisation of production facilities*. The production capacity and the operative manpower (the number of operators) are well balanced, and the utilisation of production facilities is high and stable.
(3) *Small work-in-process inventories*. The smooth work flow results in the minimal work-in-process inventories between workstations.
(4) *Flexibility and adaptability of production*. Various kinds and amounts of products can be manufactured, adapting to the dynamically changing market; further, the plant holds the possibility for future expansion.
(5) *Economy of production*. It is possible to produce products economically, at the optimum minimum cost.

Fundamental Patterns of Plant Layout

The pattern of the plant layout is basically decided by the relationship between the number of products P and the production quantity Q. As a result of this $P-Q$

analysis (or *volume–variety analysis*) a *P–Q chart* is constructed by arranging products in descending order of Q, as illustrated in Figure 9.1. This chart suggests the basic search for alternative layouts, which include the following three types.

(I) *Product* (or *flow-line* or *production-line*) *layout*. In the case of a large ratio of Q/P, continuous mass production is justified, and production facilities and auxiliary services are located according to the process route for producing the product.
(II) *Process* (or *functional*) *layout*. In the case of a small ratio of Q/P, or jobbing or small-lot production, machines and services of like types are located together as work centres in one area of the plant.
(III) *Group*(*-technology*)(or *cellular*) *layout*. In the case of a medium-sized ratio of Q/P where the various products can be grouped into several group cells, these grouped items are produced as apparent lots, and machines and services are arranged to meet this type of production, based upon the group-technology (GT) principles. This layout leads to *cellular manufacturing*.

—— EXAMPLE 9.1 ——

The above three types of layout pattern are illustrated in Figure 9.2.   □

In these layout patterns workpieces move through the plant. Another type of layout is where production equipment and tools are brought to one place, where operations and assembly are implemented, such as ship building in a yard or building construction on a land space.

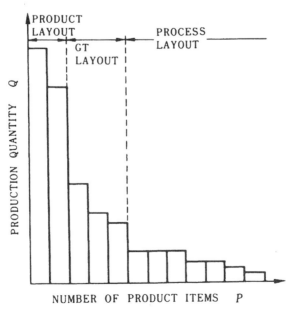

Figure 9.1 An illustration of the P–Q chart. This is constructed by arranging products P in descending order of production quantity Q. The basic search for three alternative layouts (product layout, process layout, and group-technology (GT) layout) is suggested by this chart.

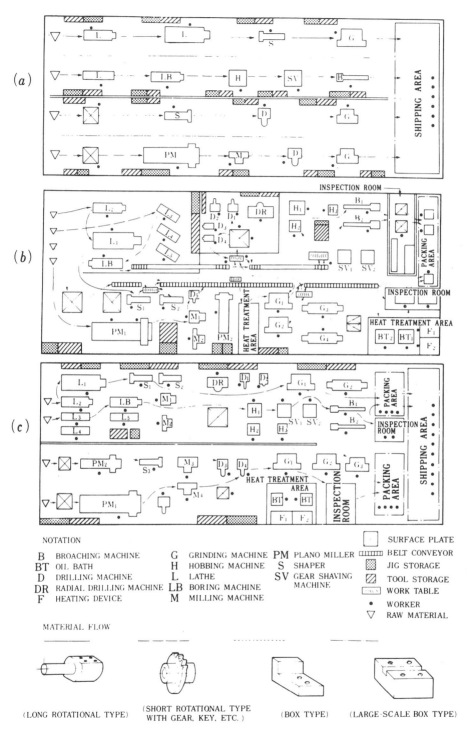

Figure 9.2 Typical illustration of three types of layout pattern: (a) product or flow-line layout for a large ratio of quantity Q over number of product P is the flow-type; (b) process or functional layout for a small ratio of Q/P locates machines of like types as a work centre; and (c) group-technology (GT) or cellular layout for medium-sized Q/P is a pattern between (a) and (b).

Plant Layout Design

The design of plant layout is mostly made in a heuristic way by a choice of one layout from among the above alternatives such that appropriate criteria may be 'satisfied'. Some of the objectives for this layout planning are (Francis *et al.*, 1992):

(1) minimise the overall production time;
(2) minimise the overall production cost;
(3) minimise the material-handling time and cost;
(4) minimise variation in types of material-handling equipment;
(5) minimise investment in equipment;
(6) utilise existing space most effectively;
(7) maintain flexibility of arrangement and operation.

9.2 Systematic Layout Planning (SLP)

Basics of SLP

Various procedures for layout planning have been proposed and developed. Most of them use some sort of heuristic approach, since optimisation analysis is rather difficult for both process and layout planning. An organised approach, referred to as *systematic layout planning* or simply *SLP*, developed by Muther (1973), has received considerable publicity due to its practical application in determining an appropriate 'best' layout plan.

Layout planning proceeds in the following four phases:

(1) *location*—determining the plant site to be laid out;
(2) *general overall layout*—establishing the general arrangement of the area to be laid out from the basic flow patterns;
(3) *detailed layout plans*—establishing the detailed actual placement of each specific physical machine and equipment;
(4) *installation*—executing the layout plan.

Among these phases, general overall layout and detailed layout plans are especially important phases of layout planning which will be made in sequence in SLP, both according to essentially the same SLP procedures.

The SLP procedure is depicted in Figure 9.3. The following five key factors are considered in this layout design process.

- Product P: What is to be produced?
- Quantity Q: How much of each item will be made?
- Routing R: How will each item be produced?
- Supporting services S: With what support will production be backed?
- Time T: When will each item be produced?

The layout procedure is based directly on three fundamentals, which are always at the heart of any layout project:

(1) *relationships*—the relative degree of closeness desired or required among things;
(2) *space*—the amount, kind, and shape or configuration of the things being laid out;
(3) *adjustment*—the arrangement of things into a realistic best fit.

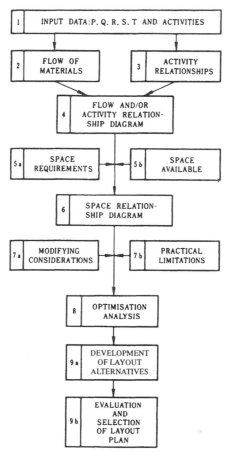

Figure 9.3 The systematic layout planning (SLP) procedures, after Muther (1973), with modification for determining an appropriate 'best' layout plan.

The SLP Procedure

Basic steps involved in SLP are as follows (refer to Figure 9.3).

(1) *Input data.* The preliminary planning step of SLP is to input five key factors: P, Q, R, S, and T. An analysis over a moderately large term of P and Q individually and in their relationship is especially important.

(2) *Flow of materials.* This determines the most effective process or routing R by selecting the operations and sequences that will optimally produce P and Q in the specified T, which is process planning. For this flow analysis, techniques of process charting for a single item and for multi-product output are useful. Flow-of-materials analysis also includes the frequency or magnitude of materials movement. When the products or parts under consideration are very numerous, the *from–to* (or *cross* or *travel*) *chart* is convenient to employ. This chart lists the distance or cost data, as well as the frequency of transportation between each pair of activities or work centres concerned with the flow of those materials. This is illustrated in Table 24.2.

(3) *Activity relationships.* The flow of materials is a common basis for layout arrangements, but it is idealistic in most circumstances; in practice, the related

supporting services must integrate with the layout to ensure efficient productive activities. Further, in some cases operations are dangerous, noisy, etc. and must be isolated from the main flow of materials. Consequently a systematic way of relating the main activities concerning the flow of materials to those concerning the supporting services is necessary. The *activity relationship (REL) diagram* is often used for this purpose and indicates the relative importance and closeness between the two for every activity and every supporting service with reasons coded on a form, as illustrated in Figure 9.4.

(4) *Flow and/or activity relationship diagram*. After carrying out a flow-of-material analysis and an activity relationship charting, a combination of the two is diagrammed in the form of a *flow and/or activity relationship diagram*. This diagram relates the various activities or departments geographically with their relative closeness and intensity to each other without any regard to the actual spaces required. Use is made of operation symbols for the activities, identification numbers or letters for those activities, and a number of lines or colour codes, which express the relationship and the frequency or closeness value between two activities.

(5) *Space determination*. The space or area of each piece of machinery, equipment, and service facilities required for producing the products is determined. This is processed by first determining the standard area needed for installing each facility, then multiplying this by the number of facilities required, and, finally, adding any extra space which may be required, e.g. gangways. The number of facilities required are calculated by dividing the number of pieces per hour to meet production requirements by production rate produced by each facility per hour, or alternatively by dividing the available capacity of a facility by the capacity required to meet production demand. The total sum of space requirements must be balanced against the space available.

(6) *Space relationship diagram*. The area to be allowed for each activity is shown on the flow and/or activity relationship diagram, resulting in the *space relationship diagram*. This is essentially a rough layout plan and is illustrated in Figure 9.5, which was generated from the REL diagram shown in Figure 9.4.

(7) *Adjusting the diagram*. The space relationship diagram represents a theoretically ideal arrangement and is seldom usable. It is adjusted and manipulated to incorporate modifying considerations such as the material-handling methods, storage facilities, site conditions and surroundings, personnel requirements, building features, utilities and auxiliaries, procedures and controls, shape of detailed activities layouts, etc. and practical limitations like cost, safety, and employee preference.

(8) *Optimisation analysis*. Where a certain evaluation criterion is decided for layout planning, optimisation analysis is made such that the relative location of areas for the activities involved is optimised or satisfied when the optimisation is impossible, e.g. by repeatedly replacing the relative allocation of two or three departments until no further decreases in transportation costs are generated.

(9) *Evaluating and determining the best layout*. After abandoning those plans which do not seem worthy, relatively few layout proposals will remain. Each has its particular advantages and disadvantages. Which of these alternative plans to select as a 'best' plan is now decided by means of balancing advantages against disadvantages, the factor-analysis method, or cost comparison including both capital expenditures and operating expenses. The *factor-analysis method* is said to be the most effective general method of

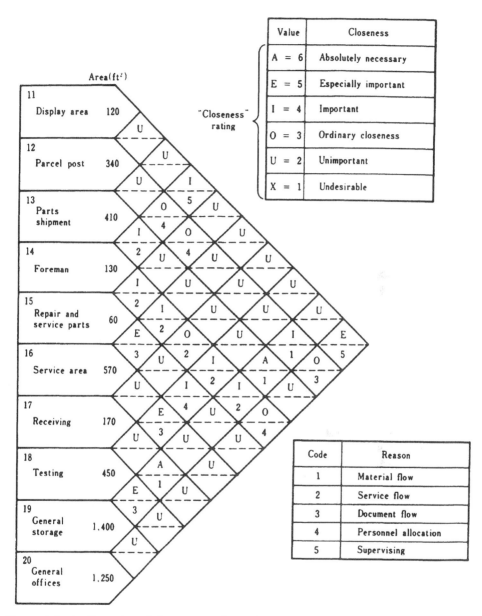

Figure 9.4 An illustration of the activity relationship (REL) diagram. This indicates the relative importance and closeness between two activities involved in the productive work (after Moore, 1971, reproduced by permission of the Institute of Industrial Engineers).

evaluating layout alternatives; its procedure is:
(a) List all of the factors to be considered important.
(b) Weight the relative importance of each of these factors to each other.
(c) Rate the alternative plans against one factor at a time.
(d) Calculate the weighted rating values and sum up those values to obtain the total value for each of the alternatives.
(e) Select the alternative with the highest total value.

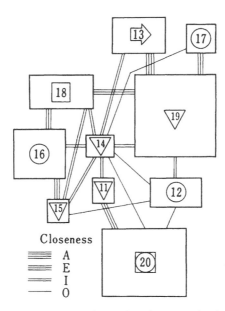

Figure 9.5 An illustration of the space relationship diagram. This is essentially a rough layout plan and is generated from the activity relationship diagram shown in Figure 9.4.

Table 9.1 Mechanism of the factor-analysis method for evaluating several alternatives. A plan with the largest total of weighted rating values is selected (after Muther, 1973, courtesy of Cahners Books).

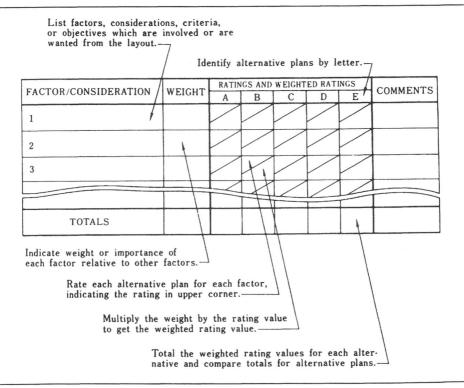

LAYOUT PLANNING AND DESIGN

This procedure is set out in Table 9.1. Finally a scale model or a virtual reality model of the selected layout may be used to aid visualisation of the selected layout.

9.3 Mathematical Layout Design

9.3.1 Machine Assignment

Machine Assignment Problem

In the case that it is predecided that N machine tools are to be placed in N positions on a shop floor, an optimal layout can be determined to minimise the total cost z required to allocate all the machines.

Introducing the binary variable x_{ij} which is 1 if machine $i(M_i)$ is placed in position $j(P_j)$ with a cost of c_{ij} and 0 otherwise, the machine assignment problem is formulated as follows:

$$\text{minimise } z = \sum_{i=1}^{N} \sum_{j=1}^{N} c_{ij} x_{ij} \tag{9.1}$$

$$\text{subject to } \sum_{j=1}^{N} x_{ij} = 1 \quad (i = 1, 2, \ldots, N) \tag{9.2}$$

$$\sum_{i=1}^{N} x_{ij} = 1 \quad (j = 1, 2, \ldots, N) \tag{9.3}$$

$$x_{ij} = 0 \text{ or } 1 \quad (i, j = 1, 2, \ldots, N). \tag{9.4}$$

Constraint (9.2) implies that machine M_i is placed in one of the positions $\{1, 2, \ldots, N\}$, and (9.3) implies that one of the machines $\{1, 2, \ldots, N\}$ is allocated to position i.

Solving the Machine Assignment Problem

The following proposition is useful to solving the current problem (Sasieni et al., 1959).

── **PROPOSITION 9.1** ──────────────────────────

The optimal solution is invariable, even if a constant is added to or subtracted from every element of a row or a column in the cost matrix (c_{ij}). □

──────────────────────────

Using the above proposition, first we construct a revised cost matrix with every row and every column having at least one zero element. Then the machines are assigned to such elements whenever possible.

── **EXAMPLE 9.2** ──────────────────────────

(Machine assignment) A cost matrix is given by Table 9.2 for assigning five machines in five positions. Following the proposition, 1, 3, 6, 10, and 2 are subtracted from every element of rows 1 to 5 respectively, and subsequently 2 and 5 from every element of columns 3 and 5, resulting in Table 9.3(a). In this table a unique element having 0 in every row or column is marked with □; all these positions are selected. ☒ means positions eliminated during the search process. ✓ means rows and columns which have not yet been considered for assignment.

In this example, subtracting 3 from each element in row 2, an element having 0 in row 2 and column 5 results and completes the machine assignment (b)—the optimal solution is obtained; that is, place M_1 to P_2, M_2 to P_5, M_3 to P_3, M_4 to P_1, and M_5 to P_4. □

9.3.2 Heuristic Algorithms for Layout Design

Otimisation of Plant Layout

Denoting the distance between the positions of facilities (machines) i and j by d_{ij}, the number of workpieces between these places by v_{ij}, and the cost of transferring a workpiece a unit distance by u_{ij}, an objective is to minimise the total transportation cost z. This is formulated as follows:

$$\text{minimise } z = \sum_{i=1}^{N} \sum_{j=1}^{N} d_{ij} w_{ij} \tag{9.5}$$

where $w_{ij} = u_{ij} v_{ij}$.

Table 9.2 Cost matrix for machine assignment.

Machine \ Position	P_1	P_2	P_3	P_4	P_5
M_1	3	1	16	11	17
M_2	3	9	20	3	11
M_3	14	13	8	6	11
M_4	10	16	16	15	22
M_5	3	7	11	2	16

Table 9.3 Solution process for machine assignment, based upon the basic data indicated in Table 9.2.

2̶	0̶	1̶3̶	1̶0̶	1̶1̶	
☒	6	15	☒	3	✓
8̶	7̶	0̶	☒	☒	
⓪	6	4	5	7	✓
	5	7	⓪	9	✓
✓			✓		

(a)

5	⓪	13	13	11
☒	3	12	☒	⓪
11	7	⓪	3	☒
⓪	3	1	5	4
1	2	4	⓪	6

(b)

LAYOUT PLANNING AND DESIGN

For example, denoting the coordinates of the places of facilities i and j by (x_i, y_i), and (x_j, y_j), two alternative distances d_{ij} are expressed by:

$$d_{ij} = \begin{cases} \sqrt{(x_i - x_j)^2 + (y_i - y_j)^2} \\ |x_i - x_j| + |y_i - y_j| \end{cases}. \quad (9.6)$$

Heuristic Approach to Deciding Near-optimal Machine Layout

A heuristic approach—a sort of 'pattern search'—is taken to decide near-optimal solutions for machine layout, minimising the objective function (9.5) (Buffa *et al.*, 1964).

Heuristic algorithm for layout for minimising the total transportation cost:

Step 1: Decide an initial feasible layout pattern.
Step 2: Calculate matrices (d_{ij}) and (w_{ij}).
Step 3: Obtain the amounts of cost reduction when any two machines are exchanged. If no positive value exists, stop (a near-optimal solution has been obtained). Otherwise, proceed.
Step 4: Exchange two machines—p and q—with the largest positive value, recalculate matrix (d_{ij}), and return to step 3. □

The following cost reduction is expected in step 4 above:

$$\Delta z = \sum_{j=1}^{N} d_{pj} w_{pj} + \sum_{j=1}^{N} d_{qj} w_{qj} - \sum_{j=1}^{N} d_{pj} w_{qj} - \sum_{j=1}^{N} d_{qj} w_{pj} - 2 d_{pq} w_{pq}. \quad (9.7)$$

The above algorithm is employed in computerised layout planning—CRAFT—as will be discussed in Section 24.5.3.

9.4 Production Flow Analysis

Cellular Layout by Production Flow Analysis

A method of constructing a cellular (or GT) layout is *production flow analysis* originated by Burbidge (1989). This classifies work flows logically and arranges production facilities to several cells, using, e.g. cluster analysis which deals with 'similarity'.

Table 9.4 Job work flow.

Job	M_1	M_2	M_3	M_4	M_5	M_6
J_1	①			③	②	
J_2		①				
J_3			②		①	
J_4	①		③	②		
J_5		②				①

Table 9.5 Rearranging job–machine combination, constructing two groups, based upon the basic data indicated in Table 9.4.

Machine Job	M_1	M_5	M_4	M_3	M_6	M_2
J_1	①	②	③			
J_3		①		②		
J_4	①		②	③		
J_2						①
J_5					①	②

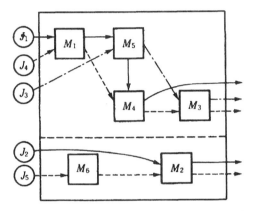

Figure 9.6 Construction of a cellular layout by production flow analysis, based on the basic data indicated in Table 9.4. This is 'cellular manufacturing'.

── EXAMPLE 9.3 ──

Work flows for five jobs—the technological processing order being indicated by ①, ②, ③ for each job—are given by Table 9.4. Combinations of jobs and machines are rearranged as shown in Table 9.5; we obtain two cells—machine groups. Two patterns of cellular layout are constructed as depicted in Figure 9.6. □

References

BUFFA, E.S., ARMOUR, G.C. and VOLLMAN, T.E. (1964) Allocating facilities with CRAFT, *Harvard Business Review*, **42** (2).
BURBIDGE, J.L. (1989) *Production Planning* (Oxford: Oxford University Press).
FRANCIS, R.L., MCGINNIS, L.F. Jr. and WHITE, J.A. (1992) *Facility Layout and Location* (Englewood Cliffs, NJ: Prentice-Hall), pp. 30–31.
MOORE, J.M. (1971) Computer program evaluates plant layout alternatives, *Industrial Engineering*, **3** (8).
MUTHER, R. (1973) *Systematic Layout Planning* (2nd edn) (Boston, Mass: Cahners Books).
SASIENI, M., YASPAN, A. and FRIEDMAN, L. (1959) *Operations Research* (New York: Wiley), p. 186.

Supplementary reading

FUJIMOTO, Y. (1993) *Management and Locations for Production* (in Japanese) (Tokyo: Koyo-shubo).

CHAPTER TEN

Logistic Planning and Design

10.1 Transportation Problems

10.1.1 Transportation Problems Defined

Problems such as the allocation of the raw materials purchased from various suppliers to various manufacturing divisions and delivery of the finished goods produced in various factories to various distribution depots/markets are called *transportation problems*.

Transportation-type Linear Programming Models

A basic transportation problem is formulated as follows: M sources of supply (e.g. factories) deliver quantities of a product to N delivery points (markets) (henceforth called 'sinks'), as depicted in Figure 10.1. The quantities available at the sources, the quantities required at the sinks, and the unit cost of transportation from each source to each sink are shown on the matrix represented in Table 10.1.

Table 10.1 Cost matrix with supply and demand for the transportation problem. This problem determines the optimal quantity to be transported from each factory to each market such that the total transportation cost is minimised.

Factory from which transported	Destination market				Supply
	1	2	...	N	
1	c_{11}	c_{12}	...	c_{1N}	a_1
2	c_{21}	c_{22}	...	c_{2N}	a_2
⋮		...	...	...	⋮
M	c_{M1}	c_{M2}	...	c_{MN}	a_M
Demand	b_1	b_2	...	b_N	

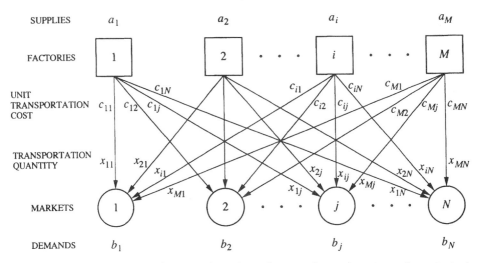

Figure 10.1 Transportation planning of products from M factories to N markets. A single product is manufactured at M factories, and delivered to N markets with transportation costs. The optimum transportation quantities are decided by transportation-type linear programming.

When supply and demand are in balance,

$$\sum_{i=1}^{M} a_i = \sum_{j=1}^{N} b_j \qquad (10.1)$$

where a_i is the supply which is available at factory i ($=1, 2, \ldots, M$) and b_j is the demand needed at market j ($=1, 2, \ldots, N$).

If supply and demand are not in balance a dummy source or dummy sink is introduced to account for excess demand or excess production respectively.

Designating the quantity to be transported from source i to sink j by x_{ij}, we can then formulate a transportation problem for finding x_{ij} so as to achieve the least cost z:

$$\text{minimise } z = \sum_{i=1}^{M}\sum_{j=1}^{N} c_{ij} x_{ij} \qquad (10.2)$$

$$\text{subject to } \sum_{j=1}^{N} x_{ij} = a_i \quad (i = 1, 2, \ldots, M) \qquad (10.3)$$

$$\sum_{i=1}^{M} x_{ij} = b_j \quad (j = 1, 2, \ldots, N) \qquad (10.4)$$

$$x_{ij} \geq 0 \quad (i = 1, 2, \ldots, M; \ j = 1, 2, \ldots, N) \qquad (10.5)$$

and (10.1), where c_{ij} is the transportation cost per unit product from factory i to market j.

The above model is called a *transportation-type linear program* and it is a special-type linear program. (The general linear programming (LP) model will be explained in detail in Chapter 13.) In the following this special LP is solved with a numerical example.

10.1.2 Solution Procedure for Transportation-type Linear Program

Basic Proposition for Solving Transportation-type Linear Program

Since (10.1) is equal to $\sum_{i=1}^{M} \sum_{j=1}^{N} x_{ij}$, we have $(M + N - 1)$ independent variables in total. Among $M \times N$ x_{ij}s, $[MN - (M + N - 1)]$ variables (called *non-basic variables*) are set at zero, and the remaining $(M + N - 1)$ variables (called *basic variables*) are nonzero, satisfying (10.3) and (10.4). Hence, where $a_i \neq 0$ $(i = 1, 2, ..., M)$ and $b_j \neq 0$ $(j = 1, 2, ..., N)$, at least one nonzero element should exist in each row and in each column.

Introducing new variables, u_i $(i = 1, 2, ..., M)$ and v_j $(j = 1, 2, ..., N)$, as

$$z = \sum_{i,j} c_{ij} x_{ij} - \sum_i u_i \left(\sum_j x_{ij} - a_i \right) - \sum_j v_j \left(\sum_i x_{ij} - b_j \right)$$

$$= \sum_{i,j} (c_{ij} - u_i - v_j) x_{ij} + \sum_i u_i a_i + \sum_j v_j b_j \quad (10.6)$$

we obtain the following conclusion.

THEOREM 10.1

At the optimal solution for the transportation-type linear program,

(a) $c_{ij} = u_i + v_j$ for basic variables, $\quad$ (10.7)
(b) $s_{ij} = c_{ij} - (u_i + v_j) \geq 0$ for non-basic variables. $\quad$ (10.8)

□

$(M + N)$ new variables, $u_i (i = 1, 2, ..., M)$ and v_j $(j = 1, 2, ..., N)$, can be uniquely determined by (10.7), setting an arbitrary u_i or v_j as zero. Hence, as far as condition (10.8) is not satisfied, we can further reduce the total transportation cost by successively exchanging such a non-basic variable having the largest s_{ij} with a certain basic variable.

Obtaining an Initial Feasible Solution for the Transportation-type Linear Program

There are three ways to obtain an initial (starting) feasible solution as follows (Hitomi, 1978).

(1) *Northwest corner rule.* We begin in the northwestern corner of the routing table composed of x_{ij}s, allocating to the (1,1) cell as much as we need for sink 1 or as much as we can supply from source 1, whichever is less, namely, $\min(a_1, b_1)$. If the need of sink 1 is satisfied, no assignment to any other cell in the first column would be appropriate, so we move to the (1, 2) cell of the first row and allocate $\min(a_1 - b_1, b_2)$ to this cell; and so on. Continuing in this way, we reach the southeastern corner with an assignment that meets all requirements and exhausts the supplies.
(2) *Least unit transportation cost rule.* Starting with a combination of factory and market having least unit transportation cost, we allocate to this cell as much as is needed; no assignment is needed to any other cell in this row when all products are exhausted or in this column when all demands are satisfied. The

same process is continued further for combinations of factory and market with the second, third, ..., least unit transportation cost.

(3) *VAM (Vogel's approximation method)*. In each row or in each column, the difference between the second least and the least unit transportation costs is calculated; we assign a cell with the least unit transportation cost in the row or in the column having the greatest difference. Deleting this row or column, the same procedure is continued. This method is said to generate an initial feasible solution relatively near the optimal one.

Transportation Algorithm

The general solution procedure for solving the transportation-type linear programming problem is obtained as follows:

Step 1: Generate an initial feasible starting solution.
Step 2: Select the row (or column) with the largest number of basic variables. Assign that row (or column) $u_i = 0$ or $v_j = 0$. Compute all other u_i and v_j values to the basic variables using (10.7).
Step 3: Calculate s_{ij} for all non-basic variables using (10.8).
Step 4: If $s_{ij} \geq 0$ for all non-basic variables, the current solution is optimal (stop). Otherwise, introduce into the basis that cell with the largest negative s_{ij} value, draw a stepping stone path from that cell, and determine the maximum to which it can be raised. Update the tableau and return to step 2. □

The *stepping stone path method* employed in step 4 is a procedure to enter a cell with the largest negative s_{ij} value into the basis and assign the maximum amount to which it can be raised. In compensation for this, the values of basic variables in the row and in the column containing the cell are decreased, and further, the values of basic variables in the row and in the column containing the decreased basic variables are increased, and so forth. Finally, a stepping stone path is created so as not to violate the constraints — (10.3) and (10.4).

── **EXAMPLE 10.1** ──

(Transportation problem) Three factories produce a product to deliver to four markets. Quantities of supply and demand and unit transportation cost are shown in Table 10.2, where supply and demand are in balance.

Table 10.2 Basic data of cost matrix with supply and demand for the transportation problem.

Factory from which transported	Destination market				Supply
	1	2	3	4	
1	2	3	5	1	7
2	7	3	4	6	9
3	4	1	7	2	18
Demand	5	8	7	14	

Table 10.3 Application of the transportation algorithm to a problem indicated in Table 10.2. The initial tableau (a) is obtained by the northwest corner rule. This is updated as (b), and finally an optimal solution is obtained as (c). Dashed lines in (a) and (b) indicate the stepping stone path. The total transportation cost reduces from 102 for (a) to 82 for (b), and the minimum cost is obtained in (c) as 76.

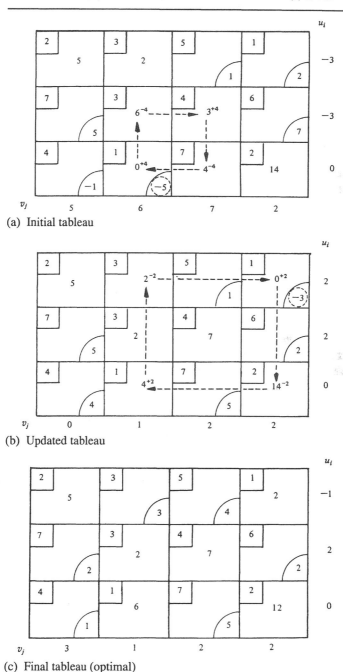

(a) Initial tableau

(b) Updated tableau

(c) Final tableau (optimal)

An initial feasible starting solution ($z = 102$) is obtained by the northwest corner rule (step 1). u_is and v_js and then s_{ij}s are determined by step 2 and step 3, as shown in Table 10.3(a).

Following step 4, cell (3,2) which has the largest negative s_{ij} value ($= -5$), is now introduced into the basis. It can be raised to 4 by drawing a stepping stone path: $(3,2) \to (2,2) \to (2,3) \to (3,3) \to (3,2)$.

Now we obtain the updated tableau ($z = 82$), as shown in Table 10.3(b).

Returning to step 2 and repeating the same procedure, the updated tableau is further updated ($z = 76$), as shown in Table 10.3(c), which is the final tableau indicating an optimal solution, since all $s_{ij} > 0$. The optimal solution is $z^* = 76$ (minimum); $x_{11} = 5$, $x_{14} = 2$, $x_{22} = 2$, $x_{23} = 7$, $x_{32} = 6$, $x_{34} = 12$, and the other x_{ij}s are all 0. □

10.2 Distribution Problems

Travelling Salesperson Problems

A distribution for minimising the total transportation distances (or times) needed to circulate the products made in a factory among several places (markets) is called the *travelling salesperson problem*.

Solving Travelling Salesperson Problems

Travelling salesperson problems can be solved by the following methods (Conway *et al.*, 1967).

- The *branch-and-bound method* for obtaining the optimal solution. However, this method requires a great deal of computation time as the size of the problem becomes large (NP-hard problems).[1] This algorithm will be explained in Section 14.2.4, being applied to solving large-scale flow-shop scheduling problems.
- The *dynamic programming approach* for obtaining the optimal solution, by constructing the recursive functional equation based upon the principle of optimality, which was discussed in Section 8.4.2.
- The *heuristic approach* by starting from the factory and, first, taking a route to the nearest market from the factory, then to the nearest market from the current market, and so on, ... until all the places have been visited.

Table 10.4 Distances between factory F, markets M_1, M_2, M_3, and M_4 (km).

To From	F	M_1	M_2	M_3	M_4
F	—	50	(40)	120	50
M_1	50	—	80	100	(80)
M_2	40	(80)	—	160	90
M_3	(120)	100	160	—	110
M_4	50	80	90	(110)	—

― **EXAMPLE 10.2** ―

(Heuristic approach to solving travelling salesperson problems) The factory (F) has four markets (M_1 to M_4) where the products are demanded, and the distances between F, M_1, ..., and M_4 are given by Table 10.4. A circulation route starting from F, visiting all markets, and coming back to F is obtained by the heuristic approach: $F \to M_2 \to M_1 \to M_4 \to M_3 \to F$ with a total travel distance of 430 km.[2] □

Notes

1. Non-deterministic polynomial; NP-hard problems cannot be solved within polynomial time.
2. The optimal transportation route obtained by the branch-and-bound method is: $F \to M_4 \to M_3 \to M_1 \to M_2 \to F$, with a minimum distance of 380 km.

References

CONWAY, R.W., MAXWELL, W.L. and MILLER, L.W. (1967) *Theory of Scheduling* (Reading, MA: Addison-Wesley), Chap. 4.

HITOMI, K. (1978) *Production Management Engineering* (in Japanese) (Tokyo: Corona Publishing), pp. 73–74.

Supplementary reading

BALLOW, R.H. (1992) *Business Logistics Management* (3rd edn) (London: Prentice-Hall).

BLANCHARD, B.S. (1992) *Logistics Engineering and Management* (Englewood Cliffs, NJ: Prentice-Hall).

GOPAL, C. and CAHILL, G. (1992) *Logistics in Manufacturing* (Homewood, IL: Irwin).

CHAPTER ELEVEN

Manufacturing Optimisation

11.1 Evaluation Criteria for Manufacturing Optimisation

Short History of Economics of Machining

The optimisation analysis of manufacturing has been studied since Gilbert's first work on the economics of machining (Gilbert, 1950). He introduced the 'maximum production rate' and the 'minimum production cost' criteria, under which optimal machining speeds were analysed by developing mathematical models for single-stage manufacturing. In 1964 a new criterion for manufacturing—the 'maximum profit' in a limited time interval—was proposed (Okushima and Hitomi, 1964); the optimal machining speed was derived under this criterion by break-even analysis and by marginal analysis. Since 1966, this third criterion has been named the 'maximum profit rate' (Armarego and Russell, 1966). The term 'profit rate' usually means return on investment in the field of economics and engineering economy. In this book, *profit rate* means profit per unit time interval.

Manufacturing Criteria

Summarising, the following three basic criteria (or principles) are utilised in manufacturing optimisation.

(I) *Maximum-production-rate or minimum-time criterion*. This maximises the number of products produced in a unit time interval; hence, it minimises the production time per unit piece. It is the criterion to be adopted when an increase in physical productivity or productive efficiency is desired, neglecting the production cost needed and/or profit obtained.

(II) *Minimum-cost criterion*. This criterion refers to producing a product at the least cost, and coincides with the *maximum-profit criterion*, if the unit revenue is constant. It is the criterion to be adopted when there is ample time for production.

(III) *Maximum-profit-rate criterion*. This maximises the profit in a given time interval. It is the criterion to be recommended when there is insufficient capacity for a specific time interval.

Where market demands are large compared with productive capacity, a larger total profit is obtained by producing and selling a large amount of products resulting from reducing unit production time and sacrificing unit production cost based upon the minimum-cost criterion. This is the fundamental principle of profit rate.

Usually one of the above three criteria is employed according to the manufacturing objective. In some cases the time minimum and the cost minimum are required simultaneously. This multiple(two)-objective optimisation generates the maximum-profit-rate solution as a preferred solution.

Manufacturing Optimisation

Ideally, optimisation should be applied to a total manufacturing system such as material fabrication (casting, forging, etc.)—part machining (cutting, grinding, etc.)—product assembly; however, the present theory has not been extended to this kind of analysis for total optimisation of a manufacturing system. In this chapter, then, the theory of optimisation for part machining will be explained, first, for single-stage manufacturing, namely, on a single machine tool, and then, for a multiple-stage manufacturing system where many machine tools are sequenced in an order decided by technology. Practically 'machining' is a complex physical phenomenon; accordingly, in an attempt to express the machining state with an abstract mathematical model for optimisation analysis, it is important to make the formulation as simple as possible.

For mathematical optimisation analysis, as mentioned in Section 2.4, there are 'unconstrained optimisation' and 'constrained optimisation'. Both cases are discussed in this chapter for manufacturing optimisation.

Decision Variables for Manufacturing Optimisation

Once it has been reasonably determined which work materials are to be machined, which machine tool utilised, which cutting tool employed, and who is to carry out the work, these are the uncontrollable parameters for manufacturing. Then typical controllable variables are machining conditions; namely,

- depth of cut;
- feed rate;
- machining speed.

Depth of cut can often be determined mainly by the sizes of the work material and the product; hence, it is assumed constant in this chapter. The other two conditions can be arbitrarily determined within ranges set in the machine tool. In unconstrained optimisation of manufacturing, as will be mentioned later, optimality can never be achieved when determining optimal values for both conditions. Hence, in this case, setting a fixed value for feed rate, *optimal machining speed* will be determined. In constrained optimisation, in addition to optimal machining speed, *optimal feed rate* is simultaneously decided. As long as the machine tool is not equipped with a variable speed drive, feed rate and machining speed are both discrete; however, for the sake of simplicity of analysis, these decision variables are often treated as continuous.

Manufacturing optimisation for multistage systems is done by constrained optimisation for the maximum production rate, minimum production cost, and maximum profit rate, generating the optimal machining speeds to be utilised on multiple stages. The maximum-profit-rate principle is more meaningful in this case than in a single-stage case.

11.2 Optimisation of Single-stage Manufacturing

11.2.1 Basic Mathematical Models

Basic Factors in Machining Operation

Optimisation analysis of single-stage manufacturing is the fundamental of optimisation of integrated manufacturing systems. For constructing basic mathematical models based upon three evaluation criteria mentioned in the previous section the following three important factors are formulated:

- unit production time;
- unit production cost;
- profit rate.

Unit Production Time

Unit production time is the time needed to manufacture a unit of product. The shorter this time, the higher the productivity; and the machining conditions for the least unit production time are based upon the minimum-time criterion. It is generally assumed that unit production time comprises the following three time elements:

(1) *Preparation* (or *setup*) *time* t_p (min/pc)—time necessary to prepare for machining. This includes the loading and unloading time of workpieces to a machine tool, the approaching time of a cutting tool to the workpiece, etc.
(2) *Machining time* t_m (min/pc)—time necessary to handle the machine tool for cutting.
(3) *Tool-replacement time* t_r (min/pc)—time required to exchange a worn cutting tool with a new edge or insert a new throw-away tip.

Denoting the time required to replace a worn cutting tool with a new one by t_c (min/edge), and the *tool life*, which is the time from the beginning of using a new edge until its replacement, by T (min/edge), the tool-replacement time per unit piece is:

$$t_r = t_c t_m / T, \tag{11.1}$$

since a cutting edge can machine T/t_m pieces on average during its lifetime.

Accordingly, the unit production time, t (min/pc), is given by:

$$t = t_p + t_m [+t_r] = t_p + t_m [+t_c t_m / T]. \tag{11.2}$$

Automatic machine tools can replace the worn tool during setup of a workpiece; hence the last term of the above equation is eliminated. Brackets in equation (11.2) and those which follow mean this situation.

Production Rate

Production rate is the number of pieces produced per unit time, and is the reciprocal of the unit production time given by (11.2); hence, denoting the production rate by q (pc/min),

$$q = 1/t = 1/(t_p + t_m[+t_ct_m/T]). \tag{11.3}$$

Unit Production Cost

Unit production cost is the cost required to manufacture a unit piece. The machining conditions for the least unit production cost result from the minimum-cost criterion. Unit production cost comprises the following three kinds of cost elements.

(1) *Capacity utilisation cost* u_o ($/pc)—cost of utilising the production facility (machine tool) for making a part. It includes operation and depreciation costs of the production facility, labour cost and overhead.
(2) *Machining cost* u_m ($/pc)—cost needed for the machining time. It includes expenses for electricity, cutting oil, etc.
(3) *Tool cost* u_t ($/pc)—cost of a cutting edge required to produce a piece of product. It includes purchasing cost[s] of the cutting tool (and the grinding wheel, depreciation cost for the tool grinder, direct labour and overhead costs for grinding worn cutting edge), etc.

Denoting the overhead cost per unit time by k_o ($/min), the capacity utilisation cost is expressed by multiplying the unit production time by k_o:

$$u_o = k_o t = k_o(t_p + t_m[+t_c t_m/T]). \tag{11.4}$$

Denoting the machining overhead, such as cost of cutting oil, electricity charge, etc., during actual cutting operation, by k_m ($/min),

$$u_m = k_m t_m. \tag{11.5}$$

Denoting the cost of a cutting edge by k_t ($/edge),

$$u_t = k_t t_m/T, \tag{11.6}$$

since with a cutting edge T/t_m pieces can be produced during its life.

Hence, the unit production cost, u ($/pc), is given by:

$$u = u_o + u_m + u_t = k_o t_p + (k_o + k_m)t_m + ([k_o t_c +]k_t)t_m/T. \tag{11.7}$$

Unit Profit

Unit profit is the gain obtained by producing a unit piece of product. The gross profit per unit piece produced, g ($/pc), is the unit revenue or selling price r_u ($/pc) minus the material cost m_c and the unit production cost u ($/pc):

$$g = r_u - m_c - u = r_0 - (k_o t_p + (k_o + k_m)t_m + ([k_o t_c +]k_t)t_m/T), \tag{11.8}$$

where $r_0 = r_u - m_c$, the unit net revenue ($/pc)—value added.

Profit Rate

According to its definition mentioned previously, *profit rate* p ($/min) is obtained by dividing the unit profit g ($/pc) by the production time t (min/pc), neglecting the law of diminishing returns.

$$p = \frac{g}{t} = \frac{r_0 - (k_m t_m[+ k_t(t_m/T)])}{t_p + t_m[+ t_c(t_m/T)]} - k_o. \tag{11.9}$$

The machining conditions for maximising this profit rate are based upon the maximum-profit-rate criterion.

11.2.2 Single-stage Manufacturing Models as a Function of Machining Speed

Dependent Factors of Machining Speed

Among the three machining variables—machining speed, feed rate, and depth of cut the optimal decision for just one variable can easily be made. In this section, machining speed is chosen as a representative decision variable, and an analysis is developed to determine the optimal machining speed to be utilised on the machine tool. For this purpose basic mathematical models, especially unit production time (11.2), unit production cost (11.7), and profit rate (11.9) are to be expressed as a function of machining speed. The tool life value T (min) and the machining time t_m (min/pc) are the only factors in the basic models which vary with change of machining speed.

Machining Time as a Function of Machining Speed

The machining time is inversely proportional to machining speed v (m/min); hence,

$$t_m = \lambda/v \tag{11.10}$$

where λ is a machining constant, and it is determined, depending upon machining patterns, as follows:

- Turning, boring, drilling, and reaming operations:

$$\lambda = \pi DL/1000s \tag{11.11}$$

  because

$$t_m = L/Ns \tag{11.12}$$

  and

$$v = \pi DN/1000. \tag{11.13}$$

- Milling operation

$$\lambda = \pi D(l+L)/1000sZ \tag{11.14}$$

  because

$$t_m = (l+L)/sNZ \tag{11.15}$$

  and because of (11.13).

In the above equations D is the machining diameter (mm) for turning and boring operations and tool diameter (mm) for milling, drilling, and reaming operations; L is the machining length (mm) for turning and milling operations and length of the hole to be machined (mm) for boring, drilling, and reaming operations; l is the additional feed length (mm) for milling operation; N is the main spindle speed (rev/min); s is the feed rate per tooth (mm/tooth) for milling operation and the feed rate per revolution (mm/rev) for other operations; and Z is the number of teeth of the milling cutter (see Figure 11.1).

Tool-life Equation

The life of a cutting tool is the time until the width of the flank wear (or the depth of the crater wear on the rake face) reaches a certain amount (e.g. 0.5–1 mm for rough machining/0.2–0.3 mm for finish). The amount of wear increases with time, depending on the cutting speed v, as depicted in Figure 11.2.

The relationship between the tool life and machining conditions is expressed by the tool-life equation. The most typical model is the *Taylor tool-life equation*, which shows the relationship between tool life T (min) and machining speed v (m/min) as follows:

$$vT^n = C, \tag{11.16}$$

where C and n are constants depending upon kinds of cutting tool used and work material to be machined, with or without cutting fluids, machine tools used, other

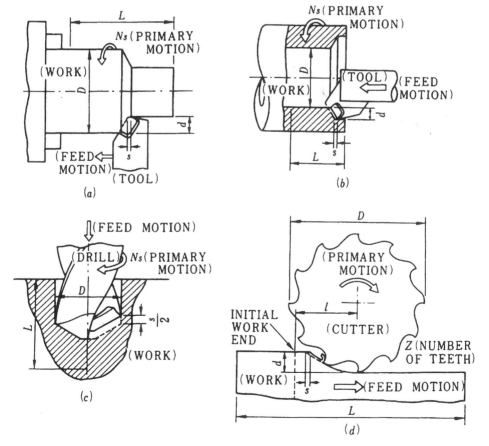

Figure 11.1 Explanation and dimensions of (a) turning, (b) boring, (c) drilling and (d) milling operations.

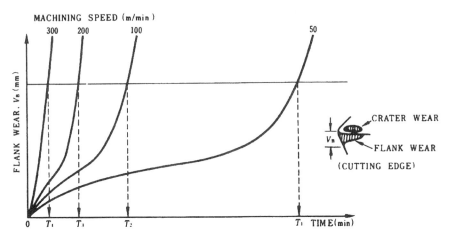

Figure 11.2 Illustration of wear-out failure. A cutting tool wears out with time. Its life is in general described as the time at which flank or crater wear reaches a certain amount, and is greatly affected by machining conditions, especially machining speed; the tool life decreases as the machining speed increases. (Also see Figure 11.3).

machining conditions, etc. C is the 1-minute tool-life machining speed, and n is the slope of the tool-life curve (straight line)—in general between 0 and 1, usually 0.1–0.4.

The above *tool-life* curve is a straight line on bilogarithmic paper, as indicated in Figure 11.3.

If the tool life is expressed as the cutting length T'(mm) during the lifetime T, as is often done for milling, drilling, and reaming operations, with constants C' and n' in place of C and n, transform C' and n' into C and n as follows:

$$C = \begin{cases} (C'/(1000s/\pi D)^{n'})^{1/(1+n')} & \text{for drilling and reaming operations} \\ (C'/(1000sZ/\pi D)^{n'})^{1/(1+n')} & \text{for milling operation} \end{cases} \quad (11.17)$$

$$n = 1/(1/n' + 1) \quad (11.18)$$

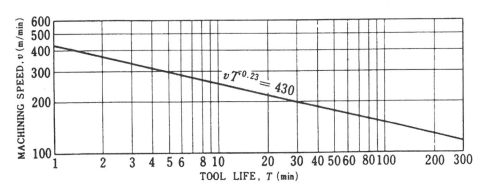

Figure 11.3 The Taylor tool-life curve. It shows a linear relationship between tool life and machining speed on a bilogarithmic graph.

MANUFACTURING OPTIMISATION   161

since there is a relationship between T and T':

$$T' = \begin{cases} 1000svT/\pi D \text{ for drilling and reaming operations} \\ 1000sZvT/\pi D \text{ for milling operation.} \end{cases} \quad (11.19)$$

Production Time, Cost, and Profit Rate as a Function of Machining Speed

Substituting t_m and T, expressed as a function of v given by (11.10) and (11.16), into (11.2), (11.7), and (11.9), we obtain unit production time and cost, and profit rate expressed as a function of v as follows:

$$t = t_p + \lambda/v \, [+ t_c \lambda v^{1/n-1}/C^{1/n}] \quad (11.20)$$

$$u = k_o t_p + (k_o + k_m)\lambda/v + ([k_o t_c +]k_t)\lambda v^{1/n-1}/C^{1/n} \quad (11.21)$$

$$p = \frac{r_0 - k_m \lambda/v - k_t \lambda v^{1/n-1}/C^{1/n}}{t_p + \lambda/v[+ t_c \lambda v^{1/n-1}/C^{1/n}]} - k_o. \quad (11.22)$$

11.2.3 Determining Optimal Machining Speeds—Unconstrained Optimisation

Optimal Machining Speeds

Optimal machining speeds under three kinds of evaluation criteria for a fixed depth of cut and a fixed feed rate can be easily determined by differentiating (11.20), (11.21) and (11.22) with respect to machining speed v and setting them equal to zero. To minimise the production time when an automatic machine is used, however, the optimal machining speed is limited to the maximum speed, v_{max}, available to that machine.

(I) The *minimum-time* or *maximum-production-rate machining speed*:

$$v_t = \begin{cases} C/[(1/n - 1)t_c]^n \text{ for ordinary machines} \\ v_{max} \text{ for automatic machines.} \end{cases} \quad (11.23)$$

The *minimum-time* or *maximum-production-rate tool life*:

$$T_t = \begin{cases} (1/n - 1)t_c \text{ for ordinary machines} \\ (C/v_{max})^n \text{ for automatic machines.} \end{cases} \quad (11.24)$$

(II) The *minimum-cost machining speed*:

$$v_c = C((k_o + k_m)/(1/n - 1)([k_o t_c +] k_t))^n. \quad (11.25)$$

The *minimum-cost tool life*:

$$T_c = (1/n - 1)([k_o t_c +]k_t)/(k_o + k_m). \quad (11.26)$$

(III) The *maximum-profit-rate machining speed*:

$$v_p = (1-n)(k_t t_p[+rt_c])v_p^{1/n} + \lambda(k_t[-k_m t_c])v_p^{1/n-1} - nC^{1/n}(k_m t_p + r_0) = 0. \quad (11.27)$$

The maximum-profit-rate machining speed v_p can be explicitly expressed from (11.27) for particular values of n, the constant showing the slope of the Taylor tool-life diagram, that is, 1, $\frac{1}{2}$, $\frac{1}{3}$ and $\frac{1}{4}$, since this equation becomes algebraic equations of 1st, 2nd, 3rd, and 4th degree, respectively. For example, if $n = \frac{1}{2}$,

$$v_{p:n=1/2} = \{\lambda^2(k_t[-k_m t_c])^2 - C^2(k_m t_p + r)(k_t t_p[+r_0 t_c])\}^{1/2} - \lambda(k_t[-k_m t_c])/ k_t t_p[+rt_c]. \tag{11.28}$$

High-efficiency Range and Efficiency Machining Speeds

What quantitative relationship exists between the minimum-time machining speed v_t, the minimum-cost machining speed v_c, and the maximum-profit-rate machining speed v_p, which were derived above, for a fixed depth of cut and a fixed feed rate?

It is quite natural from the practical standpoint to assume that the machining speed for the maximum production rate (minimum time) is greater than that for the minimum production cost per piece produced; hence,

$$v_c < v_t. \tag{11.29}$$

A speed range between these two optimal machining speeds is called the *high-efficiency* (or merely *efficiency*) [*speed*] *range*, or *non-inferior range*, which means that any machining speed in this range is preferable from the managerial standpoint.

It can be proved mathematically that the maximum-profit-rate machining speed exists in the high-efficiency range (for proof, see Hitomi, 1976), which is an interesting and noteworthy relationship.

--- **PROPOSITION 11.1** ---

$$v_c < v_p < v_t. \tag{11.30}$$

□

--- **EXAMPLE 11.1** ---

(Calculating optimal speeds) Compute optimal machining speeds for production data as given in the following:

- Tool-life parameters
  - —slope constant n — 0.23
  - —1-min tool-life machining speed C (m/min) — 430
- Machining parameters
  - —depth of cut d (mm)[1] — 1.00
  - —feed rate s (mm/rev) — 0.20
- Work parameters[2]
  - —work diameter D (mm) — 50.00
  - —work length L (mm) — 200.00
- Time parameters
  - —setup time t_p (min/pc) — 0.75
  - —tool-replacement time t_c (min/edge) — 1.50
- Cost parameters
  - —direct labour cost and overhead k_o ($/min) — 0.50
  - —machining overhead cost k_m ($/min) — 0.05
  - —tool cost k_t ($/edge) — 2.50
  - —material cost m_c ($/pc) — 20.00
  - —gross revenue r_u ($/pc) — 32.00

Based upon the above data, optimal machining speeds are calculated using (11.23), (11.25), and (11.27), and we obtain the optimal machining speeds, v_c^*, v_p^*, v_t^*, as follows:

- 216, 279, 297 for ordinary machines
- 230, 404, 450 for automatic machines

These numerical results indicate that proposition (11.1) is right. □

The unit production time, the unit production cost, and the profit rate given by equations (11.20) to (11.22) are a function of machining speed with minimal or maximal points at the optimal machining speeds.

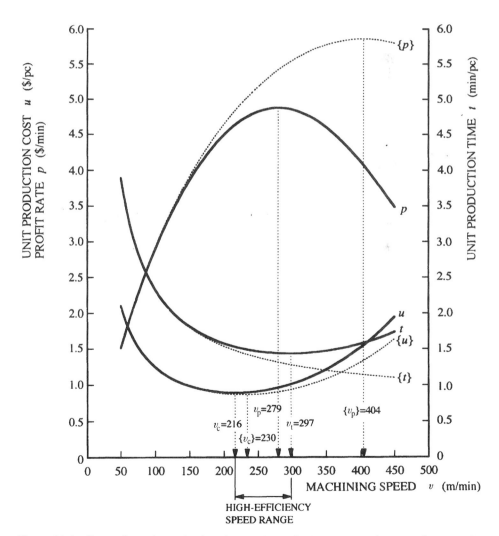

Figure 11.4 Curves for unit production time, unit production cost, and unit profit rate (solid lines for ordinary machines/broken lines for automatic machines).

--- **EXAMPLE 11.2** ---

Figure 11.4 shows time, cost, and profit-rate curves versus machining speed for production data given in the previous example. □

As seen in Figure 11.4, the time and cost curves are fairly flat at their minimal points. Hence, the increase in unit production time or cost is small even if the machining speed deviates from the optimal values. However, deviation of machining speed should be directed towards the inside of the [high-]efficiency range so that the increase in unit production cost or time can be kept less. This is why the high-efficiency range may be called the non-inferior range.

Machining speeds in this range are the *high-efficiency* (or *non-inferior*) *machining speeds* and are to be employed in preference to machining speeds outside the range. The maximum-profit-rate machining speed, which falls within this range, is the preferred machining speed.

Efficiency Machining Speed as a Pareto Optimum

In the above, the high-efficiency speed range is the *Pareto optimum*, where no objective can be improved without impairing another objective; the maximum-profit-rate machining speed is a typical Pareto-optimum solution.

--- **EXAMPLE 11.3** ---

(Pareto-optimum speeds) The relationship between the unit production time and cost for conventional machining is shown in Figure 11.5. A solid line indicates the range of Pareto optimum; the ends are the minimum-time and minimum-cost speeds, and the maximum-profit-rate speed lies on this line. □

11.2.4 Determining the Optimal Machining Speed—Constrained Optimisation for Practical Cases

Practical Speed Range

In practice, the machining speed employed should be in a certain allowable range. In addition, in order to save manpower for replacing a worn tool, especially for automatic manufacturing, the length of tool life should be greater than an allowable minimum life. In this section optimal machining speed will be determined under such practical restrictions.

Denoting the maximum and minimum spindle speeds (rev/min) set in the machine tool by N_{min} and N_{max}, respectively, and the diameter for the work material or cutter by D (mm), the minimum and maximum machining speeds are given by:

$$v_{min} = \pi D N_{min}/1000, \qquad v_{max} = \pi D N_{max}/1000. \tag{11.31}$$

Hence, we have the following constraint:

$$v_{min} \leq v \leq v_{max}. \tag{11.32}$$

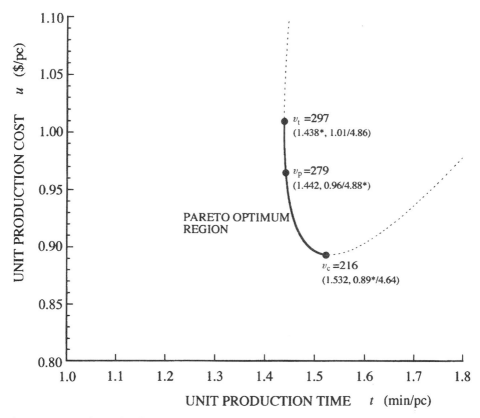

Figure 11.5 Relationship between unit production time and unit production cost—existence of the 'Pareto optimum' region indicated by the solid line between the minimum-time (v_t) and the minimum-cost (v_c) speeds for ordinary machines. The maximum profit-rate (v_p) lies in this region. Three figures in parentheses indicate unit production time (min/pc), unit production cost ($/pc)/profit rate ($/min), and * stands for optimal values.

Non-replacement Machining Speed

In order to machine for a specified time r without replacing the cutting tool, the following condition must hold:

$$(T/t_m)(t_p + t_m) \geq r. \tag{11.33}$$

Substituting equations (11.10) and (11.16) into the above equation,

$$f(v) \equiv v^{1/n} - (t_p C^{1/n}/\lambda r)v - C^{1/n}/r \leq 0. \tag{11.34}$$

The v-value satisfying this inequality condition is *r-time non-replacement machining speed*, which is denoted by v_r.

Practical Optimal Machining Speed

Taking the two constraints (11.32) and (11.34) into consideration, the optimal machining speed under constrained optimisation is given by:

$$v^* = \min(v_o, v_r, v_{max}) \tag{11.35}$$

where $v_o = v_c$, v_p or v_t, according to one of the three evaluation criteria employed. If $v_r < v_{min}$, there exists no optimal machining speed.

EXAMPLE 11.4

Basic data are given, the same as in Example 11.1; in addition, spindle speeds set in the lathe are:

N_s = 63, 85, 120, 160, 210, 290, 390, 510, 640, 790,
950, 1120, 1300, 1490, 1700, 1930, 2200 (rev/min).

Then, the minimum and maximum machining speeds, from equation (11.31), are:

$v_{min} = 9.9$, $v_{max} = 345.6$.

Accordingly, the machining speed must exist in the range

$10 \leqslant v \leqslant 345$.

If non-replacement time is settled as $r = 30$ (min), then the machining speed satisfying (11.34) with the equality sign is:

$v_r = 233.7$.

Machining speeds less than this value guarantee a tool life of more than 30 min.

Hence, we obtain the optimal machining speeds from equation (11.35) as follows (refer to the results in Example 11.1):

- The maximum-production-rate and the maximum-profit-rate machining speeds = 233 m/min.
- The minimum-cost machining speed = 216 m/min.

Converting these values into suitable speeds of the main spindle (values for N_s above), from equation (11.13) we obtain 1490 rev/min for the maximum-production-rate and the maximum-profit-rate criteria, and 1300 rev/min for the minimum-cost criterion.

If $r = 180$ min, then for all criteria the optimal machining speed is 147 m/min and the corresponding speed of the main spindle is 950 rev/min. □

Important Practical Fact

As non-replacement time r increases, non-replacement machining speed v_r decreases. The relationship between r and v_r is shown in Figure 11.6. As seen in this figure, non-replacement time of the cutting tool increases significantly even with a small reduction of machining speed. This is an important fact from the practical viewpoint.

In-life Machined Quantity

Practically it is more important to know the number of pieces to be machined during the lifetime of a cutting tool than the length of tool life. Let us call this quantity *in-life machined quantity*; denoting it by c (pcs/edge), from equations (11.10) and (11.16),

$$c = T/t_m = C^{1/n}/\lambda v^{1/n-1}. \tag{11.36}$$

Converting v into a suitable rotation speed of the main spindle, N_s, using (11.13) and rounding up, we obtain:

$$\hat{c} = \lfloor C^{1/n}(1000/\pi D N_s)^{1/n-1}/\lambda \rfloor \tag{11.37}$$

where $\lfloor A \rfloor$ stands for the Gaussian symbol, meaning the greatest integer equal to or less than A.

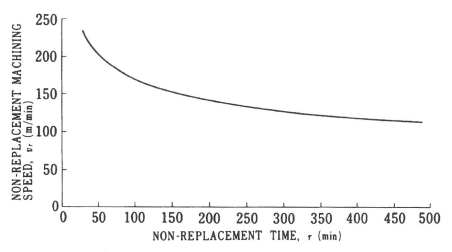

Figure 11.6 Non-replacement time increases rapidly as non-replacement machining speed decreases.

— **EXAMPLE 11.5** —

Calculate the in-life machined pieces under the minimum-cost criterion for basic data given in Example 11.1. In this case, the optimal machining speed is:

$v_c = 216$ (m/min).

A suitable spindle speed for this machining speed is:

$N_s = 1300$ (rev/min).

Hence, from equation (11.37),

$\hat{c} = 33$.

That is, a tool edge should be replaced after machining 33 pieces. □

11.2.5 Determining Optimal Machining Speed and Feed Rate—Constrained Optimisation

Generalised Tool-life Equation

In practice, determination of both optimal machining speed and optimal feed rate is required. For this purpose, first let us construct a basic mathematical model for single-stage manufacturing as a function of feed rate and machining speed.

Unit production time and cost, which are two basic measures of performance for the minimum-time and the minimum-cost criteria, are expressed in general form as in (11.2) and (11.7). In these formulae the tool life T (min/edge) and the machining time t_m (min/pc) are dependent upon feed rate s (mm/rev) and machining speed v (m/min). To calculate T, the following *generalised Taylor tool-life equation* is employed.

$$Ts^{1/m_o}v^{1/n_o} = C_o \quad (11.38)$$

where m_o, n_o, and C_o are constants.

Generalised Machining Objective Function

The machining time, t_m, is now expressed as a function of s and v as follows, similar to (11.10):

$$t_m = \lambda_o/sv, \qquad \lambda_o = \lambda s \tag{11.39}$$

where λ_o is a machining constant.

Substituting (11.38) and (11.39) into (11.2) and (11.7), unit production time and cost are expressed as:

$$t = t_p + \frac{\lambda_o}{sv} + t_c \frac{\lambda_o}{C_o} s^{1/m_o - 1} v^{1/n_o - 1} \tag{11.40}$$

$$u = k_o t_p + (k_o + k_m)\frac{\lambda_o}{sv} + (k_o t_c + k_t)\frac{\lambda_o}{C_o} s^{1/m_o - 1} v^{1/n_o - 1}. \tag{11.41}$$

These are to be minimised. Substituting m for $1/m_o - 1$, n for $1/n_o - 1$, a for t_c/C_o or $(k_o t_c + k_t)/(k_o + k_m)C_o$ and z for t/λ_o or $u/(k_o + k_m)$ and omitting the constant term t_p/λ_o or $k_o t_p/(k_o + k_m)\lambda_o$, we obtain the general goal function as

$$z = \frac{1}{sv} + as^m v^n \tag{11.42}$$

where index constants m and n are usually related as $0 < m < n$, since usually $0 < n_o < m_o < 1$ from the standpoint of metal-cutting practice.

Unconstrained Optimisation of the Generalised Machining Objective

Unconstrained optimisation of (11.42) is carried out by setting partial derivatives of this equation with respect to each of these decision variables—feed rate s and machining speed v—equal to 0, as follows.

$$\frac{\partial z}{\partial s} = -\frac{1}{s^2 v} + ams^{m-1}v^n = 0; \quad \text{i.e.} \quad ams^{m+1}v^{n+1} = 1 \tag{11.43}$$

$$\frac{\partial z}{\partial v} = -\frac{1}{sv^2} + ans^m v^{n-1} = 0; \quad \text{i.e.} \quad ans^{m+1}v^{n+1} = 1. \tag{11.44}$$

From these equations, the s- or v-minimum point for a fixed v or s is obtained as follows:

s-minimum point:

$$s^* = \frac{1}{(amv^{n+1})^{1/(m+1)}}\bigg|_{v:\text{fixed}} \tag{11.45}$$

v-minimum point:

$$v^* = \frac{1}{(ans^{m+1})^{1/(n+1)}}\bigg|_{s:\text{fixed}} \tag{11.46}$$

Let the loci of s and v, when changing v and s in equations (11.45) and (11.46), be called the *s-minimum line* and *v-minimum line* respectively; these are given by:

s-minimum line:

$$sv^{(n+1)/(m+1)} = \frac{1}{(am)^{1/(m+1)}} \tag{11.47}$$

v-minimum line:

$$sv^{(n+1)/(m+1)} = \frac{1}{(an)^{1/(m+1)}}. \tag{11.48}$$

Both possess the equal tangential gradient for the same s- or v-value, that is, they are parallel on bilogarithmic paper. For $m < n$, the s-minimum line lies above the v-minimum line on a regular graph as shown in Figure 11.7.

When v is fixed at a certain value v_1 (point A), z_1—the minimum value of z—clearly occurs at B—the point of intersection of AA', a line drawn parallel to the s-axis from the point A, and the s-minimum line. Fixing the s-value at s_1 corresponding to the point B, the minimum of z occurs at C—the point of intersection of BB', a line drawn parallel to the v-axis from B, and the v-minimum line. This minimum value z_2 is clearly less than z_1—the z-value at point B. Again fixing the v-value at v_2 corresponding to the point C, the z-minimum point reaches a point D, at

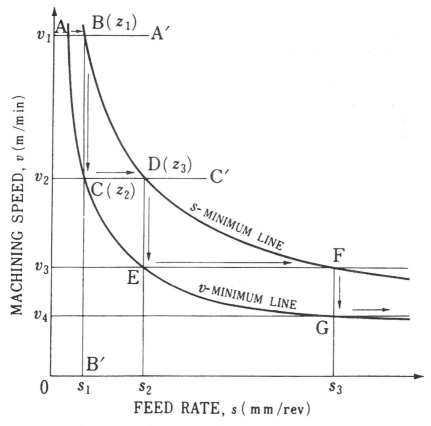

Figure 11.7 s- and v-minimum lines for $z = 1/sv + as^m v^n$ (for $m < n$); the z-value decreases infinitely as $s \to \infty$ and $v \to 0$.

which the z-value, z_3, becomes less than z_2 at C. Repeating this process, the z-minimum point moves successively from B to C, from C to D, from D to E, from E to F, from F to G, etc., and finally reaches $s \to \infty$ and $v \to 0$. Thus the following rule is concluded.

--- **PROPOSITION 11.2** ---

The unconstrained minimum of (11.42) is non-existent in the range of $s, v > 0$. It decreases infinitely as $s \to \infty$ and $v \to 0$ when $m < n$. □

Practical Machining Constraints

From the above proposition a unique optimal feed rate and machining speed cannot be determined for unconstrained minimisation of production time and cost. The optimisation is then performed by taking practical manufacturing constraints into account.

The following are typical constraints considered in practical machining:

(1) *Feed-rate constraint*:

$$s_{min} \leq s \leq s_{max} \tag{11.49}$$

where s_{min} and s_{max} are the lowest and highest feed rates set in the machine tool employed.

(2) *Machining-speed constraint*:

$$v_{min} \leq v \leq v_{max} \tag{11.50}$$

where v_{min} and v_{max} are the lowest and highest machining speeds, respectively. These are decided by the work diameter and the lowest and highest speeds of the main spindle set in the machine tool employed.

(3) *Roughness constraint.* The surface roughness should be less than a certain amount to ensure good product accuracy; hence,

$$\frac{s^2}{8R} \leq R_{max} \tag{11.51}$$

where R is the nose radius of the cutting tool and R_{max} is the required surface roughness. From this equation,

$$s \leq \sqrt{(8RR_{max})} = s_a \tag{11.51'}$$

where s_a is constant. This is a sort of feed constraint, as mentioned in (1). Combining (11.49) and (11.51'), we get:

$$s_{min} \leq s \leq \min(s_{max}, s_a). \tag{11.52}$$

(4) *Power constraint.* The machining resistance is in general given by the power function of machining speed and feed rate, and it must not exceed the motor power of the machine tool employed; hence

$$\frac{\gamma d^\alpha s^\beta v}{60} \leq P \tag{11.53}$$

where d is depth of cut (mm), α and β are positive values near 1, γ is the specific machining resistance (GPa), and P is the motor power (kW).

Constrained Optimisation of the Generalised Machining Objective

Summarising the above, a problem of the constrained optimisation determining optimal values for two decision variables—machining speed and feed rate—is formulated as follows.

Minimise (11.42) subject to:

$$0 \leq s \leq b \tag{11.54}$$

$$0 \leq v \leq c \tag{11.55}$$

$$sv^k \leq d \tag{11.56}$$

where b, c, d, and k are constants For convenience sake, minimal feed and speed values are both set at zero. This setting is acceptable, since the optimal solution is always generated on a line $s = b$, as clarified later.

This is a typical nonlinear-programming problem.

Simplified Optimisation Algorithmic Procedure for Determining Optimum Machining Speed and Feed Rate

Suppose that $m < n$ as in the practical situation. To obtain the slope of the contour of (11.42) in the sv-plane, differentiating this equation at $z = h$, where h is constant, with respect to s, the slope of the contour at the point (s, v) is:

$$\left.\frac{dv}{ds}\right|_z = -\frac{v - ams^{m+1}v^{n+2}}{s - ans^{m+2}v^{n+1}}. \tag{11.57}$$

On the other hand, the slope of the sv-constraint line given by (11.56) is, by differentiating this equation with respect to s,

$$\left.\frac{dv}{ds}\right|_{sv} = -\frac{v}{ks}. \tag{11.58}$$

From the above two equations, in general,

$$\left.\frac{dv}{ds}\right|_z \neq \left.\frac{dv}{ds}\right|_{sv}. \tag{11.59}$$

Thus, where the sv-constraint line and the z-contour hold a point (s_0, v_0) in common, their slopes at this point are not equal, which results in the following.

--- **PROPOSITION 11.3** ---

For $m \neq n$, the sv-constraint line and the z-contour intersect each other. □

From Propositions 11.2 and 11.3, we obtain the following proposition relating to the constrained minimisation for (11.42).

— PROPOSITION 11.4 —

For $m < n$, the constrained minimisation for (11.42) occurs on the s-constraint line ($s = b$). There are two cases (Figure 11.8):

(a) it occurs at P, the point of intersection of its line and the sv-constraint line; or
(b) it occurs at Q, a point inside its point of intersection, at which unconstrained minimisation occurs for a fixed s-value ($= b$). □

With use of the above proposition, an algorithmic procedure to obtain the constrained minimisation is proposed as follows (Hitomi, 1975):

Algorithm for determining optimal feed rate and machining speed:

Step 1: The optimal feed rate is:

$$s^* = b.$$

Step 2: Calculate

$$v_1 = \frac{1}{(ab^{m+1}n)^{1/(n+1)}} \qquad (11.60)$$

and

$$v_2 = \left(\frac{d}{b}\right)^{1/k}. \qquad (11.61)$$

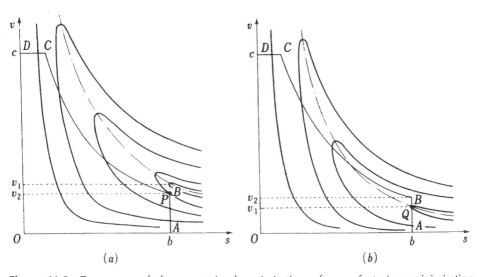

Figure 11.8 Two cases of the constrained optimisation of manufacturing: minimisation occurs (a) at P, the point of intersection of the s- and the sv-constraint lines, or (b) at Q, a point inside P on the s-constraint line. In each figure, OABCD is a feasible region satisfying constraints (11.54)–(11.56); a chain line stands for the v-minimum line (equation (11.48)), which expresses the locus of the minimal value of v at a fixed s-value, and other curves are contour lines of z.

Step 3: The optimal machining speed is:

$$v^* = \begin{cases} v_1, & \text{if } v_1 \leq v_2 \\ v_2, & \text{if } v_1 > v_2. \end{cases} \qquad (11.62)$$

Stop. □

— EXAMPLE 11.6 —

Determine the optimal machining speed and feed rate for the basic data given in Example 11.1, except that the life equation of the cutting tool with a nose radius of $R = 0.8$ mm is assumed as:

$$T s^{1/0.55} v^{1/0.23} = 1.51 \times 10^{10}$$

namely, $m_0 = 0.55$, $n_0 = 0.23$, $C_0 = 1.51 \times 10^{10}$.

The following machining constraints are taken into account.

(a) Feed-rate constraint: the maximum surface roughness allowed is 10 μm; then, from equation (11.51'),

$$s \leq 0.252 = s_a.$$

(b) Machining-speed constraint: since the maximum spindle speed set in the lathe is 2200 rev/min,

$$v \leq 345.$$

(c) Power constraint: this is given by:

$$s^{0.85} v \leq 48$$

when machining the work material with a specific cutting resistance of 5 GPa by a lathe with the motor power of 4 kW, assuming $\beta = 0.85$.

Based upon the above machining information, the total production cost in $/pc is, from equation (11.41),

$$u = 0.375 + \frac{17.2788}{sv} + 0.676\,17 \times 10^{-8} s^{0.82} v^{3.35}.$$

Hence, the cost minimisation is expressed as follows: find optimal values of s and v such that

$$z = \frac{1}{sv} + 0.391 \times 10^{-9} s^{0.82} v^{3.35}$$

is a minimum subject to

$$0 \leq s \leq 0.252$$
$$0 \leq v \leq 345$$
$$s^{0.85} v \leq 48.$$

In this case, $a = 0.391 \times 10^{-9}$, $b = 0.252$, $d = 48^{1/0.85}$, $k = 1/0.85$, $m = 0.82$, and $n = 3.35$.
From step 1, the optimal feed rate is $s^* = 0.252$. Then from step 2, $v_1 = 196.1$ and $v_2 = 154.9$; hence, from step 3 the optimal machining speed is $v^* = 155$ m/min.
If the motor power is 8 kW, then $d = 96^{1/0.85}$. In this case, $v_2 = 309.8$; hence, $v^* = 196$ m/min. □

Where the feed rates and the speeds of the main spindle of the machine tool are discrete, we can still use the above algorithm with slight modification.

--- **EXAMPLE 11.7** ---

The speeds of the main spindle are given for N_s in Example 11.4; and the feed rates are:

$s_d = 0.025, 0.050, 0.075, 0.100, 0.130, 0.160, 0.190, 0.220, 0.250, 0.280, 0.320,$
$\qquad 0.360, 0.400, 0.450, 0.500, 0.560, 0.630, 0.700$ (mm/rev).

In this case $s_{min} = 0.025$, $s_{max} = 0.700$, and $s_a = 0.252$; hence, $0.025 \leq s \leq 0.252$. The optimal feed rate is, from step 1 and the values of s_d, $s^* = 0.250$ mm/rev (this corresponds to b). In step 2, for the motor power of 4 kW, from (11.60) and (11.61), $v_1 = 196.7$ and $v_2 = 156.0$. The speed of the main spindle for v_2, which is less than v_1, is $N_s = 93.1$. From values of N_s the optimal speed is $N_s^* = 950$ rev/mm; hence, the optimal machining speed is $v^* = 149$ m/min. □

11.3 Optimisation of Multistage Manufacturing Systems

11.3.1 Scope of Multistage Manufacturing Systems

Optimisation Analysis for Multistage Manufacturing Systems

In practice, the conversion of a raw material into the finished part or product is rarely done in single-stage manufacturing; rather it is usually completed by being processed successively on several machine tools or workstations that constitute a multistage manufacturing system. Therefore, optimisation analysis should be made on such a manufacturing system (Hitomi, 1991), utilising several analytical results for single-stage manufacturing developed in the previous section.

It is enough to set the feed rate as the maximum determined from the feed-rate and/or roughness constraints, when the motor power is large enough. Then 'optimal machining speeds' at multiple stages are decision variables. In this section, optimisation analysis of a multistage manufacturing system will be made, determining

- the optimal cycle time;
- optimal machining speeds to be utilised at multiple stages.

Here *cycle time* means the time interval between successive completions of the machined product through the manufacturing system.

For this purpose basic mathematical models for processing work materials through a flow-type multistage manufacturing system, which consists of N machine tools or workstations sequenced in the technological order, as shown in Figure 11.9, are constructed.

At each stage the workstation is automatically loaded with incoming work material which has been transferred from the previous stage, and a specified operation is conducted by the assigned machine tool. The finished material is then unloaded from the workstation, and transferred to the following stage. Thus, in passing through the manufacturing system of N stages, the raw material of cost m_c is converted into the finished product which produces revenue r_u.

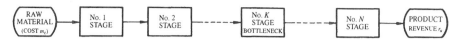

Figure 11.9 Schematic model of a flow-type multistage manufacturing system, through which a piece of raw material of cost m_c is converted sequentially into the finished product of revenue r_u.

11.3.2 Basic Mathematical Models

Preconditions for an Automated Manufacturing System

In an automated manufacturing system loading and unloading of workpieces at all stages are run automatically during the same cycle time. Each setup time is denoted by a (min/pc). The replacement of a worn cutting edge with a new one is done automatically within this setup time.

Regarding tool life, T_j (min/edge), at stage j (S_j), follows the Taylor tool-life equation:

$$v_j T_j^{n_j} = C_j \qquad (j = 1, 2, \ldots, N) \tag{11.63}$$

where n_j and C_j are constants (slope constant and 1-min tool-life machining speed).

Production time and production cost on S_j are dependent on the machining speed, v_j (m/min), utilised at that stage.

Stage Production Time

Machining time (min/pc) at S_j is λ_j / v_j, where λ_j is a machining constant given by:

$$\lambda_j \equiv b_j = \begin{cases} \pi D_j L_j / 1000 s_j & \text{for turning, boring, drilling and reaming} \\ \pi D_j (l_j + L_j)/1000 s_j Z_j & \text{for milling} \quad (j = 1, 2, \ldots, N) \end{cases} \tag{11.64}$$

where D_j is the machining diameter (mm) for turning and boring or the tool diameter (mm) for milling, drilling and reaming; L_j is the machining length (mm) for turning and milling or the length of the hole to be machined (mm) for boring, drilling and reaming; l_j is the additional feed length (mm) for milling; s_j is the feed rate per tooth (mm/tooth) for milling or the feed rate per revolution (mm/rev) for the other operations; and Z_j is the number of teeth of the milling cutter ($j = 1, 2, \ldots, N$).

Hence, the stage production time (min/pc) on S_j to manufacture a workpiece is

$$t_j(v_j) = a + b_j / v_j \qquad (j = 1, 2, \ldots, N), \tag{11.65}$$

since the setup time is a.

Total Production Time

The total production time (min/pc) to complete a piece of product through N-stage manufacture is the sum of (11.65) for all stages.

$$t(v) = \sum_{j=1}^{N} t_j(v_j) = aN + \sum_{j=1}^{N} b_j / v_j \tag{11.66}$$

where $v = (v_1, v_2, \ldots, v_N)$.

Cycle Time

The most important factor regarding a multistage manufacturing system is the cycle time which is a time interval of the successive flowout of products completed by passing through the system.

The cycle time is a maximum among stage production times given by (11.65).

$$\tau = \max_{1 \leq j \leq N} t_j(v_j) \equiv a + b_K/v_K \tag{11.67}$$

where K signifies a bottleneck stage, which has the largest stage production time.

Stage Production Cost

The production cost of a workpiece on S_j consists of the machining cost and the tool cost. Denoting machining overhead on S_j such as motor power, cutting fluid, and others by c_j (\$/min), the machining cost required to manufacture a part of a workpiece on S_j is:

$$u_{jc}(v_j) = b_j c_j / v_j = \beta_j / v_j \quad (j = 1, 2, \ldots, N) \tag{11.68}$$

where $\beta_j = b_j c_j$.

Each tool can produce an amount, $T_j/(b_j/v_j)$, of products during its life time. Hence, by denoting the cost of a new tool utilised on S_j by c_{tj} (\$/edge) and by using (11.63), the tool cost required to manufacture a workpiece is:

$$u_{jt}(v_j) = b_j c_{tj} v_j^{1/n_j - 1}/C_j^{1/n_j} = \gamma_j v_j^{k_j} \quad (j = 1, 2, \ldots, N) \tag{11.69}$$

where $k_j = 1/n_j - 1$ and $\gamma_j = b_j c_{tj}/C_j^{1/n_j}$. Since it is usual that $0 < n_j \leq 0.5$, $k_j \geq 1$.

Then the stage production cost required to manufacture a workpiece on S_j is given by the sum of (11.68) and (11.69):

$$u_j(v_j) = \beta_j/v_j + \gamma_j v_j^{k_j} \quad (j = 1, 2, \ldots, N). \tag{11.70}$$

This is a convex function with respect to v_j. Hence, there exists a minimum-cost machining speed, v_{jc}, to minimise the stage production cost as follows:

$$v_{jc} = (\beta_j/\gamma_j k_j)^{1/(k_j+1)} \quad (j = 1, 2, \ldots, N). \tag{11.71}$$

Total Production Cost

Invested capital for an automated manufacturing system and direct and indirect costs for workers on this system are collectively denoted by α (\$/min). Since a workpiece is produced in each cycle time, the overhead required to manufacture a part is

$$u_0(v) = \alpha \tau. \tag{11.72}$$

Then the total production cost required to manufacture a unit of product is given by

$$u(v) = u_0(v) + \sum_{j=1}^{N} u_j(v_j) = \alpha \tau + \sum_{j=1}^{N} (\beta_j/v_j + \gamma_j v_j^{k_j}). \tag{11.73}$$

Profit Rate

Denoting the revenue (selling price minus material cost) obtained by selling a unit of product by r_0, the profit obtained is

$$r(v) = r_0 - u(v) = r_0 - \alpha\tau - \sum_{j=1}^{N}(\beta_j/v_j + \gamma_j v_j^{k_j}). \tag{11.74}$$

Profit rate, which is the profit obtained in a unit time, is expressed by dividing the above by the cycle time:

$$p(v) = r(v)/\tau = \left\{ r_0 - \sum_{j=1}^{N}(\beta_j/v_j + \gamma_j v_j^{k_j}) \right\}/\tau - \alpha. \tag{11.75}$$

Constraints

The machining speed v_j to be utilised is technologically limited within a suitable 'allowable speed range':

$$\underline{v}_j \leq v_j \leq \bar{v}_j \quad (j = 1, 2, \ldots, N) \tag{11.76}$$

where $\underline{v}_j$ and $\bar{v}_j$ are the lower and upper limits of the machining speed on S_j.

If $v_{jc} < \bar{v}_j$, the speed range E_j between v_{jc} and $\bar{v}_j$ is the efficiency speed range:

$$E_j = [v_{jc}, \bar{v}_j] \quad (j = 1, 2, \ldots, N). \tag{11.77}$$

Any speed in this range is called the 'efficiency speed'.

11.3.3 Determining the Optimal Machining Speeds under the Maximum Production-rate Criterion

Analysis for the Minimum Production Time

The maximum production rate is attained by minimising the cycle time τ, given by (11.67), subject to the technological constraints given by (11.76). Then we have a problem:

(P_1) minimise $\max_{1 \leq j \leq N} b_j/v_j$ subject to (11.76).

To solve this problem, first, calculate $b_j/\bar{v}_j$ for $j = 1, 2, \ldots, N$, and then find K such that this amount be a maximum. (If there exist several such Ks, find $K_1, K_2, \ldots, K_L$, $L \leq N$; a set of K_ks ($k = 1, 2, \ldots, L$) is denoted by $\{K\}$.) This is the bottleneck stage.

The optimal machining speed to be set on the bottleneck stage is

$$v_K^* = \bar{v}_K. \tag{11.78}$$

Then the minimum (or optimal) cycle time, which produces the maximum number of products in a unit time, is

$$\tau^* = a + b_K/\bar{v}_K. \tag{11.79}$$

Regarding machining speeds on the slack stages it is reasonable to decide optimal ones within the technological ranges given by equation (11.76), so as to minimise the stage production costs under the constraints that the stage production times on the slack stages do not exceed the minimum cycle time obtained above. Then we have a problem:

(P_2) minimise (11.70) subject to (11.76) and

$$a + b_j/v_j \leq \tau^* \tag{11.80}$$

for $j \in \{N\} - \{K\}$, where $\{N\}$ signifies a set of N stages.

This is a non-linear program. Applying the Kuhn–Tucker necessary condition (see Section 11.3.6) to the above problem, the following equations are obtained:

$$-\beta_j/v_j^{*2} + k_j \gamma_j v_j^{*k_j-1} + \lambda_1(-b_j/v_j^{*2}) + \lambda_2 - \lambda_3 = 0, \tag{11.81}$$

$$a + b_j/v_j^* - \tau^* \leq 0, \quad \lambda_1(a + b_j/v_j^* - \tau^*) = 0, \tag{11.82}$$

$$\underline{v}_j \leq v_j^* \leq \bar{v}_j \quad (a), \quad \lambda_2(v_j^* - \bar{v}_j) = 0 \quad (b), \quad \lambda_3(v_j^* - \underline{v}_j) = 0 \quad (c), \tag{11.83}$$

where λ_1, λ_2, and λ_3 are Lagrange multipliers corresponding to (11.80), the right-hand and left-hand sides of inequalities (11.76). The v_j^*s are the optimal machining speeds to be obtained.

Typically four solutions are obtained from the above three equations for the optimal machining speeds to be utilised as follows:

(1) In the case that $\lambda_1 = \lambda_2 = \lambda_3 = 0$: if $a + b_j/v_j^* \leq \tau^*$ and $\underline{v}_j \leq v_j \leq \bar{v}_j$, then from (11.81),

$$v_j^* = (\beta_j/\gamma_j k_j)^{1/(k_j+1)} \equiv v_{jc}. \tag{11.84}$$

That is to say, the optimal machining speed is the minimum-cost machining speed.

(2) In the case that $\lambda_1 > 0$: from (11.82),

$$v_j^* = b_j/(\tau^* - a) \equiv v_{je}. \tag{11.85}$$

(3) In the case that $\lambda_2 > 0$: from (11.83(b)),

$$v_j^* = \bar{v}_j. \tag{11.86}$$

(4) In the case that $\lambda_3 > 0$: from (11.83 (c)),

$$v_j^* = \underline{v}_j. \tag{11.87}$$

Decision Factors for Multistage Manufacturing Systems

The optimising algorithm determining the optimal machining speeds includes the following three items:

(1) determining the bottleneck stage[s];
(2) determining the optimal cycle time;
(3) determining optimal machining speeds on all stages of the manufacturing system.

Optimising Algorithm under the Minimum Production Time

Step 1: Calculate $B_j = b_j/\bar{v}_j$ for $j = 1, 2, \ldots, N$.
Step 2: (Determining the bottleneck stage) Set j for $\max_{j \in \{N\}} B_j$ to K (the bottleneck stage). (If there are more than two such Ks, denote them by $K_1, K_2, \ldots, K_L$: $\{K\} = \{K_1, K_2, \ldots, K_L\}$.)

Step 3: (Determining the optimal cycle time) $\tau^* = a + b_K/\bar{v}_K$.
Step 4: (Determining the optimal machining speeds on the bottleneck stage) $v_K^* = \bar{v}_K$.
Step 5: Calculate v_{jc} for $j \in \{N\} - \{K\}$ using (11.71). If $v_{jc} < \underline{v}_j$, set $v_{jc} \leftarrow \underline{v}_j$. If $v_{jc} > \bar{v}_j$, set $v_{jc} \leftarrow \bar{v}_j$.
Step 6: Calculate $t_{jc} = a + b_j/v_{jc}$ for $j \in \{N\} - \{K\}$.
Step 7: If $t_{jc} \leq \tau^*$, then set $v_j^* = v_{jc}$ for such js: $\{J\}$, and go to step 9. Otherwise proceed.
Step 8: Calculate v_{je} for $j \in \{N\} - \{K\} - \{J\}$ using (11.85). If $v_{je} < \underline{v}_j$, set $v_{je} \leftarrow \underline{v}_j$. If $v_{je} > \bar{v}_j$, set $v_{je} \leftarrow \bar{v}_j$.
Step 9: (Determining the optimal machining speeds on slack stages) $v_j^* = v_{jc}$ for $j \in \{N\} - \{K\}$ and v_{je} for $j \in \{N\} - \{K\} - \{J\}$. Stop. □

11.3.4 Determining the Optimal Machining Speeds under the Minimum Production Cost Criterion

Analysis for the Minimum Production Cost

The minimum total production cost is attained by solving the following problem: (P_3) minimise (11.73) subject to (11.76) and

$$a + b_j/v_j - \tau \leq 0 \tag{11.88}$$

for $j \in \{N\} - \{K\}$.

Relaxing the constraint given by (11.76) (this is considered in developing the algorithm) and applying the Kuhn–Tucker condition, there exist $\lambda_j \geq 0$, $j \in \{N\} - \{K\}$, for the optimal v_j^* such that

$$-\beta_K/v_K^{*2} + k_K \lambda_K v_K^{*k_K - 1} - (a - \lambda_K)b_K/v_K^{*2} = 0, \quad \lambda_K = \sum_{j \in \{N\} - \{K\}} \lambda_j \tag{11.89}$$

$$-\beta_j/v_j^{*2} + k_j \gamma_j v_j^{*k_j - 1} - \lambda_j b_j/v_j^{*2} = 0, \quad j \in \{N\} - \{K\} \tag{11.90}$$

$$\lambda_j s_j^* = 0, \quad s_j^* = b_K/v_K^* - b_j/v_j^* \geq 0. \tag{11.91}$$

Hence,

$$v_K^* = [\{\beta_K + (a - \lambda_K)b_K\}\gamma_K k_K]^{1/(k_K + 1)}, \tag{11.92}$$

$$v_j^* = [(\beta_j + \lambda_j b_j)/\gamma_j k_j]^{1/(k_j + 1)}, \quad j \in \{N\} - \{K\}. \tag{11.93}$$

Then the optimal cycle time is:

$$\tau^* = a + b_K/v_K^*. \tag{11.94}$$

As above, the optimal machining speeds vary with Lagrange multipliers, λ_j, $j = 1, 2, \ldots, K-1, K+1, \ldots, N$. We have the following results.

(1) The optimal machining speed on the slack stage, v_j^*, is the minimum-cost speed, v_{jc}, if $\lambda_j = 0$. v_j^* increases with an increase in λ_j, passing through the efficiency speed range. This decreases the production time on the slack stage.
(2) The optimal machining speed on the bottleneck stage, v_K^*, is the efficiency speed:

$$v_{Ke} = [(\beta_K + ab_K)/\gamma_K k_K]^{1/(k_K + 1)} \tag{11.95}$$

if λ_js are all zero and $v_{Ke} < \bar{v}_K$. The speed v_K^* decreases with an increase in λ_js, reaches the minimum-cost speed, v_{Kc}, when the sum of λ_js becomes a, and further decreases to zero. This increases the production time on the bottleneck stage.

(3) Hence, by setting the appropriate values for λ_js it is possible to let the production time on the bottleneck stage be equal to or greater than the production times on the slack stages.

As in the previous section, the optimal machining speed on the slack stage is the minimum-cost speed if the production time on this stage is less than the optimal cycle time, τ^*; it is the efficiency speed if the production time is equal to τ^*.

Appropriate values for λ_js are set such that for the optimal machining speeds given by (11.92) and (11.93) and the optimal cycle time given by (11.94), (11.84) holds for stages having $\lambda_j = 0$, and (11.85) for stages having $\lambda_j > 0$.

Optimising Algorithm under the Minimum Production Cost

Step 1: Calculate v_{jc} for all $j \in \{N\}$ using (11.71). If $v_{jc} < \underline{v}_j$, set $v_{jc} \leftarrow \underline{v}_j$. If $v_{jc} > \bar{v}_j$, set $v_{jc} \leftarrow \bar{v}_j$, and denote such js by $\{J\}$.
Step 2: Calculate $t_{jc} = b_j/v_{jc}$ for all $j \in \{N\}$.
Step 3: Calculate v_{Ke} for all $K \in \{N\}$ using (11.95). If $v_{Ke} < \underline{v}_K$, set $v_{Ke} \leftarrow \underline{v}_K$. If $v_{Ke} > \bar{v}_K$, set $v_{Ke} \leftarrow \bar{v}_K$.
Step 4: Calculate $t_{Ke} = b_K/v_{Ke}$ for all $K \in \{N\}$.
Step 5: Proceed to the following by sequentially setting $K = 1, 2, \ldots, N$.
  (5.1) If $t_{Ke} \geq t_{jc}$ for all $j \in \{N\} - \{K\}$, set $t_K \leftarrow t_{Ke}$ and go to Step 6.
  (5.2) Denote by $\{\hat{J}\}$ a set of j such that $t_{jc} > t_{Ke}$.
  (5.3) Set appropriate positive values for λ_js for $j \in \{\hat{J}\}$ such that $t_K = b_K/v_K^* \geq b_j/v_j^*$ using (11.92) and (11.93), and set $v_{Ke} \leftarrow v_K^*$ and $v_{je} \leftarrow v_j^*$. Go to Step 6.
Step 6: Calculate $u_K = u(v)$ using (11.73) for $K = 1, 2, \ldots, N$.
Step 7: (Determining the bottleneck stage) Set K for $\min_{K \in \{N\}} u_K$ as K^* (the bottleneck stage).
Step 8: (Determining the optimal cycle time) $\tau^* = t_{K^*} + a$.
Step 9: (Determining the optimal machining speeds) $v_{K^*}^* = v_{K^*e}$, $v_j^* = v_{je}$ for $j \in \{N\} - \{K^*\}$. Stop. □

11.3.5 Determining the Optimal Machining Speeds under the Maximum Profit-rate Criterion

Analysis for the Maximum Profit Rate

The maximum profit rate is attained by solving the following problem.
(P$_4$) minimise

$$\left\{ \sum_{j=1}^{N} (\beta_j/v_j + \gamma_j v_j^{k_j}) - r_0 \right\}/\tau$$

subject to (11.76) and (11.88).

Applying the Kuhn–Tucker condition, for $\lambda_j \geq 0$, $j = 1, 2, \ldots, K-1, K+1, \ldots, N$.

$$(k_K \gamma_K v_K^{*k_K-1} - \beta_K / v_K^{*2})/\tau^* + \left\{ \sum_{j=1}^{N} (\beta_j/v_j^* + \gamma_j v_j^{*k_j}) - r_0 \right\} b_K / \tau^{*2} v_K^{*2}$$

$$+ \lambda_K b_K / v_K^{*2} = 0, \qquad \lambda_K = \sum_{j \in \{N\} - \{K\}} \lambda_j \quad (11.96)$$

$$(k_j \gamma_j v_j^{*k_j-1} - \beta_j / v_j^{*2})/\tau^* - \lambda_j b_j / v_j^{*2} = 0 \quad (11.97)$$

$$\lambda_j s_j^* = 0, \qquad s_j^* = \tau^* - (a + b_j / v_j^*) \geq 0. \quad (11.98)$$

Hence, from (11.97),

$$v_j^* = \{(\beta_j + \lambda_j b_j \tau^*)/\gamma_j k_j\}^{1/(k_j+1)}. \quad (11.99)$$

v_K^* can be obtained from the following equation:

$$\gamma_K a k_K v_K^{*k_K+3} + (k_K + 1)\gamma_K b_K v_K^{*k_K+2}$$
$$- \{a\beta_K + r_K b_K - \lambda_K b_K a^2\} v_K^{*2} + 2\lambda_K b_K^2 a v_K^* + \lambda_K b_K^3 = 0 \quad (11.100)$$

where

$$r_K = r_0 - \sum_{j \in \{N\}-\{K\}} (\beta_j/v_j^* + \gamma_j v_j^{*k_j}). \quad (11.101)$$

If λ_js are all zero,

$$v_K^* = v_{Kp} \text{ for } \gamma_K a k_K v_{Kp}^{k_K+1} + (k_K + 1)\gamma_K b_K v_{Kp}^{k_K} - (a\beta_K + r_K b_K) = 0 \quad (11.102)$$

where v_{Kp} is the maximum-profit-rate speed.

The optimal cycle time is

$$\tau^* = a + b_K/v_K^*. \quad (11.103)$$

The algorithm under the maximum-profit-rate criterion is similar to the one under the minimum-production-cost criterion; hence it is omitted.

--- **EXAMPLE 11.8** ---

(Optimisation of automated multistage manufacturing systems) Basic data for manufacturing a product on a three-stage manufacturing system are shown in Table 11.1.

Following the algorithms developed in the previous sections, the results obtained are indicated in Table 11.2.

The results show that No. 3 stage has slack for all the criteria and the optimal machining speed at this stage is 94 m/min which is the minimum cost speed obtained by (11.71).

Under the maximum-production-rate criterion No. 2 stage is a bottleneck, and the optimal machining speed there is the upper limit as given by (11.78). The optimal machining speed at No. 1 stage is the efficiency speed obtained by (11.85); this stage has no slack at all.

Under the minimum-cost and the maximum-profit-rate criteria the optimal machining speeds at 1 and 2 stages were determined through the Lagrange multipliers by using (11.92) and (11.93) and (11.99) and (11.100).

It is a noteworthy fact that an important relationship holds for the optimal machining speeds under the three criteria:

$$v^*(\text{min. cost}) < v^*(\text{max. profit rate}) < v^*(\text{min. time}) \quad (11.104)$$

Table 11.1 Basic data for the numerical example for optimisation of an automated flow-type multistage manufacturing system.

Stage j	Machining parameters					Tool parameters			Setup time a	Cost Parameters				
	s_j	D_j	L_j	L'_j	$\bar{v}_j$	Z_j	C_j	n_j		c_j	c_{tj}	m_o	r_p	α
1	0.20	104	200	—	400	1	450	0.25	0.5	0.10	7.50	25	—	⎫
2	0.05	100	300	100	350	8	500	0.33	0.5	0.15	15	—	—	⎬ 1.3
3	0.15	10	95	—	250	2	400	0.33	0.5	0.15	6	—	40	⎭

$j = 1$: turning; 2: milling; 3: drilling.

Table 11.2 The optimal solution for an automated flow-type multistage manufacturing system.

Stage	Minimum time		Minimum cost		Maximum profit rate	
	Optimal speed (m/min)	Unit time (min/pc)	Optimal speed (m/min)	Unit time (min/pc)	Optimal speed (m/min)	Unit time (min/pc)
1	364	1.40	172	2.40	265	1.73
2	350	1.40	166	2.40	255	1.73
3	94	0.71	94	0.71	94	0.71
Cycle time (min/pc)	1.40 (minimum)		2.40		1.73	
Total cost ($/pc)	34.54		29.94 (minimum)		31.12	
Profit rate ($/min)	3.91		4.19		5.13 (maximum)	

for both No. 1 and No. 2 stages as in the case of single-machine manufacturing, discussed in Section 11.2.3, Proposition 11.1.

Needless to say, the cycle time is minimum under the maximum-production-rate criterion, the total cost is minimum under the minimum-cost criterion, and the profit rate is maximum for the maximum-profit-rate criterion. □

11.3.6 The Kuhn–Tucker Theorem for Optimisation with Inequality Constraints

Optimisation Problem with Inequality Constraints

Problems P_2 and P_3 in the previous sections are optimisation with inequality constraints. A typical formulation is:

minimise $g(\mathbf{x})$ (11.105)

subject to $f_i(\mathbf{x}) \leq 0$ ($i = 1, 2, \ldots, M$) (11.106) where $\mathbf{x} = (x_1, x_2, ..., x_N)$. We assume that $g(\mathbf{x})$ and $f_i(\mathbf{x}) (i = 1, 2, ..., M)$ are differentiable convex functions.

The Kuhn–Tucker Necessary Conditions

First we define a Lagrangian-like function:

$$L(\mathbf{x}, \boldsymbol{\lambda}) = g(\mathbf{x}) + \sum_{i=1}^{M} \lambda_i f_i(\mathbf{x}) \tag{11.107}$$

where $\boldsymbol{\lambda} = (\lambda_1, \lambda_2, ..., \lambda_M)$ are Lagrange multipliers.

Then for optimal solutions, $\mathbf{x}^*$ and $\boldsymbol{\lambda}^*$, for the constrained optimisation problem—(11.105) and (11.106)—the following proposition must hold (Wismer and Chattergy, 1978).

— **PROPOSITION 11.5** —

The Kuhn–Tucker necessary (or stationary) conditions

$$\frac{\partial g(\mathbf{x}^*)}{\partial x_j} + \sum_{i=1}^{M} \lambda_i^* \frac{\partial f(\mathbf{x}^*)}{\partial x_j} = 0 \quad (j = 1, 2, ..., N) \tag{11.108}$$

$$f_i(\mathbf{x}^*) \leq 0 \quad (i = 1, 2, ..., M) \tag{11.109}$$

$$\lambda_i^* f_i(\mathbf{x}^*) = 0 \quad (i = 1, 2, ..., M) \tag{11.110}$$

$$\lambda_i^* \geq 0 \quad (i = 1, 2, ..., M). \tag{11.111}$$

□

Notes

1. This information is not required in the computation process.
2. By the use of these data the machining constant λ is calculated from (11.11).

References

ARMAREGO, E.J.A. and RUSSELL, J.K. (1966) Maximum profit rates as a criterion for the selection of machining conditions, *International Journal of Machine Tool Design & Research*, **6** (1).

GILBERT, W.W. (1950) Economics of machining, in *Machining—Theory and Practice* (Cleveland, OH: American Society for Metals), pp. 465–485.

HITOMI, K. (1975) *Theory of Planning for Production* (in Japanese) (Tokyo: Yuhikaku), Sec. 14.5.

HITOMI, K. (1976) Analysis of production models—Part I: The optimal decision of production speeds, *AIIE Transactions*, **8** (1).

HITOMI, K. (1991) Analysis of optimal machining conditions for flow-type automated manufacturing systems, *International Journal of Production Research*, **29** (12).

OKUSHIMA, K. and HITOMI, K. (1964) A study of economical machining—An analysis of the maximum-profit cutting speed, *International Journal of Production Research*, **3** (1).

WISMER, D.A. and CHATTERGY, R. (1978) *Introduction to Nonlinear Optimization* (New York: North-Holland), p. 82.

Supplementary reading

Data Files for Machining Operations, 12 volumes (in Japanese) (Tokyo: Machine Promotion Association).

FIELD, M., ZLATIN, N., WILLIAMS, R. and KRONENBERG, M. (1968, 1969) Computerized determination and analysis of cost and production rates for machining operations, Part 1/Part 2, *Journal of Engineering for Industry, Trans. ASME*, Series B, **90** (3), **91** (3).

Machinability Data Handbook (2nd edn) (1972) (Cincinnatti, OH: Metcut Research Associates).

PART THREE

Management Systems for Manufacturing

This part describes basic principles of production management technology—managerial information flow. The meaning of this flow and the functions and decision problems of strategic production planning and operational production management are introduced (Chapter 12). Basic theories and solution algorithms of production planning (Chapter 13)—short- and long-terms, product mix, economic lot size, MRP/CRP, production forecasting—and production scheduling (Chapter 14)—operations (flow- and job-shop) scheduling and project scheduling (PERT/CPM)—are discussed.

Inventory is 'stock' in the material flow: its models—fixed order quantity (Wilson model), replenishment, and (S, s) systems, together with two (or double) bin system and ABC classification—are explained (Chapter 15).

In the field of production control (Chapter 16) process control including line of balance, just-in-time (JIT) production or *kanban* (instruction card) system, productive maintenance, and replacement or renewal theories are mentioned.

Methods of quality control (QC), total quality control/management (TQC/TQM), quality function deployment (QFD), and quality engineering or Taguchi method are discussed (Chapter 17).

CHAPTER TWELVE

Managerial Information Flow in Manufacturing Systems

12.1 Managerial Information Flow

Information Flow in Manufacturing

As discussed in Section 1.2, the 'flow of materials' is a basic indispensable function in manufacturing. This flow is accompanied by the 'flow of costs' and the raw materials are converted into products with increased value. The driving force of this function is the 'flow of information'. The flow of materials which accompanies the flow of costs proceeds according to the instructions from the flow of information based on market needs. This synergistic action realises high quality (Q), low costs (C) and just-in-time or quick-response delivery (D). Realisation of this ideal production mode is now observed in automated or CIM-related factories.

Information-flow Process

The information-flow process in the manufacturing system is usually called *manufacturing* (or *production*) *management*; it proceeds with the following steps (Hitomi, 1975), as explained in Section 3.4 (see Figure 3.5).

(I) strategic production planning;
(II) operational production management;
    (1) aggregate production planning;
    (2) production process planning;
    (3) production scheduling;
    (4) production implementation;
    (5) production control.

II-4: production implementation (the production function) is an actual activity of production/fabrication in a workshop; it is the 'flow of materials' through production operations and processes. II-2: production process planning (the design function) is concerned with intrinsic (or pure) production technology as well; it is called the 'flow of technological information'.

188　MANAGEMENT SYSTEMS FOR MANUFACTURING

Other activities—I: strategic production planning, II-1: aggregate production planning, II-3: production scheduling, and II-5: production control (the [strategic] management function)—constitute a series of management [systems]; i.e. the 'flow of managerial information,' as mentioned in Section 5.3. In particular phase I treats the 'flow of administrative, or strategic management information'. The following section lists decision problems of these activities (Hitomi, 1994).

The structure and operating procedures of the above information flow in an on-line, real-time basis with the use of computer facilities and information technology are called *manufacturing information systems*.

This part is concerned with this type of information flow.

12.2 Decision Problems in Managerial Information Flow

Issues in Strategic Production Planning

Strategic issues treat matters concerning the relationship between the system and its environment, usually in the long term (5 to 10 years). The main problems of *strategic production planning* are:

- new product planning: Chapter 7;
- long-term profit planning: Chapter 20;
- capital investment for new plant construction: Chapter 21;
- international manufacturing: Chapter 33.

Decision Problems in Aggregate Production Planning

Aggregate production planning determines the kinds of product items and the quantities to be produced in the specified time periods. The important problems involved in this field are:

- short-term production planning;
- long-term production planning;
- optimal product mix;
- lot-size analysis;
- material requirements planning (MRP);
- production smoothing;
- production forecasting.

These decision problems are discussed in Chapter 13.

Decision Problems in Production Scheduling

Production scheduling determines an actual implementation plan as to the time schedule for every job contained in the process route adopted. The major problems are:

- job sequencing;
- operations scheduling for flow and job shops;
- project scheduling.

These decision problems are solved in Chapter 14.

Decision Problems in Production Control

Production control reduces or eliminates the deviations of actual production performances from the production standards (plans and schedules). Important problems to be solved are:

- process control: Chapter 16;
- inventory control: Chapter 15;
- quality control: Chapter 17;
- cost control: Chapter 19.

References

HITOMI, K. (1975) *Theory of Planning for Production* (in Japanese) (Tokyo: Yuhikaku).

HITOMI, K. (1994) Manufacturing systems: past, present and for the future, *International Journal of Manufacturing System Design*, **1** (1), 1–17.

CHAPTER THIRTEEN

Aggregate Production Planning

13.1 Production Planning Defined

Meaning of [aggregate] Production Planning

Since available resources for production, such as raw materials, machines, labour forces, funds, etc. are limited, it is desirable to allocate effectively and utilise those production resources to determine optimal kinds and quantities of products to manufacture. This is [*aggregate*] *production planning* in a specified time period (usually a comparatively long range, such as month or year units).

Short- and Long-term Production Planning

If the time range is short, usually less than a year, this is *short-term* (or *range*) (or *static*) *production planning*; the time factor is not considered in this case. On the other hand, if the time range is large, such as several years, we need inclusion of the time factor in the analysis. This is *long-term* (or *dynamic*) *production planning*, which will be dealt with in a later section.

13.2 Short-term Production Planning

13.2.1 Linear-programming Model for Short-term Production Planning

Mathematical Model for Production Planning by Linear Programming

Suppose we want to produce N kinds of products (or parts) with M kinds of production resources. Let us assume that a_{ij} units of resource $i(=1, 2, ..., M)$ are required to produce a unit of product $j(=1, 2, ..., N)$, from which c_j units of profit are gained. If only b_i units are available for resource i, determine an optimal product mix and optimal production quantities. Here, the a_{ij}s, b_is, and c_js are all constants; a_{ij} is called a *technology coefficient*.

To solve this product-mix and requirement problem, let the production amount for product $j(=1, 2, ..., N)$ be x_j—*system variables*. By taking the total profit to be obtained through manufacture and sales of the products as a production objective, this short-range production planning is formulated so as to:

$$\text{maximise } z = \sum_{j=1}^{N} c_j x_j \tag{13.1}$$

$$\text{subject to } \sum_{j=1}^{N} a_{ij} x_j \leq b_i \quad (i = 1, 2, ..., M) \tag{13.2}$$

$$x_j \geq 0 \quad (j = 1, 2, ..., N). \tag{13.3}$$

Linear Program[ming]

In the above formulation (13.1) is the *objective function* expressing the total profit gained. Equations (13.2) and (13.3) are *constraints*; (13.2) indicates restriction of production resources and (13.3) shows non-negativity requirements, meaning zero or positive quantity of production. The problem is then to determine optimal product items j^* ($\in \{1, 2, ..., N\}$) and production amounts x_j^*, thereby earning a maximum profit z^*.

The problem presented by (13.1) to (13.3) is one of the well-known *linear programming* (LP) type, since the objective function and the constraints are all expressed in linear form. The use of linear programming techniques is now very common in aggregate production planning and the neoclassical production analysis of economic theory, often referred to as *activity analysis*.

A combination of x_js which satisfies (13.2) and (13.3) is called a *feasible solution*. Any feasible solution[s] which maximises z is an *optimal solution*.

── EXAMPLE 13.1 ──

(Short-range production planning via linear programming) When producing two kinds of products, P_1 and P_2, in a specified production period, raw materials required are 4 and 10 kg, labour, 5 and 7 hours, and production facilities, 16 and 5 units, respectively, for producing a unit of each of P_1 and P_2. A unit of product P_1 creates a profit of $4, and product P_2 $9. Given that available production resources are limited, such as 40 kg for raw materials, 35 man-hours for labour, and 80 units for production facilities, formulate production planning for total profit maximisation.

Decision variables for this problem are production amounts of products P_1 and P_2, which are denoted by x_1 and x_2, respectively. Then, we formulate the following linear programming model according to (13.1)–(13.3):

$$\text{maximise } z = 4x_1 + 9x_2 \tag{13.4}$$

$$\begin{aligned}
\text{subject to } & 4x_1 + 10x_2 \leq 40 \; ... \; (a) \\
& 5x_1 + 7x_2 \leq 35 \; ... \; (b) \\
& 16x_1 + 5x_2 \leq 80 \; ... \; (c)
\end{aligned} \tag{13.5}$$

$$x_1, x_2 \geq 0. \tag{13.6}$$

□

Graphical Solution for Linear Programs

If there are only two decision variables in the LP problem, as in the above example, it is convenient to obtain an optimal solution by the graphical method.

─── EXAMPLE 13.2 ───

As shown in Figure 13.1, the above constraints (including the non-negativity requirement)—(13.5) and (13.6)—are represented by OABCD, which is called a *feasible region*. Each point existing in this region (including boundaries) is a feasible solution.

An optimal solution—the maximum profit—is obtained by finding a point in the feasible region at which the objective function (13.4) is maximised. This is point C (*the extreme or corner point*) at which the objective function line contacts the feasible region, and the coordinates of this point indicate the optimal production amounts for products P_1 and P_2. These values can also be calculated precisely by simultaneously solving (a) and (b) of (13.5) when the equality sign is used as follows (these two equations are called *active*):

$$x_1^* = 3.18 \text{ (kg)} \quad \text{and} \quad x_2^* = 2.73 \text{ (kg)}.$$

The maximum profit (the optimal value) is then

$$z^* = 4 \times 3.91 + 9 \times 2.73 = 37.27 \text{ (\$)}.$$

Notice in this case that, at the extreme point C, a normal (CC') of the objective-function line is included inside normal lines (CC'' and CC''') of the constraint lines, BC and CD.   □

Four Categories for LP Solutions

It is found by referring to Figure 13.1 that the feasible region representing the constraints of a linear program is always convex. The optimal solution is in general

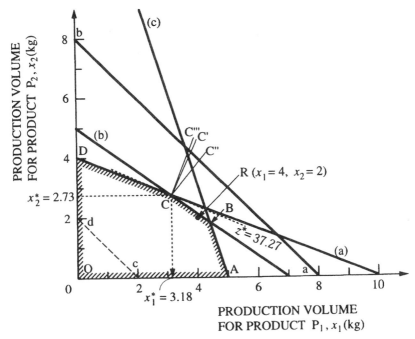

Figure 13.1 Graphical approach to solving the linear program (equations (13.4)–(13.6)). Constraints are represented by a convex feasible region (OABCD), and an optimal solution is obtained at an extreme point (C) of this feasible region.

classified into the following four categories:

(1) a unique solution, which is located on an extreme point of the feasible region (as in Example 13.2);
(2) an infinite number of optimal solutions which are located on a boundary of the feasible region;
(3) no feasible solution, hence, no optimal solution;
(4) unbounded feasible solutions, and the optimal solution is infinite.

--- **EXAMPLE 13.3** ---

(Case of category (2)) If in Example 13.1 the unit profit for product P_2 is 10, then the objective function becomes

$$z' = 4x_1 + 10x_2. \qquad (13.7)$$

In this case every point on boundary CD is optimal.

Since there are infinitely many points on a line segment, we have an infinite number of optimal solutions. In this case, point C or D—the extreme point—represents a unique optimal solution. □

--- **EXAMPLE 13.4** ---

(Case of category (3)) If we are required to produce the total amount of products greater than 8, then the following new constraint is to be added:

$$x_1 + x_2 \geq 8. \qquad (13.8)$$

This constraint indicates the region above ab in Figure 13.1. Since this region and the original feasible region contradict, there is no feasible region; hence, we have no optimal solution. □

--- **EXAMPLE 13.5** ---

(Case of category (4)) If we have ample amounts of production resources, we do not need to consider constraint (13.5); in this case, we would obtain as much profit as possible by producing as many as possible of both products P_1 and P_2 if the demands are not limited. □

13.2.2 Introduction to the Simplex Method

Basics of the Simplex Method

In cases where a linear program contains more than three decision variables, the graphical approach to solving linear programs is impossible; hence, they should be solved analytically. A most famous and representative analytical solution procedure is the simplex method[1] proposed by Dantzig (1963), which will be briefly explained in the following.

Introducing non-negative *slack variables* x_{N+i} ($i = 1, 2, \ldots, M$), which means the remains of the production resources, inequality constraint (13.2) is replaced by the equality constraint as follows:

$$\sum_{j=1}^{N} a_{ij}x_j + x_{N+i} = b_i \quad (i = 1, 2, ..., M) \tag{13.9}$$

$$x_j \geq 0 \quad (j = 1, 2, ..., N), \quad x_{N+i} \geq 0 \quad (i = 1, 2, ..., M). \tag{13.10}$$

It is seen from (13.9) that the number of unknown variables is greater than the number of equations; hence, $(N+M)$ decision variables cannot be uniquely determined. By setting N variables, which are called *non-basic variables*, at zero, the remaining M variables, which are called *basic variables*, can be determined. This is a *basic solution*. If it is a feasible solution as well, it is a *basic feasible solution*. Sets of basic and non-basic variables are called a *basis* and a *non-basis* respectively.

The *simplex method* is an iterative procedure to obtain an optimal solution for linear programming problems in a finite number of steps. It starts from an initial basic feasible solution and repeats the *pivoting process* moving iteratively to a better basic feasible solution by replacing a basic variable with a non-basic variable one by one. The algorithm below is for solving a well-structured linear program; in this solution procedure the *simplex tableau* indicated in Table 13.1 is conveniently utilised. In the algorithm, overbars stand for the values of the constants which appear in the process of pivoting.

The Simplex Algorithm

Step 1: Initialisation. As an initial basic feasible solution, set

$$x_j = 0 \quad (j = 1, 2, ..., N), \quad x_{N+i} = b_i \; (>0) \quad (i = 1, 2, ..., M). \tag{13.11}$$

Step 2: Optimality test. The current solution is optimal, if

$$\bar{c}_{jM+k} \leq 0 \quad (k = 1, 2, ..., N). \tag{13.12}$$

In this case, the first column in the simplex tableau gives the optimal solution—stop. Otherwise, proceed.

Step 3: Pivoting processes.

(3-1) Select the column q given by

$$\bar{c}_q = \max_{1 \leq k \leq N} \bar{c}_{jM+k} > 0 \tag{13.13}$$

as the column of the pivot element.

(3-2) Select $\bar{a}_{pq}$ given by

$$\frac{\bar{b}_p}{\bar{a}_{pq}} = \max_{-a_{iq} < 0, \, 1 \leq i \leq M} \frac{\bar{b}_i}{\bar{a}_{iq}} \tag{13.14}$$

as the pivot element.

(3-3) Pivoting rules.

(i) Divide the row p by $\bar{a}_{pq}$, except the pivot element, which is set at $-1/\bar{a}_{pq}$.

AGGREGATE PRODUCTION PLANNING

Table 13.1 The simplex tableau for solving a linear program. (a) is the initial basic feasible solution. (b) is the tableau obtained by the pivoting process. A process from (b) to (c) shows the pivoting process by pivoting rules, where an element with $-\bar{a}_{pq}$ is a pivot element.

(a) Initial tableau

			x_1	x_2	...	x_N
Goal function	{	z	0	c_1	c_2	...	c_N
Basic variables	{	x_{N+1}	b_1	$-a_{11}$	$-a_{12}$	...	$-a_{1N}$
		x_{N+2}	b_2	$-a_{21}$	$-a_{22}$	...	$-a_{2N}$
		$\vdots$	...	...	...	...	...
		x_{N+M}	b_M	$-a_{M1}$	$-a_{M2}$	...	$-a_{MN}$

Non-basic variables (=0)

(b)

		x_{jM+1}	x_{jM+2}	...	x_q	...	x_{jM+N}
z	$\bar{z}_0$	$\bar{c}_{jM+1}$	$\bar{c}_{jM+2}$	...	$\bar{c}_q$	...	$\bar{c}_{jM+N}$
x_{j_1}	$\bar{b}_1$	$-\bar{a}_{1jM+1}$	$-\bar{a}_{1jM+2}$	...	$-\bar{a}_{1q}$	...	$-\bar{a}_{1jM+N}$
x_{j_2}	$\bar{b}_2$	$-\bar{a}_{2jM+1}$	$-\bar{a}_{2jM+2}$	...	$-\bar{a}_{2q}$	...	$-\bar{a}_{2jM+N}$
$\vdots$	...	...	...	...	...	...	...
x_p	$\bar{b}_p$	$-\bar{a}_{pjM+1}$	$-\bar{a}_{pjM+2}$	...	$-\bar{a}_{pq}$	...	$-\bar{a}_{pjM+N}$
$\vdots$	...	...	...	...	...	...	...
x_{jM}	$\bar{b}_M$	$-\bar{a}_{MjM+1}$	$-\bar{a}_{MjM+2}$	...	$-\bar{a}_{Mq}$	...	$-\bar{a}_{MjM+N}$

If $\bar{c}_{jM+k} \leq 0$, $k = 1, 2, ..., q-1, q+1, ..., N$, $\bar{c}_q \leq 0$, optimal solution is:

$z^* = \bar{z}_0$, $x_{ji}^* = \bar{b}_i$, $i = 1, 2, ..., p-1, p+1, ..., M$, $x_p^* = \bar{b}_p$

$x_{jM+k}^* = 0$, $k = 1, 2, ..., q-1, q+1, ..., N$, $x_q^* = 0$.

Otherwise, proceed to pivoting process from (b) to (c): pivoting element, $-\bar{a}_{pq}$.

(c)

		x_{jM+1}	x_{jM+2}	...	x_p	...	x_{jM+N}
z	$\hat{z}_0$	$\hat{c}_{jM+1}$	$\hat{c}_{jM+2}$	...	$\hat{c}_q$	...	$\hat{c}_{jM+N}$
x_{j_1}	$\hat{b}_1$	$-\hat{a}_{1jM+1}$	$-\hat{a}_{1jM+2}$	...	$-\hat{a}_{1q}$	...	$-\hat{a}_{1jM+N}$
x_{j_2}	$\hat{b}_2$	$-\hat{a}_{2jM+1}$	$-\hat{a}_{2jM+2}$	...	$-\hat{a}_{2q}$	...	$-\hat{a}_{2jM+N}$
$\vdots$	...	...	...	...	...	...	...
x_q	$\hat{b}_p$	$-\hat{a}_{pjM+1}$	$-\hat{a}_{pjM+2}$	...	$-\hat{a}_{pq}$	...	$-\hat{a}_{pjM+N}$
$\vdots$	...	...	...	...	...	...	...
x_{jM}	$\hat{b}_M$	$-\hat{a}_{MjM+1}$	$-\hat{a}_{MjM+2}$	...	$-\hat{a}_{Mq}$	...	$-\hat{a}_{MjM+N}$

$$\hat{a}_{pq} = \frac{1}{\bar{a}_{pq}}, \quad \hat{a}_{pjM+k} = \frac{\bar{a}_{pjM+k}}{\bar{a}_{pq}}, \quad \hat{a}_{iq} = -\frac{\bar{a}_{iq}}{\bar{a}_{pq}}, \quad \hat{a}_{ijM+k} = \bar{a}_{iqM+k} - \bar{a}_{iq}\hat{a}_{pjM+k}$$

$$\hat{b}_p = \frac{\bar{b}_q}{\bar{a}_{pq}} \geq 0, \quad \hat{b}_i = \bar{b}_q - \bar{a}_{iq}\hat{b}_p \geq 0$$

$$\hat{c}_q = -\frac{\bar{c}_q}{\bar{a}_{pq}}, \quad \hat{c}_{jM+k} = \bar{c}_{jM+k} - \hat{a}_{pjM+k}\bar{c}_q$$

$$\hat{z}_0 = \bar{z}_0 + \bar{b}_p \frac{\bar{c}_q}{\bar{a}_{pq}}$$

$i = 1, 2, ..., p-1, p+1, ..., M; \quad k = 1, 2, ..., q-1, q+1, ..., N$

Table 13.2 Solving by the simplex method for a linear program given by equations (13.4)–(13.6). In these simplex tableaux $\theta = b_j/\bar{a}_{iq}$ in (13.14).

(a) Initial tableau

		x_1	x_2	θ
z	0	4	9	
x_3	40	−4	−10*	4
x_4	35	−5	−7	5
x_5	80	−16	−5	16

(b) Updated tableau

		x_1	x_3	θ
z	36	0.4	−0.9	
x_2	4	−0.4	−0.1	10
x_4	7	−2.2*	0.7	3.2
x_5	60	−14	0.5	4.3

(c) Final tableau (optimal)

		x_4	x_3
z	37.273	−0.182	−0.773
x_2	2.727	0.182	−0.227
x_1	3.182	−0.455	0.318
x_5	15.455	6.364	−3.955

(ii) Divide the column q, excepting the row p, by $-\bar{a}_{pq}$.
(iii) Add the amount of (i) multiplied by $-\bar{a}_{pq}$ or $\bar{c}_q$ to all the corresponding elements for the other rows except the column q.
Return to step 2. □

--- **EXAMPLE 13.6** ---

(Solving by the simplex tableau) Table 13.2 shows an analytical solution process by the simplex method for solving a linear program given by (13.4) to (13.6) in Example 13.1. In this table, x_3, x_4 and x_5 are slack variables:

$$\begin{aligned} 4x_1 + 10x_2 + x_3 &= 40 \quad \text{(a)} \\ 5x_1 + 7x_2 + x_4 &= 35 \quad \text{(b)} \\ 16x_1 + 5x_2 + x_5 &= 80 \quad \text{(c)} \end{aligned} \quad (13.15)$$

$$x_1, x_2, x_3, x_4, x_5 \geq 0. \quad (13.16)$$

In Table 13.2 the entries indicated with asterisks are *pivot elements*. The initial tableau (a) is obtained directly by step 1 from (13.4) and (13.15), indicating an initial feasible solution. In this table, elements in the first row excepting the first element are all positive, hence this tableau is not optimal by step 2. The pivot element is selected as (2,3) according to substeps (3-1) and (3-2). Then, an updated tableau (b) is generated according to the pivoting rules mentioned in substep (3-3). This tableau is not yet optimal; hence it is updated and (c) is generated. This tableau is proved optimal by the optimality test in step 2. The optimal solution (optimal production plan) is indicated in the first column on this final tableau (c). This solution is completely coincident with the one obtained previously by the graphical method.

One basic feasible solution or the associated simplex tableau corresponds exactly to a particular extreme point on the graph. The pivoting process means the transfer of an extreme point to a neighbouring extreme point along the boundary of the feasible region. In the example, the pivoting process from the initial tableau (a) to the updated tableau (b) corresponds to the transfer from point O (origin) to point D in Figure 13.1, and the pivoting process from (b) to the final tableau (c) (optimal solution), the transfer from point D to point C (optimal point). □

Shadow Price and Reduced Profit

From the final tableau (c) above, which indicates the optimal solution obtained by solving the linear program via the simplex method, the following interesting and noteworthy facts can be further clarified.

(1) There is a remainder of 15.45 units of production facilities in the optimal production schedule, which is indicated as the value for x_5 on the final simplex tableau.
(2) Both raw materials and labour force are completely exhausted up to their available amounts. However, if there existed one unit more of the raw material or the labour force, the total profit to be obtained would increase $0.77 or $0.18 more (see the entries of the first row corresponding to x_3 and x_4 on the final simplex tableau, respectively—Question 13.12). This augmentation of the profit is called the *shadow price* of the production resources.

In the above example, all system variables (production volumes) were basic variables (positive) in the optimal solution. If, however, any system variable belongs to a non-basis, this variable is zero (no production) in the optimal solution. In this case, the corresponding entry in the first row of the final simplex tableau means *reduced profit* (in the case of minimisation problems this is called the *reduced cost*), and is a decreased profit associated with producing a unit of the corresponding product.

— EXAMPLE 13.7

(Reduced profit) If in Example 3.1 the unit profit of product P_2 is increased to 20 the objective function is:

$$z'' = 4x_1 + 20x_2. \tag{13.17}$$

The final (optimal) simplex tableau for this problem is indicated in Table 13.3. In this case

Table 13.3 The final simplex tableau for an objective function, $4x_1 + 20x_2$.

		x_1	x_3
z''	80	−4	−2
x_2	4	−0.4	−0.1
x_4	7	−2.2	0.7
x_5	60	−14	0.5

only 4 units of product P_2 should be produced. That is, product P_1 is not produced; if a unit of this product was manufactured and sold, the total profit would be decreased by $4. □

13.2.3 Duality in Linear Programming

The Dual Problem

The above-mentioned shadow price is closely related to the dual problem for the original problem indicated by (13.1)–(13.3). In the case that this maximising problem is given as a *primal problem*, then the corresponding minimising problem is called the *dual problem*, which is expressed as:

$$\text{minimise } w = \sum_{i=1}^{M} b_i y_i \tag{13.18}$$

$$\text{subject to } \sum_{i=1}^{M} a_{ij} y_i \geq c_j \quad (j = 1, 2, ..., N) \tag{13.19}$$

$$y_i \geq 0 \quad (i = 1, 2, ..., M) \tag{13.20}$$

where y_i is called the *dual variable* (corresponding to the ith constraint of the primal problem), and w is called the *dual objective*, whilst x_j and z are the *primal variable* (corresponding to the jth constraint of the dual problem) and the *primal objective*, respectively.

Notice that, if the ith (jth) constraint of the primal (dual) problem holds with equality, the corresponding dual (primal) variable is no longer non-negative; that is, unrestricted in sign.

—— EXAMPLE 13.8 ——

The dual problem corresponding to the primal problem given by (13.4) to (13.6) in Example 13.1 is expressed as:

$$\text{minimise } w = 40y_1 + 35y_2 + 80y_3 \tag{13.21}$$

$$\text{subject to } \begin{array}{l} 4y_1 + 5y_2 + 16y_3 \geq 4 \quad (a) \\ 10y_1 + 7y_2 + 5y_3 \geq 9 \quad (b) \end{array} \tag{13.22}$$

$$y_1, y_2, y_3 \geq 0. \tag{13.23}$$

□

Duality Theorems

The significance of *duality* viewed as the primal and dual models expressed in (13.1) to (13.3) and (13.18) to (13.20) is mentioned by the following two theorems (Wagner, 1975).

—— THEOREM 13.1 ——

Duality. In the event that both the primal and dual problems possess feasible solutions, the primal problem has an optimal solution x_j^* ($j = 1, 2, ..., N$), the dual problem has an optimal solution y_i^* ($i = 1, 2, ..., M$), and

$$\sum_{j=1}^{N} c_j x_j^* = \sum_{i=1}^{M} b_i y_i^*, \text{ namely max } z = \min w. \quad (13.24)$$

□

— **THEOREM 13.2** —

Complementary slackness. x_j^* ($j = 1, 2, ..., N$) and y_i^* ($i = 1, 2, ..., M$) are optimal if and only if

$$y_i^* \left(\sum_{j=1}^{N} a_{ij} x_j^* - b_i \right) = 0 \quad (i = 1, 2, ..., M) \quad (13.25)$$

$$x_j^* \left(\sum_{i=1}^{M} a_{ij} y_i^* - c_j \right) = 0 \quad (j = 1, 2, ..., N). \quad (13.26)$$

□

Theorem 13.2 means that, whenever a constant in one of the problems holds with strict inequality, so that there is slack in the constraint, the corresponding variable in the other problem equals 0.

Solving the Dual Problem

It should be noted that the optimal solution to the dual problem has also been calculated when the final tableau has been obtained for solving the primal problem. The first value in the first row is the optimal (minimum) value for the dual objective, which is equal to the optimal (maximum) value for the primal objective. The value for the slack variable x_{N+i} in the first row is the optimal value for the corresponding dual variable y_i, and the value for the primal variable x_j in the first row represents the difference between the left- and right-hand sides of the jth constraint (slack variable y_{M+j}) for the dual problem for the associated optimal solution.

— **EXAMPLE 13.9** —

(Solving the dual problem) By introducing slack variables, y_4 and y_5, (13.21) is replaced by the following equality constraints:

$$\begin{array}{ll} 4y_1 + 5y_2 + 16y_3 - y_4 = 4 & \text{(a)} \\ 10y_1 + 7y_2 + 5y_3 - y_5 = 9 & \text{(b)} \end{array} \quad (13.27)$$

$$y_1, y_2, y_3, y_4, y_5 \geq 0. \quad (13.28)$$

The optimal solution for this dual problem is given in the final tableau in Table 13.2 as follows:

$w^* = 37.27 \, (= z^*)$
$y_1^* = 0.77$, $y_2^* = 0.18$, $y_3^* = 0$; $y_4^* = y_5^* = 0$.

Notice that y_1^*, y_2^*, and y_3^* are shadow prices.

Proven as the theorem of complementary slackness, the corresponding first and second constraints for the primal problem—(13.5)(a) and (b)—hold with equality sign for the optimal solution ((a) and (b) are *active*), since y_1^* and y_2^* are positive. Similarly, it is reasonable that y_4^* and y_5^* are both zero, since both x_1^* and x_2^* are positive. The third constraint for the primal problem—(13.5) (c)—holds with strict inequality for the optimal solution ((c) is *inactive*), which means that y_3^* should be zero (no contribution to the further profit increase). □

As seen in the above discussion, the optimal values for the dual variables are shadow prices that imply marginal profit associated with resource increase, that is, an increase in the total profit caused by an increase of one more unit of the production resource. Accordingly, the duality has a connection with sensitivity analysis.

13.2.4 Limitations of Application of Linear Programming to Production Planning

In applying linear programming to optimisation analysis of production planning, the decision-makers should keep in mind the following notable remarks:

(1) Two crucial technical and economic assumptions are made in formulating a production planning model by linear programming:
   (a) *proportionality*: for each activity, the total amounts of each input (production resource) and the associated profit (objective) are directly proportional to the level of output.
   (b) *additivity*: the total amounts of each input and the associated profit are the sums of the inputs and profit for each individual activity.
   If such linear axioms do not hold, the model is of a *non-linear programming* type, which is much more difficult to solve (e.g. see Section 11.3).
(2) Frequently an initial basic feasible solution is not easily found out. In such cases, the *two-phase method* or *big-M (penalty) method* is effective in the search for the initial feasible solution (see Section 13.2.5).
(3) When decision variables are defined as unit product (integers), and fractional values are meaningless (unrealisable) in practice, such as a car, an oceangoing oil tanker, a building, etc., rounded figures obtained from an optimal solution by the above simplex method are not always optimal in the true sense. In such a case, *integer programming* is applicable to finding an integer-valued solution (see Section 13.2.6).
(4) Production information is often fuzzy; e.g. the future demands are not always deterministic. If such information can be treated as probabilistic, *stochastic* or *chance-constrained programming* techniques are applicable (see Section 13.2.7).
(5) The optimal number of product items determined by means of linear programming is limited to that equal to or less than M—the amount of the constraints (13.2). This is often contrary to reality, thereby indicating a serious limitation of the optimal decision-making by a mathematical approach such as linear programming.
(6) Aggregate production planning often aims at multiple objectives such as maximum profit, minimum cost, maximum market share, maximum growth rate, etc. at the same time. In this case, ordinary linear (mathematical) programming is not effective; *multiple-objective optimisation* techniques are applied to this multiple-objective production planning (see Section 13.3).

13.2.5 Two-phase Method in Linear Programming

Difficult Case of Finding an Initial Basic Solution

The following example demonstrates a case where an initial basic feasible solution is not easily obtained.

AGGREGATE PRODUCTION PLANNING

--- **EXAMPLE 13.10** ---

A stable utilisation of the workshop is also required in Example 13.1 such that the total of production quantities for both products P_1 and P_2 exceeds at least 2. In this case the following additional constraint must be added to (13.5):

$$x_1 + x_2 \geq 2. \tag{13.29}$$

This constraint eliminates a portion of Ocd from the original feasible region shown in Figure 13.1; this does not affect the optimal solution at all. However, it is not easy to find an initial basic feasible solution when solving this problem analytically, because, if we choose slack variables x_3, x_4, x_5 and x_6 as an initial set of basic variables, as we have done previously in Example 13.6, from (13.15) and the following equation derived from (13.29) by introducing a new variable x_6, called a *surplus variable*, we obtain:

$$x_1 + x_2 - x_6 = 2, \qquad x_6 \geq 0. \tag{13.30}$$

In this case an initial basic solution is not feasible, since $x_6 = -2$. □

Basics of the Two-phase Method

To overcome the above difficulty, we introduce the two-phase method as follows (Hadley, 1962).

If some of the constraints, e.g. first K ($<M$) constraints of (13.2), hold with equality, artificial variables, $x'_{N+1}, x'_{N+2}, \ldots$, and x'_{N+K} are added to these constraints such that

$$\sum_{j=1}^{N} a_{ij} x_j + x'_{N+i} = b_i \qquad (i = 1, 2, \ldots, K) \tag{13.31}$$

$$x'_{N+i} \geq 0 \qquad (i = 1, 2, \ldots, K). \tag{13.32}$$

The idea of introducing artificial variables is that, if the original problem is to be solved, the artificial variables must be reduced to zero. The *two-phase method* solves the problem in two parts: phase I and phase II.

Phase I   Instead of considering the actual objective function, (13.1), we try to maximise the auxiliary objective function:

$$y = -\sum_{i=1}^{K} x'_{N+i} \tag{13.33}$$

subject to (13.31) and the last $(M - K)$ constraints of (13.9) as well as non-negativity requirements for all original decision variables and the associated slack and artificial variables. If the maximum value of y does not vanish, the artificial variables cannot be reduced to zero, and the original problem has no feasible solution; hence stop here. If the maximum value of y is zero, all the artificial variables vanish, then we end phase I and proceed to phase II.

Phase II   Assigning its actual profit value to each original decision value and a profit of 0 to any artificial variables which may appear in the basis at a zero level, we maximise the original function z—(13.1). The first tableau for phase II is the last tableau for phase I, which is an initial basic feasible solution for the original problem.

--- EXAMPLE 13.11 ---

(Applying the two-phase method) Upon solving Example 13.1 with a new constraint given by (13.29) or (13.30), we add an artificial variable x_7 to (13.30):

$$x_1 + x_2 - x_6 + x_7 = 2; \qquad x_6, x_7 \geq 0. \tag{13.34}$$

Applying the two-phase method to this problem, first in phase I we maximise

$$y = -x_7 = -2 + x_1 + x_2 - x_6 \tag{13.35}$$

subject to (13.15), (13.16), and (13.34). This is solved by the simplex method as shown in Table 13.4, and we obtain zero for the maximum value of y, in one pivoting process yielding (b) from (a). At the end of phase I there is no artificial variable in the basis.

Next, in phase II, by substituting x_2 obtained from (13.30) into (13.4), the original objective function z to be maximised is:

$$z = 4x_1 + 9(2 - x_1 + x_6) = 18 - 5x_1 + 9x_6. \tag{13.36}$$

The initial tableau for phase II is expressed as indicated in (a) of Table 13.5 by employing the final tableau for phase I—Table 13.4(b) with the column of the artificial variable x_7 removed. This initial basic feasible solution corresponds to point b in Figure 13.1 (the origin no longer gives us an initial basic feasible solution in this case).

Following the simplex algorithm, from this initial tableau the updated tableau (b) of Table 13.5 is generated, and finally we obtain an optimal solution in the final tableau (c) of Table 13.5, which is the same as in Examples 13.1 and 13.6, as expected. □

Table 13.4 Phase I of the two-phase method for obtaining an initial basic feasible solution to a linear program in Example 13.10. At the end of this phase, zero value for the auxiliary objective function y, which is the negative of the sum of the artificial variables, means the original problem has an optimal solution.

(a) Initial tableau

		x_1	x_2	x_6	θ
y	-2	1	1	-1	
x_3	40	-4	-10	0	4
x_4	35	-5	-7	0	5
x_5	80	-16	-5	0	16
x_7	2	-1	-1*	1	2

(b) Final tableau

		x_1	x_7	x_6
y	0	0	1	0
x_3	20	6	10	-10
x_4	21	2	7	-7
x_5	70	-11	5	-5
x_2	2	-1	-1	1

Table 13.5 Phase II of the two-phase method for obtaining an optimal solution for the original linear programming problem. The initial tableau is the same as the final tableau of phase I in Table 13.4.

(a) Initial tableau

		x_1	x_6	θ
z	18	−5	9	
x_3	20	6	−10*	2
x_4	21	2	−7	3
x_5	70	−11	−5	14
x_2	2	−1	1	—

(b) Updated tableau

		x_1	x_3	θ
z	36	0.4	−0.9	
x_6	2	0.6	−0.1	—
x_4	7	−2.2*	0.7	3.2
x_5	60	−14	0.5	4.3
x_2	4	−0.4	−0.1	10

(c) Final tableau (optimal)

		x_4	x_3
z	37.273	−0.182	−0.773
x_6	3.909	−0.273	−0.091
x_1	3.182	−0.455	0.318
x_5	15.455	6.364	−3.955
x_2	2.727	0.182	−0.227

Big-M (or Penalty) Method

In place of the two-phase method above, the *big-M* (or *penalty*) *method* produces the same result. The objective function of this method is expressed as:

$$\text{maximise } y = z - M \sum_{i=1}^{K} x'_{N+i} \tag{13.37}$$

where M signifies a very large value.

13.2.6 Integer-valued Production

Integer Programming

In the case that production volumes are integer-valued, decision variables (x_js) should hold the following constraint in place of (13.3):

$$x_j = 0, 1, 2, \ldots \quad \text{(integer)}. \tag{13.38}$$

This is *integer programming*. Rounded figures of the optimal solution obtained by ordinary linear programming often give us a wrong solution, as described in the following example. The representative solution techniques are:

- branch-and-bound method, and
- cutting plane method.

--- **EXAMPLE 13.12** ---

Suppose decision variables, x_1 and x_2—production volumes for P_1 and P_2—to be integer-valued, rounding off the optimal solution obtained in Examples 13.2 and 13.6 generates $x_1 = 4$ and $x_2 = 3$, which does not lie in the feasible region. Point R($x_1 = 4$, $x_2 = 2$) near point C in Figure 13.1 results in the total profit of 34.

The exact optimal solution, however, is obtained at point D in the case of integer-valued variables:

$x_1^* = 0$, $x_2^* = 4$; $z^* = 36$. □

13.2.7 Stochastic Production Planning

Chance-constrained Programming

If the demand for product P_j is d_j, the following constraint is further added to (3.1)–(3.3) for production planning:

$$x_j \leq d_j \quad (j = 1, 2, \ldots, N). \tag{13.39}$$

It is usual that d_j is not deterministic; if the demand has the probability density function $f_j(d_j)$, this probabilistic state creates *chance-constrained* (or *stochastic*) *programming*.

Stochastic Production Planning by Chance-constrained Programming Approach

The management policy decides that the probability of the production x_j exceeding the demand d_j should be less than a specified value α_j. Then we obtain the following stochastic constraint (Johnson and Montgomery, 1974):

$$P[x_j \geq d_j] \leq \alpha_j \tag{13.40}$$

namely

$$\int_{-\infty}^{x_j} f_j(d_j) \mathrm{d}d_j \equiv F_j(x_j) \leq \alpha_j \tag{13.41}$$

where $F_j(d_j)$ is the distribution function of demand P_j.

If the demand follows the normal distribution with mean μ_j and variance σ_j^2, (13.40) can be transformed to:

$$\Phi\left(\frac{x_j - \mu_j}{\sigma_j}\right) \leq \alpha_j \tag{13.42}$$

where $\Phi(u)$ is the standard normal distribution. Denoting the standard normal variable such that $\Phi(\theta_a) = \alpha$ by θ_a, the above equation becomes:

$$\frac{x_j - \mu_j}{\sigma_j} \leq \theta_{aj} \quad \text{i.e.} \quad x_j \leq \mu_j + \theta_{aj}\sigma_j. \tag{13.43}$$

This is a linear inequality constraint and is employed in place of (13.39).

--- **EXAMPLE 13.13** ---

(Stochastic production planning) The demand of P_1 follows the normal distribution with mean 4 and variance 1. The management policy decides $\sigma = 0.1$ in order to reduce the unsold risk.

x_1—production of P_1 is incurred with the following constraint; from (13.43),

$$x_1 \leq 4 + \theta_{0.10} = 4 - 1.28 = 2.72.$$

Graphical solution similar to Figure 13.1 generates the optimal solution:

$$x_1^* = 2.72, \ x_2^* = 2.91; \ z^* = 37.09.$$

Thus the maximum profit is reduced by addition of the stochastic constraint. □

13.3 Multiple-objective Production Planning

13.3.1 Basics of Multiple-objective Optimisation

Formulating Multiple-objective Optimisation

It is quite usual that multiple objectives are aimed in practice, e.g. at production planning. This type of decision is made by *multiple-objective* (or *multicriterion*) *optimisation*. This is to maximise (or minimise) a vector:

$$(g_1(\mathbf{x}), g_2(\mathbf{x}), \ldots, g_K(\mathbf{x})) \tag{13.44}$$

where $\mathbf{x} = (x_1, x_2, \ldots, x_N)$, N being the number of decision variables, and K, the number of objective functions.

Difficulty of Multiple-objective Optimisation

The difficulties of multiple-objective optimisation are:

- noncommensurability among objectives—difficult to measure objective values with a common dimension;
- noncompatibility among objectives—objectives are inconsistent and are in a trade-off relationship.

Hence it is very rare to be able to obtain a *supremal solution*, at which every objective reaches its optimal state, as depicted in Figure 13.2.

Pareto Optimum

Usually multiple-objective optimisation generates the *Pareto optimum* (after V. Pareto, a 19th-century Italian economist), where any objective cannot be improved

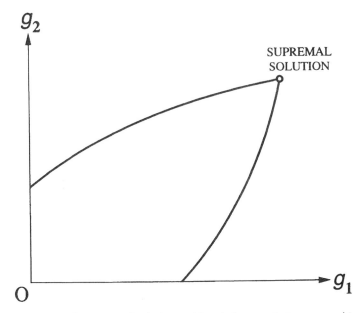

Figure 13.2 Existence of a supremal solution—this solution maximises every objective.

without incurring the inferiority of another objective, as was mentioned in Section 11.2.3. A Pareto optimum region is depicted as PQ in Figure 13.3. Any solution in the Pareto optimum region is not inferior to other solutions; hence it is called a *non-inferior solution*.

A *preferred solution* can be obtained from among Pareto optimum solutions by introducing the [social] utility function, $U(x)$. Drawing the indifference (or iso-utility) curves as in Figure 13.3, a preferred solution is decided at a contact point of the Pareto optimum and an indifference curve, as represented in the figure.

13.3.2 Techniques of Multiple-objective Optimisation

Basic Solution Methods for Multiple-objective Optimisation

Several solution methods are described below.

(1) *Parametric* (or *combined*) *method*. A single objective—the sum of all the weighted objectives established according to the relative importance of the objectives—is optimised. (This approach was employed in Example 2.13.)
(2) *Lexicographic* (or *priority*) *method*. The objectives are ranked in the order of importance, and optimisation is done simultaneously on as many of the objectives as possible, starting with the most important and going down the hierarchy.
(3) *Satisficing method*. The most important objective is optimised subject to the other objectives above their allowable satisficing levels.
(4) *Global* (or *compromise*) *evaluation method*. A single objective—the sum of the difference of each objective from its optimal state (over the optimal state), in some cases weighted—is optimised.

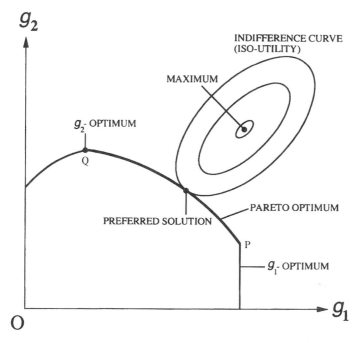

Figure 13.3 Existence of the Pareto optimum—the increase of one objective decreases the other objective. A preferred solution is decided from among this Pareto optimum region so as to maximise the utility function.

(5) *Fuzzy-programming method.* A single objective—the sum or the product of all the membership functions—is maximised.
(6) *Goal-programming* (or *goal-attainment*) *method.* All the objectives are made to approach their attainable goals as closely as possible.

13.3.3 Goal-programming Approach to Multiple-objective Production Planning

Outline of Goal Programming

Goal programming is formulated as follows (Ignizio, 1976): minimise the *achievement function* (resulting multiple-objective function):

$$\mathbf{a} = \{g_1(\mathbf{n},\mathbf{p}), g_2(\mathbf{n},\mathbf{p}), \ldots, g_K(\mathbf{n},\mathbf{p})\} \tag{13.45}$$

$$\text{subject to } f_i(\mathbf{x}) + n_i - p_i = b_i \quad (i = 1, 2, \ldots, M) \tag{13.46}$$

$$\mathbf{x}, \mathbf{n}, \mathbf{p} \geq 0. \tag{13.47}$$

The aim of (13.46) is to let the ith goal $f_i(\mathbf{x})$ approach the target b_i as closely as possible; n_i reflects negative deviation from b_i, and p_i positive deviation. Equation (13.45) is a vector function arranged in the order of descending pre-emptive priority levels, where $g_K(\mathbf{n},\mathbf{p})$ is a linear function of the *deviation variables*—$\mathbf{n} = (n_1, n_2, \ldots, n_M)$ and $\mathbf{p} = (p_1, p_2, \ldots, p_M)$, M being the number of targets. Equation (13.47) illustrates the non-negativity requirements for system and deviation variables.

Application of Goal Programming to Multiple-objective Production Planning

In this section a goal programming technique is applied to solving multiple-objective production planning.

--- **EXAMPLE 13.14** ---

(Formulating multiple-objective production planning) The following four objectives with priorities are considered for a production planning problem defined in Example 13.1, excepting that raw materials are limitlessly available and the demands for products, P_1 and P_2, are 3 and 4, respectively:

- P1 (first-priority goal): production not to exceed the demands,
- P2 (second-priority goal): minimise the excess use of labour;
- P3 (third-priority goal): maximise the total profit;
- P4 (fourth-priority goal): minimise the excess use of production facilities.

P1 is formulated as:

$$G_1 = x_1 + n_1 - p_1 = 3 \tag{13.48}$$
$$G_2 = x_2 + n_2 - p_2 = 4. \tag{13.49}$$

For P2 and P4

$$G_3 = 5x_1 + 7x_2 + n_3 - p_3 = 35 \tag{13.50}$$
$$G_4 = 16x_1 + 5x_2 + n_4 - p_4 = 80. \tag{13.51}$$

Since products 1 and 2 generate revenues, 4 and 9, respectively, the total profit expected for their demands of 3 and 4 is $4 \times 3 + 9 \times 4 = 48$. Hence P3 is formulated as:

$$G_5 = 4x_1 + 9x_2 + n_5 - p_5 = 48. \tag{13.52}$$

Under the restrictions of the above G_is, an achievement function is:

$$\text{minimise } \mathbf{a} = \{(p_1 + p_2), p_3, n_5, p_4\}. \tag{13.53}$$

This function is established by the following considerations:

- Since P1 is not to exceed the demands, overproduction—p_1 and p_2—is firstly minimised.
- Similarly P2 and P4 seek to minimise p_3 and p_4.
- Since the third priority P3 is to obtain at least 61 units of profit, n_5 is to be minimised. □

Graphical Solution of Goal Programs

The number of decision variables included in the above example is 2; hence in the following a graphical solution technique is employed to obtain the optimal solution.

--- **EXAMPLE 13.15** ---

(Graphical solution of multiple-objective production planning) Figure 13.4 is a graphical representation for (13.48)–(13.52). In this figure line segments G_1–G_5 are depicted and deviation variables to be minimised—p_1, p_2, p_3, n_5 and p_4—are circled.

In order to minimise p_1 and p_2 as the top priority P1—actually to let them be 0, the solution region obtained for satisfying P1 is OABC.

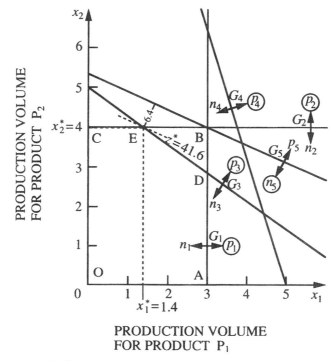

Figure 13.4 Graphical solution of a goal program—point E achieves all the goals, excepting the profit.

Similarly p_3 for the second priority P2 can be made 0 by setting the solution region as OADEC without impairing the solution for P1.

In achieving the third priority P3 it is impossible to obtain 0 for n_5 without degrading the first and the second priorities. Point E where n_5 shows the minimum value is marked. This point happens to satisfy the last priority P4, that is, to make p_4 zero as well.

As a result we obtain the optimal solution for multiple-objective production planning:

$x_1^* = 1.4$
$x_2^* = 4$
$\mathbf{a}^* = \{0, 0, 6.4, 0\}$

The total profit obtained is: $z^* = \$41.6$.

13.4 Product Mix Analysis

Optimal Product Mix

The *product mix* determines the proper combination of the kinds of items to be produced with the existing production capacity. If there is insufficient capacity to produce the entire amount of products demanded, such a demand cannot be sufficiently fulfilled. In such a case, the optimal product mix—optimal kinds of products and their production quantities—should be properly selected. Some of the policies employed to solve this problem are to:

- maximise the total profit obtained by production under the capacity constraint;
- maximise the total amount of products produced even at the expense of the profit gained; etc.

Formulating Product-mix Problems via 0–1 Linear Program

A product-mix problem can be formulated similarly to (13.1) and (13.2); (13.3) is replaced by:

$$x_j = 0 \text{ or } 1 \quad (j = 1, 2, \ldots, N). \tag{13.54}$$

If product j is accepted to be produced, $x_j = 1$; otherwise $x_j = 0$. Hence, decision variables are of 0–1 type. This mathematical programming is called *0–1 type linear* (or *integer*) *programming*.

Knapsack Problem

A case that $M = 1$ in (13.2) is called a *knapsack problem*, in that b_1 is the available capacity of a knapsack, a_{1j} is the capacity needed for item j to be loaded in the knapsack, and c_j is the degree of preference.

Solving Knapsack Problems by the Profit-rate Principle

A heuristic approach to solving knapsack problems is to employ the profit-rate principle. The *profit rate* is the profit per unit time, as defined in Chapter 11; in this case, the profit per unit capacity is:

$$r_j = c_j/a_{1j} \quad (j = 1, 2, \ldots, N). \tag{13.55}$$

Sequential selection of items in the decreasing order of the profit rate from the highest until the available capacity b_1 is filled generates a near-optimal solution.

— **EXAMPLE 13.16** —

(Product-mix decision by the profit-rate principle) We have three products—P_1, P_2, and P_3—to be manufactured in a week—40 hours. Production times needed for these three products are 15, 25, and 30 hours; gross revenues earned are $7000, $9000, and $15 000 respectively.

A product-mix problem aiming at the maximum profit is then formulated, using $1000 as the unit, as

maximise: $z = 7x_1 + 9x_2 + 15x_3$ (13.56)
subject to: $15x_1 + 25x_2 + 30x_3 \leq 40$ (13.57)
$x_1, x_2, x_3 = 0$ or 1.

Profit rates for P_1, P_2, and P_3 are:

$r_1 = 7/15 = 0.47$, $r_2 = 0.36$, $r_3 = 0.5$.

P_3 with the highest profit rate is first selected; the remaining time available is $40 - 30 = 10$ hours, but none of the other products can be produced within this time; hence stop. The profit obtained is $15 000.

However, the larger profit can be earned in this case by selecting P_1 and P_2 with priorities lower than P_3; the optimal solution is:

$$x_1^* = x_2^* = 1, \ x_3^* = 0; \ z^* = 16.$$

□

Obtaining Optimal Solution for 0–1 Type LP

The exact solution for the 0–1 type linear program can be obtained by:
- complete enumeration;
- branch-and-bound method;
- Balas' additive algorithm, etc.

Balas' additive algorithm (Balas, 1965) is an effective method in which one decision variable is selected at each stage and set at 1; with this partial solution infeasibility or attractiveness of the final solution is checked. Infeasible or unattractive cases are rejected and the final exact solution is derived from only attractive partial solutions.

To begin with, by setting as:

$$x_j = 1 - y_j \quad (j = 1, 2, \ldots, N), \tag{13.58}$$

the 0–1 type linear program is converted into the following:

$$\text{minimise: } w = -z + \sum_{j=1}^{N} c_j = \sum_{j=1}^{N} c_j y_j \tag{13.59}$$

$$\text{subject to: } g_i = \left(b_i - \sum_{j=1}^{N} a_{ij}\right) + \sum_{j=1}^{N} a_{ij} y_j \geq 0 \quad (i = 1, 2, \ldots, M) \tag{13.60}$$

$$\text{0–1 requirement: } y_j = 0 \text{ or } 1 \quad (j = 1, 2, \ldots, N). \tag{13.61}$$

Balas' Additive Algorithm

Step 1: Set the objective value $\bar{w} \leftarrow \infty$ and the partial solution $S \leftarrow \emptyset$.

Step 2: If any of the constraints is violated by setting all variables which are not included in S at 0, let a set of the constraints be G, and proceed. Otherwise go to Step 6.

Step 3: Denote by Y the set of variables $\notin S$, which have positive coefficients in the constraints $\in G$, and do not increase the objective value from $\bar{w}$ even if these variables are set at 1, denote by Y' otherwise.

Step 4: When generating complete solutions for S by setting the variables $\in Y$ at 1 and the variables $\in Y'$ at 0, proceed if $G = \emptyset$. Otherwise go to Step 7.

Step 5: Set the variables $\in Y$ at 1 one by one and calculate the total sum of the constraint values. Set a variable with the largest sum at 1 and include this variable in S. Return to Step 2.

Step 6: If the objective value as to the obtained feasible solution is not less than $\bar{w}$, proceed. Otherwise, set the variables which are not included in the current partial solutions S at 0, make a complete solution, and set the corresponding objective value at $\bar{w}$. Go to Step 8.

Step 7: If the variables $\in S$ are replaced with 0, stop (the optimal solution is the current $\bar{w}$-value and the corresponding variable values). Otherwise, proceed.

Step 8: Set the rightmost variable, which is included in S and not replaced with 0, at 0, exclude a set of variables which lie on the right side of this variable from S, and return to Step 2. □

— **EXAMPLE 13.17** ———————————————————————

(Optimal product mix by the additive algorithm) Two more constraints—capacity and raw material—are added to Example 13.16 as follows:

$$7x_1 + 6x_2 + 5x_3 \leq 20$$
$$16x_1 + 5x_2 + 20x_3 \leq 30.$$

The optimal solutions obtained in Example 13.16: $x_1^* = x_2^* = 1$, $x_3^* = 0$ also satisfy the above constraints; hence, the optimal solution is not changed even by adding the above two constraints. This optimal solution will be obtained by the additive algorithm as follows.

By replacing x_j with y_j, using (13.58), the problem is transformed into:

minimise: $w = -z + 31 = 7y_1 + 9y_2 + 15y_3$ (13.62)

subject to: $g_1 = -30 + 15y_1 + 25y_2 + 30y_3 \geq 0$ (a)
$\quad\quad\quad g_2 = 2 + 7y_1 + 6y_2 + 5y_3 \geq 0$ (b) (13.63)
$\quad\quad\quad g_3 = -11 + 16y_1 + 5y_2 + 20y_3 \geq 0$ (c)

0–1 requirements: $y_1, y_2, y_3 = 0$ or 1.

Now the additive algorithm is applied to the above problem as follows:

Step 1: Set $\bar{w} = \infty$ and $S = \emptyset$. Go to Step 2.
Step 2: Set $y_1, y_2,$ and $y_3 \notin S$ at 0. $g_1 = -30$ and $g_3 = -11$ violate the constraints; hence $G = \{g_1, g_3\}$. Go to Step 3.
Step 3: Currently $Y = \{y_1, y_2, y_3\}$ and $Y' = \emptyset$. Go to Step 4.
Step 4: Setting $y_1 = y_2 = y_3 = 1$ generates $g_1 = 40$, $g_2 = 20$, $g_3 = 30$; no constraints are violated: $G = \emptyset$. Go to Step 5.
Step 5: Setting $y_1, y_2, y_3 = 1$ sequentially, we obtain $g_1 + g_2 + g_3 = -1, -3, 16$. Since the effect of y_3 is maximal, set $S = \{3\}$ (meaning $y_3 = 1$). Return to Step 2.
Step 2: Setting $y_1 = y_2 = 0$ generates $g_1 = 0$, $g_2 = 7$, and $g_3 = 9$; no constraints are violated. Hence go to Step 6.
Step 6: $w = 15$ which is less than $\bar{w} = \infty$. We obtained a complete solution: $y_1 = y_2 = 0$, and $y_3 = 1$. Setting $\bar{w} = 15$, go to Step 8.
Step 8: Only one variable belongs to S, that is, y_3. Setting $y_3 = 0$ ($S = \{\bar{3}\}$), return to Step 2.
Step 2: Setting $y_1 = y_2 = 0$ generates violated constraints: $g_1 = -30$ and $g_3 = -11$; hence $G = \{g_1, g_3\}$. Go to Step 3.
Step 3: Variables belonging to the current S: $y_3 = 0$. If $y_1 = 1$, $w = 7$, and if $y_2 = 1$, $w = 9$; both are less than $\bar{w} = 15$. Hence $Y = \{y_1, y_2\}$ and $Y' = \emptyset$. Go to Step 4.
Step 4: Setting $y_1 = y_2 = 1$ ($y_3 = 0$) generates $g_1 = 10$, $g_2 = 15$ and $g_3 = 10$; hence $G = \emptyset$. Go to Step 5.
Step 5: Setting $y_1, y_2 = 1$ sequentially generates $g_1 + g_2 + g_3 = -1, -3$. The effect of y_1 is greater; hence y_1 is added into S, that is, $S = \{\bar{3}, 1\}$. Return to Step 2.
Step 2: Setting $y_2 = 0$ generates $G = \{g_1\}$. Go to Step 3.
Step 3: Currently $y_3 = 0$ and $y_1 = 1$. If $y_2 = 1$, $w = 16 > \bar{w} = 15$; hence $Y = \emptyset$, $Y' = \{y_2\}$. Go to Step 4.

Step 4: If $y_2 = 0$ ($y_3 = 0$, $y_1 = 1$), $G \neq \emptyset$. Go to Step 7.
Step 7: Go to Step 8.
Step 8: Replace as $y_1 = 0$: $S = \{\bar{3}, \bar{1}\}$. Return to Step 2.
Step 2: If $y_2 = 0$, $G = \{g_1, g_3\}$. Go to Step 3.
Step 3: If $y_2 = 1$, $w = 9 < \bar{w}$. $Y = \{y_2\}$ and $Y' = \emptyset$. Go to Step 4.
Step 4: Setting $y_2 = 1$ ($y_1 = y_3 = 0$), $G = \{g_1, g_3\} \neq \emptyset$. Go to Step 7.
Step 7: Since the variables belonging to S have been replaced with 0, we obtain the optimal solution:

$$y_1^* = y_2^* = 0, \quad y_3^* = 1; \quad w^* = 15.$$

Hence the optimal solution for the primal problem is:

$$x_1^* = x_2^* = 1, \; x_3^* = 0; \; z^* = 16. \qquad \square$$

13.5 Lot-size Analysis

Economic Lot Size

Intermittent production, which is often called lot (or batch) production, as was mentioned in Section 4.1, is a type of production taken when the demand for a commodity is small compared with the production capacity, namely, the production rate. The product is then manufactured periodically in a quantity which will meet the demand for some time until the next production run is resumed. In the interval between two production runs the production facilities may be employed for other work in a similar fashion. This quantity, to be manufactured periodically, is called a *lot* (or *batch*) *size*. The basic problems in this sort of intermittent production are to determine the *optimal* (or *economic*) *lot* (or *batch*) *sizes* for the products and to decide the production cycles or the order of production in which those economic lot sizes are run.

Analysis of Economic Lot Sizes

Lot production in its simplest and idealised case, where product items are continuously consumed at the uniform rate, is represented in Figure 13.5, where the variations in the level of the product inventory are plotted against time. In this case the inventory level has been reduced to zero by consumption, and it has immediately increased again by the start of production.

Let production rate (pc/h) and consumption rate (pc/h) be r_p and r_c, respectively, and $r_p > r_c$. Then, the inventory is piled up at the uniform rate of $(r_p - r_c)$ during the production period T_p and reaches a point $\hat{Q}$. During non-production period T_s the inventory decreases at the uniform rate of r_c. The total production quantity in a production cycle $(T_p + T_s)$, i.e. lot size, is:

$$Q_p = r_p T_p. \tag{13.64}$$

To determine the optimal value for this value, which is an economic lot size, the total variable cost per unit piece of production with regard to the lot size is

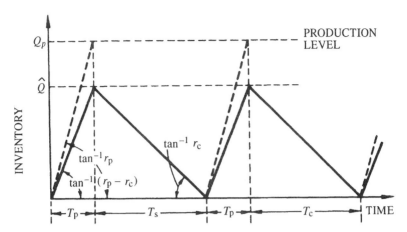

Figure 13.5 Inventory–time chart for a single-product cyclic lot production in its simplest and idealised case. The inventory is piled up at the uniform rate during the production period T_p and decreases uniformly during the non-production period T_s.

minimised. This cost is made up of:

- setup cost u_s/Q_p and
- inventory-holding cost $u_h(\hat{Q}/2)T_c/Q_p$,

where u_s is setup cost per cycle or lot, u_h is inventory-holding cost per unit piece of product per unit time, and T_c is production cycle time: $T_c = T_p + T_s$.

Since from Figure 13.5,

$$\hat{Q} = \frac{r_p - r_c}{r_p} Q_p = (1-\gamma)Q_p, \qquad 0 < \gamma = \frac{r_c}{r_p} < 1 \tag{13.65}$$

$$T_c = \frac{Q_p}{r_c} \tag{13.66}$$

the total variable cost per unit piece of product is:

$$u = \frac{u_s}{Q_p} + \frac{u_h(1-\gamma)Q_p}{2r_c}. \tag{13.67}$$

The optimal, economic or minimum-cost lot size is determined by differentiating the above equation with respect to Q_p and setting it at zero as follows:[2]

$$Q_p^* = \sqrt{\left[\frac{2r_c u_s}{(1-\gamma)u_h}\right]}. \tag{13.68}$$

The optimal production cycle time is, from (13.66) and (13.68),

$$T_c^* = \sqrt{\left[\frac{2u_s}{(1-\gamma)r_c u_h}\right]}. \tag{13.69}$$

In general, the total cost curve per unit of product is shallow near the minimum-cost (economic) lot size, which is given by a square-root function, as illustrated in Figure 13.6. Hence, it is relatively insensitive to deviations from that lot size.

--- **EXAMPLE 13.18** ---

(Calculating the economic lot size) Suppose the demand rate in the market (r_c) is 100 per day, and the production rate in the factory (r_p) is 150. When the setup cost (u_s) is $1000/cycle, and the stock-holding cost (u_h) is $1/day, from (13.68) the economic lot size is:

$$Q_p^* = \sqrt{\frac{2 \times 100 \times 1000}{(1 - 100/150) \times 1}} = 775.$$

From (13.69) the production cycle is:

$$T^* = Q_p^*/r_c = 775/100 = 8.$$

The optimal policy is, then, to manufacture 800 ($= 8r_c$) pieces of the product every 8 days, production period being 5.3 days ($= T_p$). □

Multi-product Cyclic Lot Production

In the case that consistent and continuing demands occur for certain multiple-product items, the equipment must be assigned to the multi-product cyclic lot production. If each product is manufactured once each cycle, the problem is to determine the production cycle and the economic lot sizes for all the products of M items to be produced. A subscript j is used in the previous nomenclature to indicate the association with product j.

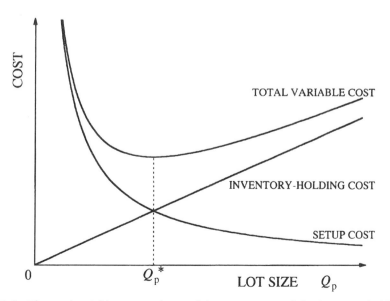

Figure 13.6 The total variable cost made up of the setup cost and the inventory-holding cost is minimal at the optimum lot size, Q_p^*.

In this analysis, time is employed as the common unit. Then, the total variable cost per unit time is expressed as:

$$U = \sum_{j=1}^{M} \left[\frac{u_{sj}}{T_c} + \frac{r_{cj}}{2} u_{hj}(1-\gamma_j)T_c \right]. \tag{13.70}$$

The optimal production cycle—the same for all M product items—is obtained from this equation as follows:

$$T_c^{**} = \sqrt{\left[2 \sum_{j=1}^{M} u_{sj} \bigg/ \sum_{j=1}^{M} (1-\gamma_j) r_{cj} u_{hj} \right]}. \tag{13.71}$$

The economic lot sizes are, from (13.67) and (13.71):

$$Q_{pj}^{**} = r_{cj} T_c^{**} = r_{cj} \sqrt{\left[2 \sum_{j=1}^{M} u_{sj} \bigg/ \sum_{j=1}^{M} (1-\gamma_j) r_{cj} u_{hj} \right]}, \quad j = 1, 2, \ldots, M. \tag{13.72}$$

The above multi-product cyclic production is possible as long as the sum of setup time S_j and optimal production period T_{pj}^{**} over all the product items does not exceed the optimum production cycle T_c^{**}; that is, if

$$\sum_{j=1}^{N} (S_j + T_{pj}^{**}) \leq T_c^{**}. \tag{13.73}$$

— **EXAMPLE 13.19** —

(Calculating multi-product economic lot sizes) Consider a cyclic lot production, in which three product items are to be produced in each cycle on specific equipment, as indicated in Table 13.6.

Table 13.6 Basic data for three-product cyclic lot production for determining optimal production cycle and economic lot sizes.

Product item	1	2	3
Production rate per day	200	250	400
Consumption rate per day	50	50	80
Inventory-holding cost ($ per unit per day)	0.01	0.01	0.005
Setup cost ($/cycle)	300	200	150

Table 13.7 Economic lot sizes, maximum inventory levels, and production times calculated for three-product cyclic lot production in the case of consistent and continuing demands. The optimal production cycle is 34.5 days (also see basic data in Table 13.6).

Product item	1	2	3
Economic lot size (pieces/cycle)	1725	1725	2760
Maximum inventory level (pieces)	1294	1380	2208
Production time (days/cycle)	8.6	6.9	6.9

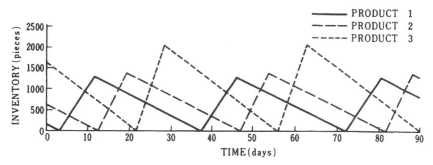

Figure 13.7 Production cycle and inventory–time chart for the optimal three-product cyclic lot production in the case of consistent and continuing demands. The optimal production cycle is 34.5 days, and the economic lot sizes are 1725 pieces for products 1 and 2, and 2760 pieces for product 3 (also see Tables 13.6 and 13.7).

From equation (13.71) the optimal production cycle is:

$T_c^{**} = \sqrt{[2(300 + 200 + 150)/(50 \times 0.01 \times 0.75 + 50 \times 0.01 \times 0.8 + 80 \times 0.005 \times 0.8)]}$
$= 34.5$ (days).

Economic lot sizes, maximum inventory levels, and production times are obtained from equations (13.72), (13.65), and (13.64), respectively, and shown in Table 13.7.

The production cycle and inventory–time chart for the above three-product cyclic lot production is shown in Figure 13.7, where setup times are assumed to be 3, 1, and 2 days for product 1, 2, and 3, respectively. Equation (13.73) also holds:

$$\sum_{j=1}^{3}(S_j + T_{pj}^{**}) = 28.4 < T_c^{**}.$$

Hence this three-product cyclic lot production is possible. □

13.6 Material Requirements Planning (MRP) and Machine Loading

13.6.1 Essentials of MRP

What is MRP?

Material requirements planning (MRP), and *manufacturing resource planning* (MRPII) (Orlicky, 1975) are widely used systematic procedures using computer-based software for production planning based upon the parts explosion, which was described in Section 7.3. They are usually applied to a large job-shop production in which multiple products are manufactured in lots through several processing stages.

First, MRP establishes a *master production schedule* (*MPS*), which is an aggregate plan showing required amounts vs. planning periods for multiple end items to be produced. Then, with the use of a bill of materials and an inventory file, production information for dispatching multiple jobs is generated on the hierarchical multiple-stage manufacturing system by taking into account commonness and possible substitution of many parts needed to assemble the multiple products.

Thus, MRP schedules and controls a total flow of materials from raw materials to finished products on a time basis (the unit of time is called a *time bucket*; usually, this is one week). The standard flow of MRP is depicted in Figure 13.8.

CRP

The possibilities of assembling products and machining parts are also examined in relation to the capacity of production facilities. This function is called *capacity requirements planning (CRP)*, as is also indicated in Figure 13.8.

13.6.2 The General Procedure of MRP

The MRP Procedure

MRP is made by the following considerations.

(1) *Calculating the requirements for products/parts*. The first step of MRP is to calculate the gross and net requirements for products/parts to be produced in the specified periods (time bucket; usually weekly basis[3]) as established in MPS. The gross requirement for parts can be obtained by multiplying the quantities of products indicated in MPS by the required number of the part shown in the bill of materials. By subtracting the inventory (including work-in-process) from the gross requirement, the net requirement is easily obtained.

(2) *Making up the lot*. In the case that a certain product or part is required to be produced in sequential periods, an appropriate lot size is decided. There are several ways to make up the lot:
   (a) *Economic-order-quantity (EOQ) method*. A method to minimise the sum of ordering cost and the inventory-holding cost. This quantity, EOQ, can

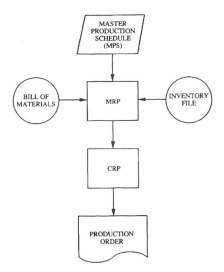

Figure 13.8 Procedure of MRP systems—based upon the master production schedule MRP proceeds with use of the bill of materials and the inventory file. After the execution of CRP the production order is issued.

be obtained through a similar analysis of production lot size, as follows:

$$Q^* = \sqrt{(2c_s D / c_h)} \qquad (13.74)$$

where D is the annual demand in units, c_s is the cost of placing an order, and c_h is the annual carrying cost per unit inventory.

(b) *Periodic-order-quantity method.* Determines the time interval of two succeeding orders by dividing the EOQ calculated using (13.74) by the average demand in the planning horizon.

(c) *Economic-part-period* (or *least-total-cost*) *method.* Determines a lot as a demand in the period such that the cumulative ordering cost equals the cumulative inventory cost. Supplementing the adjusting routine (called 'look-ahead/look-back') to this method constitutes the *part-period-balancing method*, in which the order determined by the economic-part-period method is implemented in the previous or the next period in order to cope with sudden change of demands.

(d) *Wagner–Whitin method.* Determines the optimal making-up periods for a lot so as to minimise the sum of ordering and carrying costs by dynamic programming. Incidentally the *least-unit-cost method* minimises the unit product cost, and the *Silver–Meal method* minimises the cost required in a single period.

(e) *Lot-for-lot method.* A method to simply order the amount of the demand.

(3) *Subtracting lead times.* By subtracting lead times from the period, in which the product/part is required, the start time for producing every product/part is established. For a piece part (unit) constituting a complex part (sub-assembly), first the start time for that complex part is determined, and then the start time for the piece part is calculated.

EXAMPLE 13.20

(MRP procedure) Product X with parts structure shown in Figure 13.9 is required to be produced for the next four weeks—the gross requirements are indicated by MPS as in (a) of Table 13.8. If the present inventory is 3, then the net requirements for each of four weeks can be calculated as in (c) of the table. Those for the 2nd and the 3rd weeks are added to form a lot size. Then the scheduled requirements are obtained as indicated in (d) of the table. If the

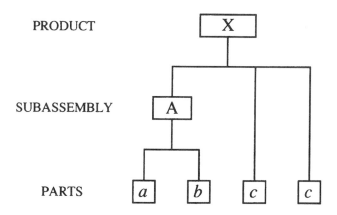

Figure 13.9 Product structure used as a sample example for executing the MRP.

Table 13.8 Requirement calculation for product X.

Week	0	1	2	3	4
(a) Gross requirement		1	3	2	4
(b) Inventory	3	2			
(c) Net requirement			1	2	4
(d) Planned requirement			3		4
(e) Scheduled order		3		4	

lead time of assembling product X is one week, then the scheduled orders are placed as in (e) of the table.

Two pieces of part c are required to assemble a unit of product X. Hence the gross requirements for the four weeks are obtained by multiplying the scheduled orders for product X by 2, as indicated in (a) of Table 13.9. If the current inventory is 5, then the net requirements can be calculated as in (c) of the table. Lot sizing is done week by week; then the net requirements are also the scheduled requirements. If the lead time of this part is one week, then the scheduled orders are determined as in (e) of the table. One piece to be ordered in week 0 is made a priority production order. □

13.6.3 Capacity Requirements Planning (CRP)

CRP Procedure

Whether or not the production facilities have enough capacity to produce multiple parts/products in every period is examined. This is done by *machine loading*, in which the work-load level is adjusted in every time period so as not to exceed the capacity of each production facility, with regard to the due dates for the products/parts to be manufactured.

—— **EXAMPLE 13.21** ——

(CRP) Part c is used in each of three types of products: X, Y, and Z. Its order quantities are: 8 in week 2 for X as found in Table 13.9, 5 in week 1 and 6 in week 4 for Y; and 4 in week 3 for Z. Then, CRP is made for c as indicated in Figure 13.10. If the normal capacity is 6 pieces a week, overload occurs by 2 in week 2. One piece can be produced in advance in the previous

Table 13.9 Requirement calculation for part c.

Week	0	1	2	3	4
(a) Gross requirement		6		8	
(b) Inventory	5				
(c) Net requirement		1		8	
(d) Planned requirement		1		8	
(e) Scheduled order	1		8		

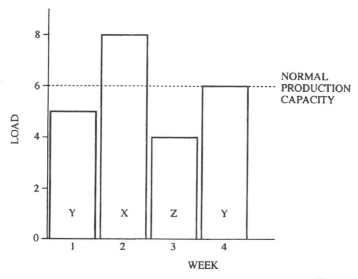

Figure 13.10 Capacity requirements planning (CRP) for part c. The overload is smoothed by employing the slacks remaining in other periods.

week and stored as inventory. The remaining one should be produced in overtime in week 2. Or, if two pieces for Y can be postponed from week 1 to week 3 by changing the due date by negotiation, two pieces for X can be produced in week 1. Thus the entire production can be made in the normal capacity. This procedure is called *flexible adjustment*, which is usually provided in the software of MRP. □

13.6.4 Machine Loading and Production Smoothing

Machine Loading

In general, the capacity of production facilities is limited; its limit is the effective work time multiplied by the number of machines, expressed in terms of machine-hours, or, in the case of operative workers, man-hours. Hence, it is necessary to adjust and optimally determine the workload for manufacturing products in every time period so as not to exceed the production capacity, in connection with due dates for those products. This decision is called *machine loading*.

Figure 13.11 illustrates the forward and backward methods for machine loading in relation to time elapsed.

Production Smoothing

It quite often happens that in some periods work load exceeds the production capacity to meet the due dates established by market demands. The function to level down the work load so that it is within the production capacity and to respond to fluctuation of the external environment of the manufacturing system is *production smoothing*.

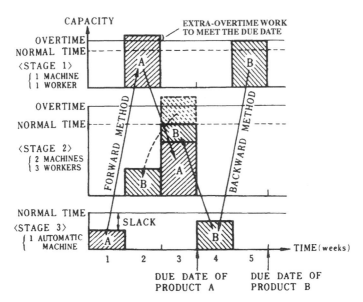

Figure 13.11 Graphical representation of the forward and backward methods for machine loading in relation to time. Overtimes are frequently required in some time periods to meet the due dates for manufacturing the products.

Policies for Production Smoothing

There are several management policies for production smoothing as follows (Gutenberg, 1955):

- *time adaptation*—a method of absorbing production fluctuation by increasing or decreasing production time, thus increasing the current production capacity; for example, by setting work overtime or undertime;
- *inventory adaptation*—a method of absorbing production fluctuation by controlling inventory level, that is, by producing and storing excessive products at periods of low demand and shipping them at periods of high demand (see Chapter 15);
- *intensive adaptation*—a method of absorbing production fluctuation by production intensity; for example, intensive utilisation of production facilities by increasing or decreasing the production speed or the speed of rotation of the main spindle of the machine tool (see Chapter 11);
- *quantitative adaptation*—a method of absorbing production fluctuation by increasing or decreasing the number of usable production facilities;
- *subcontract or outsource adaptation*—a method of absorbing production fluctuation by subcontracting production of the amounts unable to be manufactured by the existing production facilities of the manufacturing system.

A method utilising the above policies for production smoothing has been proposed for the problem of the multi-product, multistage production system when it incurs fluctuation of the demands that come from the market in the long term (Hitomi, 1975).

Optimum Machine Loading

The optimum machine loading is done by the transportation-type linear programming algorithm. Let us examine the following example, wherein the idea is extracted from Starr (1972).

--- **EXAMPLE 13.22** ---

(Optimum machine loading) Four products (or jobs), J_A, J_B, J_C, and J_D, are to be scheduled for a specified week by assigning them to each of three machines, M_1, M_2, and M_3. Production rates that characterise each job and each machine, index ratio for production rates for the three machines, available hours per week and utilisation of these machines, wage rate for running these machines, demands in this week for the four jobs, and material cost and selling price for each of these products are indicated in Table 13.10.

The problem is to select a suitable machine from among three machines and assign it for processing each of four jobs such that the total profit will be maximised.

From this table, first, the labour-cost matrix in dollars/piece is calculated, for example, for J_A–M_1, $8.00/10 = \$0.80/pc$; next, adding the material cost and subtracting the sum from the selling price, a profit matrix in dollars/piece is obtained as below:

	M_1	M_2	M_3
J_A	2.20	2.00	1.80
J_B	3.00	2.75	2.50
J_C	3.00	2.00	1.00
J_D	3.40	3.00	2.60

Table 13.10 Basic data for optimum machine loading: four products or jobs, J_A, J_B, J_C, and J_D, are to be assigned to each of three machines, M_1, M_2, and M_3 for the week's production. M_1 is the 'standard machine'.

Production rate (pc/h)	Machine			Demand (pc/week)	Material cost ($/pc)	Selling price ($/pc)
	M_1	M_2	M_3			
J_A	10	7.5	5	500	1.00	4.00
Job J_B	8	6	4	160	2.00	6.00
J_C	2	1.5	1	60	10.00	17.00
J_D	5	3.75	2.5	200	5.00	10.00
Index ratio	1	3/4	1/2			
Available capacity (h/week)	40	80	60			
Utilisation	1.0	0.9	0.8			
Wage rate ($/h)	8.00	7.50	6.00			

The above matrix is now converted into a profit-per-standard-machine-hour matrix by utilising the production rates of the standard machine M_1, i.e. by multiplying all elements in the first row by 10, which is the J_A productivity of the standard machine, etc.

We must then convert the machine hours available per week into standard machine hours available per week by multiplying the available capacity of each machine by utilisation and

Table 13.11 Matrix of profit per standard machine hour with the row and the column restrictions by demand and available machine hours in standard machine hours.

Profit ($/standard machine hour)	Machine M_1	M_2	M_3	Demand (standard machine hours)
Job J_A	22.00	20.00	18.00	50
Job J_B	24.00	22.00	20.00	20
Job J_C	6.00	4.00	2.00	30
Job J_D	17.00	15.00	13.00	40
Available standard hours	40	54	24	

Table 13.12 Final solution for the machine loading problem indicated as in Table 13.11—(a) is expressed in terms of standard machine hours, and (b) is the optimal solution indicating actual pieces to be produced by each machine assigned with the maximum total profit of $2104.

(a) Final solution in standard hour assignments

Standard machine hours	Machine M_1	M_2	M_3	Dummy	Demand
Job J_A	24	26			50
Job J_B		20			20
Job J_C		8		22	30
Job J_D	16		24		40
Available hours	40	54	24	22	140

(b) Optimal solution of actual pieces to be produced by each assignment

Actual pieces	Machine M_1	M_2	M_3	Total units	Requirements
Job J_A	240	260		500	500
Job J_B		160		160	160
Job J_C		16		16	60
Job J_D	80		120	200	200

AGGREGATE PRODUCTION PLANNING

then by index ratio. Demands for jobs are also converted into standard hours per week by dividing demand for each job by the production rate of the standard machine.

Consequently, we get the complete matrix with the row and the column restrictions, entirely in terms of standard machine hours, as shown in Table 13.11.

Since the total standard machine hours available per week are 118($= 40 + 54 + 24$) and 140($= 50 + 20 + 30 + 40$) standard hours are required to do all the jobs, a dummy machine is appointed to absorb the excess requirement of 22($= 140 - 118$) standard hours in the above matrix. The profits assigned to the dummy machine are all set at zero.

Since the problem shown in Table 13.11 is of a transportation type, the transportation algorithm of linear programming, which was introduced in Chapter 10, can be used to solve this problem.

Notice that the objective, in this case, is to maximise the total profit. Logically, therefore, all changes in assignments should be made to cells that improve the total profit. Starting from the initial solution obtained by a northwestern corner allocation, we finally obtain the solution in the number of standard hours for each assignment, as shown in Table 13.12(a). This is reconverted to actual machine hours, and then to the actual number of pieces that will be obtained for each job by multiplying actual assignment hours by the appropriate real production rates of each machine. The result is shown in Table 13.12(b). As seen from this, job J_C is the only demand that is not fully supplied as a result of demand exceeding productive capacity, and the maximum total profit for this optimal solution is $2104. □

13.7 Long-term Production Planning

Basics of Long-term Production Planning

Long-term (or *dynamic*) *production planning* establishes a production plan covering the future, usually several years, contrary to short-term (or static) production planning, which deals with a production plan for a single period (a month, three months, half a year, a year).

The following basic relation holds among inventory, production, and sales for two consecutive periods of long-term production planning (see also equation (2.1)):

(initial inventory in period $t + 1$) = (final inventory in period t)
= (initial inventory in period t) + (production volume in period t)
− (sales in period t). (13.75)

Mathematical Model for Long-term Production Planning

Long-term production planning determines the product items and their production volumes for a long term so as to minimise the total production cost including the production cost and the inventory cost or maximise the total profit under the existing production capacities, based upon the demands and/or the forecast information for a specified long term.

In the case that the demands of a product are known for H periods, long-term production planning is formulated as follows:

maximise the total profit:

$$z = \sum_{i=1}^{H}\sum_{j=1}^{H}(r_j - u_i)x_{ij} + \sum_{j=2}^{H}\sum_{i=1}^{j-1}\left(\sum_{k=i}^{j-1}c_k\right)x_{ij}, \qquad (13.76)$$

subject to the capacity requirements:

$$\sum_{j=i}^{H} x_{ij} \le a_i \quad (i = 1, 2, ..., H) \tag{13.77}$$

$$\sum_{i=1}^{j} x_{ij} \le d_j \quad (j = 1, 2, ..., H), \tag{13.78}$$

where a_i is the production capacity at period i, u_i is unit production cost at period i, d_j is the demand at period j, c_k is the inventory-carrying cost of the product at period k, r_j is the unit revenue at period j, and x_{ij} is the production volume at period i for sales at period j ($i, j, k = 1, 2, ..., H$).

In the case of multiple products—N kinds, denoting the demand of product h (P_h) at period j by $d_{(h)j}$ and unit production time of P_h by $t_{(h)}$, the total production time of P_h at period j is:

$$t_{(h)j} = t_{(h)} d_{(h)j}. \tag{13.79}$$

Then denoting the production time of product P_h for sales at period j by $x_{(h)ij}$, long-term production planning for the multiple products is formulated as:

$$\text{minimise } z = \sum_{h=1}^{N} \sum_{i=1}^{H} \sum_{j=i}^{H} u_{(h)i} x_{(h)ij} + \sum_{h=1}^{N} c_{(h)} \sum_{j=2}^{H} \sum_{i=1}^{j-1} (j-i) x_{(h)ij} \tag{13.80}$$

$$\text{subject to } \sum_{h=1}^{N} \sum_{j=i}^{H} x_{(h)ij} \le a_i \quad (i = 1, 2, ..., H) \tag{13.81}$$

$$\sum_{i=1}^{j} x_{(h)ij} = t_{(h)j} \quad (h = 1, 2, ..., N; j = 1, 2, ..., H) \tag{13.82}$$

Table 13.13 Long-term production planning by transportation-type linear programming.

Period	1	2	3	...	H	Dummy	Production capacity
1	$c_p \mid x_{11}$	$c_p + c_h \mid x_{12}$	$c_p + 2c_h \mid x_{13}$	...	$c_p + (H-1)c_h \mid x_{1H}$	$\infty \mid x_{1,H+1}$	a_1
2		$c_p \mid x_{22}$	$c_p + c_h \mid x_{23}$	...	$c_p + (H-2)c_h \mid x_{2H}$	$\infty \mid x_{2,H+1}$	a_2
3			$c_p \mid x_{33}$	...	$c_p + (H-3)c_h \mid x_{3H}$	$\infty \mid x_{3,H+1}$	a_3
⋮				...	...	...	⋮
H					$c_p \mid x_{HH}$	$\infty \mid x_{H,H+1}$	a_H
Amount of unachieved demand	$\infty \mid x_1$	$\infty \mid x_2$	$\infty \mid x_3$	...	$\infty \mid x_H$		∞
Demand	d_1	d_2	d_3	...	d_h	∞	$\sum_{i=1}^{H} a_i$ / $\sum_{j=1}^{H} d_j$

AGGREGATE PRODUCTION PLANNING

Table 13.14 Solving a three-term production planning problem.

Period	1	2	3	Dummy	Production capacity
1	3 8^1 (10)	4	5 1^5	100 1	10
2	✕	3 6^2 (2)	4 4^4 (5)	100 (3)	10
3	✕	✕	3 10^3 (10)	100	10
Amount of unachieved demand	100 (2)	100	100	✕	∞
Demand	8(12)	6(2)	15	∞	30 29

where $c_{(h)}$ is the inventory-carrying cost for P_h for one period, and $u_{(h)i}$ is the unit production cost for P_h in period i.

Solving Long-term Production Planning

The above long-term production planning problem can be solved by the transportation-type linear programming method, which was discussed in Chapter 10. The problem is tabulated as in Table 13.13, in which the slack variables for (13.77) are denoted by $x_{i,H+1}$ (≥ 0) to be produced in the dummy periods. The left and right sides of each entry indicate the unit cost and decision variable—production volume. The expenses needed, demands, and production capacity in the dummy periods and for the demands not achieved (x_j at period j) are all set to infinity; C_k is set at C_h.

--- **EXAMPLE 13.23** ---

(Long-term production planning) Let the number of periods in consideration be $H = 3$; production capacity, $a_i = 10$ ($i = 1, 2, 3$); demands, $d_1 = 8$, $d_2 = 6$ and $d_3 = 15$; unit production cost, $c_p = 3$; and inventory-carrying cost for one period, $c_h = 1$. Then we get Table 13.14 which corresponds to Table 13.13. The infinity for each period is set at 100.

Production volumes are decided in the order shown by the superscripts and are obtained by comparing both the demand and the production capacity at each entry; we get the optimal solution as follows:

$$x^*_{11} = 8, x^*_{12} = 0, x^*_{13} = 1, x^*_{22} = 6, x^*_{23} = 4, x^*_{33} = 10.$$ □

However, cases can occur that all the demands are not satisfied even if the total production capacity exceeds the entire demands.

— **EXAMPLE 13.24** —

If in Example 13.23 $d_1 = 12$ and $d_2 = 2$, not all the demands can be satisfied, as indicated by the solution shown in parentheses in Table 13.14, although the total production capacity exceeds the total demands. □

13.8 Production Forecasting

13.8.1 Time Series

Time Series, Prediction, and Forecasting

The most important information needed for production planning and scheduling is the future demand for the products; this usually varies with time. Such a varying sequence is called a *time series*.

There are two methods to estimate the future time-series values:

(1) *prediction*—subjective estimation based upon experience, intuition, and judgement, considering qualitative factors in society;
(2) *forecasting*—objective estimation by finding the time-series laws with the use of proper mathematical models and extrapolating from past data into the future.

Types of Time Series

Time series in the long run are classified into the following three types:

(1) *trend*—upward or downward (in some cases, constant) tendency in a long run;
(2) *circulation*—repeated occurrence of wavy variation such as:
   (a) 'business fluctuations' over a few years, and
   (b) 'seasonal fluctuations' in one year with the peak in a particular month or season;
(3) *irregular*—unforecastable variation such as:
   (a) short-run stochastic 'chance variation', and
   (b) unpredictable 'sudden change'.

13.8.2 Time-series Forecasting Models

Time-series Forecasting

Time-series forecasting models have been presented to predict the future value mathematically, based upon time-series historical data, as explained below. It is

AGGREGATE PRODUCTION PLANNING

assumed that time-series historical data up to the present during H periods are known, denoting the value at time t by x_t.

Total Average Method

The *total average method* is to calculate the next-period forecast value as the average of total data; hence,

$$A_t = \frac{1}{H} \sum_{j=0}^{H-1} x_{t-j}. \tag{13.83}$$

Equal weights are given to all past data in this method.

Moving-average Method

In the *moving-average method* the forecast next-period value is the average of several data near to the present; hence the $N(<H)$-term moving average is:

$$M_t = \frac{1}{N} \sum_{j=0}^{N-1} x_{t-j} = M_{t-1} + \frac{x_t - x_{t-N}}{N}. \tag{13.84}$$

This method does not adjust irregular fluctuations if N is large.

Single Exponential Smoothing Method

The *single* (or *first-order*) *exponential smoothing method* gives higher weights to the data near to the present; hence, this method performs self-correcting adjustments that regulate predicted values in the opposite direction to earlier errors. The basic model is:

$$S_t = \alpha x_t + (1 - \alpha) S_{t-1} \tag{13.85}$$

where α is the smoothing constant: $0 < \alpha < 1$; the higher this constant, the quicker the response to the new state.

---- EXAMPLE 13.25 ----

(Calculating the forecast values) Given the sales amounts for the past eight months as indicated in Table 13.15, the forecast sales amount for September is:

- total average—202
- moving average ($N = 6$)—204
- single exponential smoothing ($\alpha = 0.3$)—if the forecast sales amount for August is 205, $205 + 0.3 \times (208 - 205) = 206$. □

Table 13.15 Sales data for the past eight months.

Month	1	2	3	4	5	6	7	8
Sales	202	190	205	195	212	204	200	208

Higher-order Exponential Smoothing Method

In an attempt to forecast the high-order polynomial:

$$x_t = a_0 + a_1 t + a_2 t^2 + \cdots + a_n t^n + e_t \tag{13.86}$$

where a_i ($i = 0, 1, 2, \ldots, n$) are constants and e_t is the error at term t, kth-order *exponential smoothing* is performed:

$$S_t^{(k)} = \alpha S_t^{(k-1)} + (1 - \alpha) S_{t-1}^{(k)}. \tag{13.87}$$

--- **EXAMPLE 13.26** ---

For a linear model ($a_2 = a_3 = \cdots = a_n = 0$), 2nd-order exponential smoothing gives:

$$S_t^{(2)} = \alpha S_t + (1 - \alpha) S_{t-1}^{(2)} \tag{13.88}$$

Hence the forecast value toward τ terms into the future is:

$$\hat{x}_{t+\tau} = 2S_t - S_t^{(2)} + \frac{\alpha}{1 - \alpha}(S_t - S_t^{(2)})\tau. \tag{13.89}$$

□

Winters' Method for Seasonal Fluctuations

If time-series observations exhibit a consistent seasonal pattern from year to year, the *Winters' method* adjusts seasonal or periodic movement within the framework of linear exponential smoothing with three parameters α, β, and γ, as outlined below (Winters, 1960).

Denoting the actual value and the smoothed estimate at period t by x_t and $\tilde{x}_t$, respectively, $x_t/\tilde{x}_t$ is regarded as an index expressing the effect of seasonal variation. The seasonality index at period t is evaluated by:

$$F_t = \beta(x_t/\tilde{x}_t) + (1 - \beta)F_{t-L}, \quad 0 \le \beta \le 1, \tag{13.90}$$

where L is the length of seasonality for the time-series data.

Next, the linear trend index is estimated by:

$$R_t = \gamma(\tilde{x}_t - \tilde{x}_{t-1}) + (1 - \gamma)R_{t-1}, \quad 0 \le \gamma \le 1. \tag{13.91}$$

Finally, the following model is proposed for time-series data with seasonality and a linear trend:

$$\tilde{x}_t = \alpha(x_t/F_{t-L}) + (1 - \alpha)(\tilde{x}_{t-1} + R_{t-1}), \quad 0 \le \alpha \le 1. \tag{13.92}$$

Thus, the forecast value at τ periods ahead into the future is:

$$\tilde{x}_{t+\tau} = (\tilde{x}_t + \tau R_t)F_{t-L+\tau} \quad (\tau = 1, 2, \ldots, L). \tag{13.93}$$

In the case without trend, set R_t at 0.

--- **EXAMPLE 13.27** ---

(Applying the Winters' method for seasonal variation) Based upon seasonal data for the past two years as indicated in Table 13.16(*), the initial values for the linear trend, R_0, the

Table 13.16 Calculating the forecast demands by the Winters method.

Year	Month	Monthly No.	Actual data (*) x_t	Smoothed estimate $\tilde{x}_t$	Trend index R_t	Seasonality index F_t	Forecast value $\tilde{x}_{t,1}$	Error $x_t - \tilde{x}_{t,1}$
1994	1	1	718	789.89	1.17	0.9093	718	0
	2	2	489	794.77	1.54	0.6051	478	11
	3	3	561	798.79	1.79	0.6946	552	9
	4	4	622	796.98	1.43	0.7934	636	−14
	5	5	668	796.64	1.25	0.8453	675	−7
	6	6	721	799.38	1.40	0.8960	714	7
	7	7	813	800.61	1.38	1.0162	814	−1
	8	8	836	799.06	1.09	1.0602	852	−16
	9	9	1035	798.22	1.09	1.3080	1048	−13
	10	10	1230	802.90	1.45	1.5078	1203	27
	11	11	1062	804.78	1.50	1.3171	1059	3
	12	12	802	801.83	1.05	1.0206	825	−23
1995	1	13	733	803.54	1.12	0.9096	730	3
	2	14	476	801.04	0.76	0.6040	487	−11
	3	15	549	799.52	0.53	0.6938	557	−8
	4	16	652	804.39	1.16	0.7951	635	17
	5	17	690	807.70	1.38	0.8462	681	9
	6	18	718	807.54	1.22	0.8953	725	−7
	7	19	823	808.99	1.25	1.0162	822	1
	8	20	876	813.43	1.57	1.0619	859	17
	9	21	1079	816.98	1.76	1.3093	1066	13
	10	22	1203	814.57	1.35	1.5047	1234	−31
	11	23	1070	815.21	1.28	1.3166	1075	−5
	12	24	857	821.13	1.74	1.0229	833	24

smoothed estimate, $\tilde{x}_0$, and a set of the seasonality index, F_0, are calculated as:

$$R_0 = \left(\sum_{t=13}^{24} x_t/12 - \sum_{t=1}^{12} x_t/12\right)\bigg/[(2-1) \times 12] = 1.174$$

$$\tilde{x}_0 = \sum_{t=1}^{12} x_t/12 - (12/2 + 0.5)R_0 = 788.79$$

$F_0 = 0.9093$ (Jan.), 0.6040 (Feb.), 0.6937 (Mar.), 0.7948 (Apr.), 0.8460 (May), 0.8953 (June), 1.0163 (July), 1.0618 (Aug.), 1.3093 (Sept.), 1.5051 (Oct.), 1.3168 (Nov.), 1.0229 (Dec.).

Cyclic calculation via (13.90) through (13.93) generates Table 13.16 by setting $\alpha = 0.20$, $\beta = 0.10$, and $\gamma = 0.10$.

Table 13.17 The forecast demands obtained for 1996.

Monthly No.	1	2	3	4	5	6	7	8	9	10	11	12
Forecast value	748	498	573	658	702	744	847	887	1096	1262	1106	861

Thus the forecast values for 1996 are calculated by:

$$\tilde{x}_{t+\tau} = (821.13 + 1.74\tau)F_{t-12+\tau}$$

successively setting $\tau = 1, 2, \ldots, 12$, one by one, resulting in Table 13.17. □

13.8.3 Regression Models

Regression and Correlation

With *regression* the dependent variable to be predicted is estimated by the nature of its relationship to the independent variable, whilst *correlation* is concerned with the degree of association between the two variables.

Representative Regression Models

The value x_t at period t is estimated as follows:

- linear/polynomial: $x_t = a + bt + ct^2 + \cdots$;
- linear exponential: $x_t = ab^t$;
- double exponential: $x_t = ab^t c^{t^2}$;
- Gomperz: $x_t = ka^{b^t}$ $(0 < a, b < 1)$;
- logistic: $x_t = k/(1 + be^{-at})$,

where a, b, c, and k are constants, k being the saturation level.

13.8.4 Econometric Models

Purpose of Econometric Models

Econometrics indicates economic measurement by establishing quantitative relationships among economic variables with the aid of statistics. *Econometric models* are explicit systems which assemble and weigh economic information.

Variables which are determined from the environment are called *exogeneous*, and variables decided within the system *endogenous*.

Representative Econometric Models

The *structural equation* of the econometric model is to represent x_t, the demand at time t, by M quantitative variables, $y_{1t}, y_{2t}, \ldots, y_{Mt}$, which influence x_t and explain the nature of x_t:

$$x_t = f(y_{1t}, y_{2t}, \ldots, y_{Mt}). \tag{13.94}$$

Demands for τ terms into the future, $x_{t+\tau}$, can be forecast from this functional relationship by predicting $y_{1,t+\tau}, y_{2,t+\tau}, \ldots, y_{M,t+\tau}$ with an appropriate time-series model.

Representative models are:

- linear regression: $x_t = a_0 + a_1 y_{1t} + a_2 y_{2t} + \cdots + a_M y_{Mt}$; (13.95)
- logarithmic linear: $x_t = a_0 y_{1t}^{a_1} y_{2t}^{a_2} \ldots y_{Mt}^{a_M}$; (13.96)
- logarithmic: $x_t = a_0 + a_1 \log y_{1t} + a_2 \log y_{2t} + \cdots + a_M \log y_{Mt}$; (13.97)
- linear difference: $\Delta x_t = a_0 + a_1 \Delta y_{1t} + a_2 \Delta y_{2t} + \cdots + a_M \Delta y_{Mt}$; (13.98)
- reciprocal: $x_t = a_0 + a_1/y_{1t} + a_2/y_{2t} + \cdots + a_M/y_{Mt}$; (13.99)
- linear auto-regression: $x_t = a_0 + a_1 x_{t-1} + a_2 x_{t-2} + \cdots + a_M x_{t-M}$, (13.100)

where $a_0, a_1, a_2, \ldots, a_M$ are all constants which are determined by the method of least squares, based upon the past data.

─── EXAMPLE 13.28 ───────────────────────

(Econometric model) Suppose the demand of a product, x_t, depends upon the price y_{1t}, the advertising expenditure y_{2t}, and the consumers' income y_{3t}. Denoting the elasticities of these factors by a_1 (negative), a_2 (positive), and a_3 (positive), respectively, the econometric forecasting model is expressed:

$$x_t = a_0 y_{1t}^{a_1} y_{2t}^{a_2} y_{3t}^{a_3}. \qquad (13.101)$$

The increase of demand brought about by reducing the price by $p\%$, and raising the advertising expenditure by $q\%$, and from the income increase by $r\%$ is then, approximately: $(a_1 p + a_2 q + a_3 r) a_0 y_{1t}^{a_1} y_{2t}^{a_2} y_{3t}^{a_3} \%$. □

Notes

1 *Simplex* means the basic figure in each dimension space, such as line segment, triangle, tetrahedron, etc. The simplex method developed in 1963 seeks the optimal solution by moving along the boundaries of the feasible region, as demonstrated in Example 13.6, whilst another method—the Karmarkar method, developed in 1984—seeks optimality by iteratively modifying the solution inside the feasible region.
2 The 'one-of-a-kind' production mode, which is often performed in practice, especially in Japan, is a case that $Q_p^* = 1$. Notice that a very strict condition: $1/r_c - 1/r_p = 2u_s/u_h$ should hold for the optimality of this production mode.
3 The time bucket is recently shrinking from the 'weekly' to the 'daily' by the effect of the just-in-time (JIT) mode.

References

BALAS, E. (1965) An additive algorithm for solving linear programs with zero-one variables, *Operations Research*, **13** (4).
DANTZIG, G.B. (1963) *Linear Programming and Extensions* (Princeton, NJ: Princeton University Press).
GUTENBERG, E. (1955) *Grundlagen der Betriebswirtschaftslehre*, Bd. I—Die Produktion (Berlin: Springer), Kapitel 4.
HADLEY, G. (1962) *Linear Programming* (Reading, MA: Addison-Wesley), Chap. 5.
HITOMI, K. (1975) *Theory of Planning for Production* (in Japanese) (Tokyo: Yuhikaku), Sec. 10.

IGNIZIO, J.P. (1976) *Goal Programming and Extentions* (Lexington, MA: Heath).
ORLICKY, J. (1975) *Material Requirements Planning* (New York: McGraw-Hill).
JOHNSON, L.A. and MONTGOMERY, D.C. (1974) *Operations Research in Production Planning, Scheduling and Inventory Control* (New York: Wiley).
PROTH, J.M. and HILLION, H.P. (1990) *Mathematical Tools in Production Management* (New York: Plenum Press).
STARR, M.K. (1972) *Production Management* (2nd edn) (Englewood Cliffs, NJ: Prentice-Hall), pp. 245–253.
WAGNER, H.M. (1975) *Principles of Operations Research* (2nd edn) (Englewood Cliffs, NJ: Prentice-Hall), pp. 133–135.
WINTERS, P.R. (1960) Forecasting sales by exponentially weighted moving average, *Management Science*, **6** (3).

Supplementary reading

HAX, A.C. and CANDEA, D. (1984) *Production and Inventory Management* (Englewood Cliffs, NJ: Prentice-Hall).
JARRETT, J. (1991) *Business Forecasting Methods* (2nd edn) (Oxford: Basil Blackwell).
OZAN, T. (1986) *Applied Mathematical Programming for Engineering and Production Management* (Englewood Cliffs, NJ: Prentice-Hall).
PLOSSL, G. (1994) *Orlicky's Material Requirements Planning* (New York: McGraw-Hill).

CHAPTER FOURTEEN

Production Scheduling

14.1 Scope of Production Scheduling

Meaning of Production Scheduling

Production scheduling is a function to determine an actual (optimal or feasible) implementation plan as to the time schedule for all jobs to be executed; that is, when, with what machine, and who does what operation? This is done after the product items and the quantities to be manufactured in specified time periods have been decided by production planning and the production processes for those product items have been determined by production process planning.

Gantt Chart

This is a powerful management tool invented by H.L. Gantt in the early 20th century for planning and controlling jobs and operations. As illustrated in Figure 14.1, in this chart a time scale is placed on the horizontal axis. On the vertical axis production facilities to accomplish required jobs (or jobs to be performed) are represented. A planned time schedule for the progress of jobs to be performed on the corresponding production facilities is depicted by the horizontal bar. Performance of the actual work may be monitored in comparison with the plan and, wherever quite a large deviation between them occurs, appropriate action may be taken to improve the managing efficiency of productive work.

Job Sequencing/Operations Scheduling

A basic problem in production scheduling is to determine the order of processing of jobs waiting to be processed on a machine. This function is *job sequencing*. An optimal job sequence can be selected from among a set of 'permutation schedules' such that a certain measure of performance is optimised. Often 'dispatching (or priority) rules' are applied to obtain reasonable schedules, usually with the use of computer simulation. In the *flow shop*, where the order of machines or processes is

235

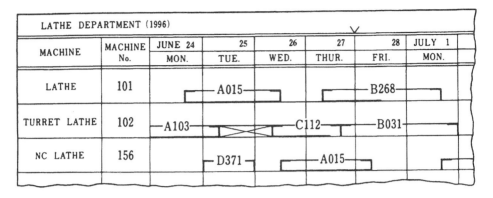

Figure 14.1 An illustration of the Gantt chart, which is a powerful management tool. Planned time schedule for progress of jobs to be run on the production facilities is depicted by the horizontal bar chart.

completely identical for all jobs to be processed, scheduling is relatively simple. On the other hand, it is rather difficult to decide an optimal schedule for *job shops* where the technological order varies depending upon types of jobs. Production scheduling for those cases, where jobs to be processed on different facilities are independent of each other, as in machining lines, is called *operations scheduling*.

Project Scheduling

In scheduling, long-range decisions sometimes have to be made, as for example in scheduling the construction of a new factory, a large ship, a building, a highway, etc. In this case, scheduling covers a relatively long term; hence, very short periods such as seconds or minutes are not pertinent to this scheduling problem. However, since many and varied kinds of work elements are contained in this type of scheduling, a reasonable schedule must be made for a smooth and effective implementation of production. Such a scheduling decision is called *project* (or *network*) *scheduling*. Typical solution techniques for solving this type of scheduling problem are the program evaluation and review technique (PERT), critical path method (CPM), and the more recent and more flexible technique known as the graphical evaluation and review technique (GERT).

14.2 Operations Scheduling

14.2.1 Scope of Operations Scheduling

Meaning of Operations Scheduling

Operations scheduling is the allocation of jobs to be processed on the corresponding machines, in a given time span, for a workshop consisting of several machines or production facilities including operative workers. A *job* is a product or part to be completed. For that a piece of raw material is converted into a finished part through a single or multiple stages, on each of which an *operation* is run, such as turning, drilling, grinding, setup, etc. on a suitable machine tool or by a skilled worker. Hence, a job is a task made up of multiple operations or work elements arranged in technological order.

Types of Operations Scheduling

There are three types of operations scheduling:

(1) *Job sequencing*—determining the order of processing jobs waiting to go on a single machine. An optimal job sequence is selected from among a set of 'permutation schedules' or by the 'dispatching (or priority) rules' established reasonably by scheduling simulation.
(2) *Flow-shop scheduling*—scheduling for a flow shop, where the sequence of machines according to multiple-stage manufacturing is completely identical for all jobs to be produced. This type of flow pattern is typical for mass production.
(3) *Job-shop scheduling*—scheduling for a job shop, where the sequence of machines differs for each job. This shop is typical for the case of varied production of most jobbing types and some batch types.

Basic Structure of Operations Scheduling

A typical example of operations scheduling is to process M jobs on N machines in a factory, as depicted in Figure 14.2. This is a combinatorial problem in that there are $(M!)^N$ alternatives, among which an optimal solution according to a certain measure of performance definitely exists and can theoretically be found in a finite number of computational iterations. However, it requires many computations, particularly as the size of the problem becomes large. That is, scheduling problems are NP-hard (see Chapter 10, Note 1).

EXAMPLE 14.1

For even a small scheduling problem such as five jobs on eight machines there exist $(5!)^8 = 4.3 \times 10^{16}$ evaluations, if a direct search by enumeration is used. Even with the use of a high-performance computer which is able to evaluate one alternative with 1 μs$(=10^{-6}$ s$)$ it takes 1363 years(!) to obtain the optimal solution. □

Importance of Scheduling Theory

Accordingly the direct search of optimal solutions by the complete enumeration

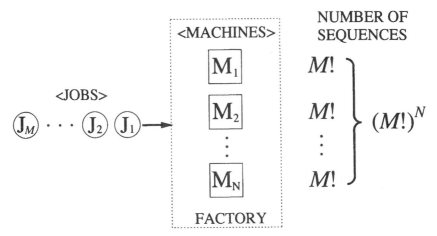

Figure 14.2 A factory processes M jobs on N machines. Each machine has $M!$ alternatives for job sequencing; hence $(M!)^N$ alternatives in total.

procedure, as above, is not practical even when using a large supercomputer, so it is wiser to use effective theorems, rules, and algorithms which have been developed through theoretical analyses based on models (see Conway *et al.*, 1967; Baker, 1974; Fench, 1982). Several important theorems and algorithms are introduced in the following sections.

14.2.2 Preliminary Analysis of Operations Scheduling

Preconditions for Operations Scheduling

For simplicity of analysis a *static* state is in most cases considered, where a set of M independent jobs is available for processing at time zero (e.g. 8:00 a.m. on Monday). If those jobs randomly enter the shop floor, this situation is *dynamic*. Some other conditions are characterised by:

- one machine is continuously available;
- each operation can be performed by only one machine;
- each machine can handle one operation at a time.

Input Data for Scheduling

Information known in advance as input to the scheduling of M jobs, J_i ($i = 1, 2, ..., M$), to be performed in the workshop, wherein an assemblage of N machines, M_k ($k = 1, 2, ..., N$), are laid out on a shopfloor, as depicted in Figure 14.2, is:

- processing time t_{ij} required by the jth operation (O_{ij}) of J_i on an associated machine ($j = 1, 2, ..., j_i$);
- ready (or release) time r_i—the time at which J_i is available for processing (=0 for static conditions);
- due (or delivery) date d_i—the time at which the processing of J_i is due to be completed.

Scheduling Criteria

A basic role of operations scheduling is to generate an optimal schedule. Such a scheduling decision should be made according to a certain measure of performance, or scheduling criterion. Important ones include:

- maximum flow time or makespan—F_{max};
- mean flow time—$\bar{F}$;
- maximum lateness or tardiness—L_{max} or D_{max};
- mean lateness or tardiness—$\bar{L}$ or $\bar{D}$;
- number of tardy jobs—m;
- average number of work-in-process jobs (or inventory);
- facility utilisation in the workshop, etc.

The last measure is maximised; others are minimised.

Basic Formula for Scheduling

In processing O_{ij} the waiting time W_{ij} preceding that operation occurs whenever the associated machine is not available owing to processing the previous job at the ready time of O_{ij}. Of course, $W_{ij} = 0$ if the associated machine is available at that time.

(I) *Flow time* or *production time* for J_i—the total time length that the job spends for processing—is:

$$F_i = W_{i1} + t_{i1} + W_{i2} + t_{i2} + \ldots + W_{ij_i} + t_{ij_i} = t_i + W_i, \tag{14.1}$$

where

$$t_i = \sum_{j=1}^{j_i} t_{ij} \quad \text{total processing time for } J_i$$

$$W_i = \sum_{j=1}^{j_i} W_{ij} \quad \text{total waiting time for } J_i.$$

For static cases F_i is also the completion time for J_i, the time at which the job is completed, since its ready time is set at zero: $r_i = 0$.

(II) *Lateness*—the difference between the completion time and the due date—is then expressed as:

$$L_i = F_i - d_i. \tag{14.2}$$

If $F_i > d_i$, a positive amount of lateness, namely *tardiness*, occurs. It is expressed as:

$$D_i = \max(0, L_i). \tag{14.3}$$

(III) *The number of tardy jobs* is then obtained as follows:

$$m = \sum_{i=1}^{M} \delta_i, \tag{14.4}$$

where

$$\delta_i = \begin{cases} 1 \text{ if } D_i > 0 \\ 0 \text{ if } D_i = 0. \end{cases}$$

(IV) *Total flow time* or *makespan*, which represents a time length from the beginning of the first operation of the first job to the end of the last operation of the last job, is then given by the maximum among the flow times for all jobs calculated from (14.1):

$$F_{max} = \max_{1 \leq i \leq M} F_i. \tag{14.5}$$

(V) *Mean flow time* is the mean time length during which all the jobs remain in the workshop, and is given by:

$$\bar{F} = \sum_{i=1}^{M} F_i / M. \tag{14.6}$$

(VI) It is important to recognise that this mean flow time is closely related to inventory in the workshop. The average number of jobs, namely, the *average work-in-process inventory* in the workshop during the interval of makespan $[0, F_{max}]$ is calculated as follows (refer to Figure 14.3):

$$\bar{M} = [MF_1 + (M-1)(F_2 - F_1) + (M-2)(F_3 - F_2) + \cdots + 1(F_M - F_{M-1})]/F_{max}$$

$$= \sum_{i=1}^{M} F_i / F_{max}. \tag{14.7}$$

Dividing both sides by M—the initial (total) number of jobs in the workshop, we get:

$$\frac{\bar{M}}{M} = \frac{\bar{F}}{F_{max}}. \tag{14.8}$$

Thus, the ratio of the average work-in-process inventory against the total number of jobs to be processed is equal to the ratio of the mean flow time against the makespan. This is true as long as all the jobs are ready at the same time period, that is, for static cases, and not in dynamic cases where jobs arrive at random time points.

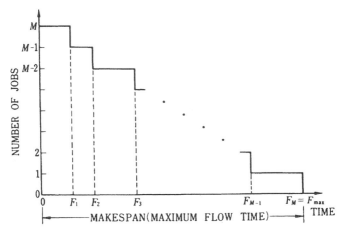

Figure 14.3 Number of jobs or work-in-process inventory in the workshop vs. time elapsed for the static case, where all the jobs are ready at time zero.

(VII) *Maximum lateness and maximum tardiness* are the maximum values among L_is and D_is, respectively; hence,

$$L_{max} = \max_{1 \leq i \leq M} L_i, \qquad (14.9)$$

$$D_{max} = \max_{1 \leq i \leq M} D_i. \qquad (14.10)$$

Then *mean lateness* and *mean tardiness* are:

$$\bar{L} = \sum_{i=1}^{M} L_i/M, \qquad (14.11)$$

$$\bar{D} = \sum_{i=1}^{M} D_i/M. \qquad (14.12)$$

Role of the Mean Flow Time

Summing (14.1) and (14.2) for all is and dividing them by the total number of jobs, M, leads to the following relations:

$$\bar{F} = \bar{t} + \bar{W}, \qquad (14.13)$$

$$\bar{L} = \bar{F} - \bar{d}, \qquad (14.14)$$

where $\bar{t} = \sum_{i=1}^{M} t_i/M$ and $\bar{d} = \sum_{i=1}^{M} d_i/M$ are constants; $\bar{W} = \sum_{i=1}^{M} W_i/M$ is the mean waiting time.

Consequently, it is concluded that an optimal schedule that achieves the minimum mean flow time has also the minimum mean lateness and the minimum mean waiting time. This schedule also minimises the mean work-in-process inventory if the makespan is fixed, as is concluded from (14.8). However, a similar conclusion cannot be drawn at all for the maximum flow time and other scheduling criteria with maximum values.

14.2.3 Single-machine Scheduling—Job Sequencing

Job Sequencing

Single-machine scheduling is a most basic problem in operations scheduling. This situation occurs when many jobs are waiting for processing prior to a machine tool, when many programs are waiting for processing in front of a computer, when an entire plant can be regarded as a single machine such as a process (chemical) industry, etc.

In the case of processing M jobs on a single machine ($N = 1$), the number of schedules to be evaluated is $M!$; hence, one can restrict attention to the class of a permutation schedule, which is completely specified by giving the order in which M jobs will be processed. In this sense this problem is called *job sequencing*.

Sequencing Rules

In the sequencing problem the total flow time is no longer a suitable scheduling criterion, since it is simply the sum of the M processing times, which is not

MANAGEMENT SYSTEMS FOR MANUFACTURING

dependent upon the order of jobs. With respect to other scheduling criteria we have the following theorems.

--- **THEOREM 14.1** ---

In the single-machine scheduling problem, mean flow time and mean lateness are minimised by sequencing the jobs in order of non-decreasing processing time—the shortest processing time (SPT) rule. □

--- **THEOREM 14.2** ---

In the single-machine scheduling problem, the maximum lateness and the maximum tardiness are minimised by sequencing the jobs in order of non-decreasing due date—the earliest due date (EDD) rule. □

--- **THEOREM 14.3** ---

In the single-machine scheduling problem, if there exists a sequence such that the maximum tardiness is zero, then there is an optimal job sequence that minimises the mean flow time by iteratively assigning job H as the last position in the optimal sequence only if

(a) $d_H \geq \sum_{i=1}^{M'} t_i$,
(b) $t_H \geq t_j$ for all j such that $d_j \geq \sum_{i=1}^{M'} t_i$,

where M' is the number of jobs which have not yet been assigned. □

The last rule for optimising single-machine scheduling is for a simple multiple (two)-objective scheduling—zero tardiness as the primary objective and the minimum mean flow time as the secondary objective.

--- **EXAMPLE 14.2** ---

(Job sequencing) Consider a five-job, single-machine scheduling problem shown in Table 14.1. In this case, the maximum flow time is the sum of processing times:

$$F_{max} = t_1 + t_2 + t_3 + t_4 + t_5 = 20.$$

This value is constant irrespective of the order of the five jobs; hence, F_{max} is not a pertinent scheduling criterion for a single-machine scheduling.

The SPT rule is one of the most useful rules applicable to the practical scheduling problems, and the job sequence: $J_3-J_1-J_5-J_4-J_2$ thus obtained minimises $\bar{F}$ (Theorem 14.1).

According to Theorem 14.2 the optimal job sequence for minimising L_{max} and D_{max} is $J_4-J_3-J_5-J_2-J_1$, which is the increasing order of due dates. In this case, it is easily shown that $D_{max} = 0$.

Table 14.1 Basic data for a single-machine scheduling problem, sequencing five jobs.

Job number	i	1	2	3	4	5
Due date	d_i	23	20	8	6	14
Processing time	t_i	3	7	1	5	4

Hence, under this condition, we can find a better job sequence minimising $\bar{F}$ as the secondary objective by Theorem 14.3; that is, J_3–J_4–J_1–J_5–J_2 thus obtained minimises $\bar{F}$ as the secondary objective. For this sequence $\bar{F} = 9.8$, while for the former sequence, $\bar{F} = 11.6$. □

Minimising the Number of Tardy Jobs

If a penalty is imposed for being tardy, the sequencing aim would be to minimise the number of tardy jobs. The following algorithm yields the desired objective (Bedworth and Bailey, 1982).

Hodgson's algorithm for minimising the number of tardy jobs:

Step 1: Order all tasks by the EDD rule; if zero or one task is tardy, then stop. Otherwise proceed.

Step 2: Starting at the beginning of the EDD sequence and working toward the end, identify the first tardy job. If no further jobs are tardy, go to step 4; otherwise go to step 3.

Step 3: Supposing that the tardy job is in the ith position in the sequence, examine the first i jobs in the sequence and identify the one with the longest processing time. Remove that job and set it aside. Revise the completion time of the other jobs to reflect the removal, and return to Step 2.

Step 4: Place all those jobs that were set aside in any order at the end of the sequence. Stop. □

--- **EXAMPLE 14.3** ---

(Minimising the number of tardy jobs) Suppose that the due dates of jobs 2 and 5 are 15 and 9, respectively, rather than 20 and 14, in Table 14.1, the EDD schedule is: J_4–J_3–J_5–J_2–J_1 (step 1). Since two jobs, J_5 and J_2, are tardy by 1 and 2 days, respectively, we proceed to step 2. The first tardy job is J_5 in the 3rd position. J_4 has the longest processing time in the first three jobs sequenced so far (step 3). Removing this job and revising the completion time of the other jobs, we find no further job is tardy (Step 2). In Step 4 the optimal sequence for the minimum tardy jobs (= 1 : J_4) is: J_3–J_5–J_2–J_1–J_4. □

14.2.4 Flow-shop Scheduling

Basic Propositions for Flow-shop Scheduling

In solving flow-shop scheduling problems, wherein all jobs are processed through an identical sequence of multiple machines arranged in series, two properties given below are dominant. Here a *regular measure of performance* is an amount to be minimised that is expressed as a function of job completion times and increases only if at least one of the completion times increases. Scheduling criteria introduced so far are mostly regular measures.

THEOREM 14.4

In the flow-shop scheduling problem minimising any regular measure of performance, it is sufficient to consider only schedules in which the same job order is prescribed on the first two machines. □

THEOREM 14.5

In the flow-shop scheduling problem minimising the makespan, it is sufficient to consider only schedules in which the same job order is prescribed on the last two machines. □

Two-stage Flow-shop Scheduling—Johnson's Method

Minimising the total flow time, F_{max}, for a two-stage (machine) flow shop is the most basic and classic problem in the field of flow-shop scheduling. S.M. Johnson developed a very useful and simple algorithm for solving this scheduling problem as follows. In this problem we have t_{i1} and t_{i2} as the processing times for the first and second operations of job J_i ($i = 1, 2, ..., M$).

Johnson's algorithm for the minimum makespan two-stage flow-shop scheduling:

Step 1: Find the minimum processing time among a set of jobs that are not yet sequenced. (In case of a tie, select arbitrarily.)

Step 2: If it requires machine 1 (or 2), place the associated job in the first (or last) position in the sequence. When all positions in the sequence are filled, stop. Otherwise, proceed.

Step 3: Remove the assigned job from consideration and return to Step 1. □

EXAMPLE 14.4

(Two-machine flow-shop scheduling) Consider a five-job, two-machine flow-shop scheduling problem for minimising the makespan, as given in Table 14.2. An optimal job sequence is the same order on machines (stages) 1 and 2 according to Theorem 14.4 or 14.5. The Johnson's algorithm is applied to obtain an optimal job sequence such as $J_2-J_4-J_3-J_5-J_1$. □

Table 14.2 Basic data for a five-job, two-machine flow-shop scheduling problem for minimising the makespan.

Job number	i	1	2	3	4	5
Processing time						
on machine 1	t_{i1}	4	1	9	3	10
on machine 2	t_{i2}	2	6	7	6	5

Calculating the Makespan

There are the following three ways to obtain the total flow time, or makespan.
(1) *Gantt charting*.

--- **EXAMPLE 14.5** ---

(Gantt charting) The Gantt chart drawn for the job sequence obtained for Example 14.4 is shown in Figure 14.4: $F_{max} = 30$ h. The series of operations marked with asterisks is called the *critical path* in that, if the completion of any operation on this path is delayed, the makespan or total production time exceeds the minimum makespan—30 h. Other operations have *slack*; they can be delayed for a certain period without prolonging the minimum makespan. For example, job 1 could be delayed for 1 h on the first stage and a series of operations—jobs 2, 4, and 3 on the second stage—could be delayed 3 h. □

(2) *Mathematical formula*. For simplicity of analysis we assume that no job overtakes any other, i.e. the job order is completely identical at all stages for all machines. The symbol $\langle \rangle$ is used to signify the order in the job sequence; $J_{\langle i \rangle} = J_\xi$ means that job ξ is processed in the ith position in the job sequence.

Referring to Figure 14.5, which shows two basic patterns[1] of processing jobs $\langle r-1 \rangle$ and $\langle r \rangle$ on stages $k-1$ and k, the completion time of $J_{\langle r \rangle}$ on stage k (M_k) is:

$$F_{\langle r \rangle k} = \max(F_{\langle r-1 \rangle k}, F_{\langle r \rangle k-1}) + t_{\langle r \rangle k}. \qquad (14.15)$$

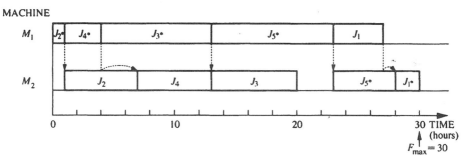

Figure 14.4 Gantt chart for an optimal schedule for a five-job, two-machine flow-shop scheduling minimising the makespan (refer to Table 14.2).

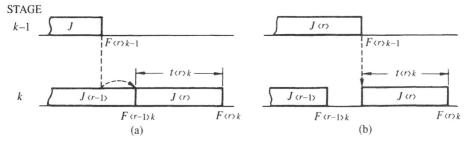

Figure 14.5 Two basic patterns of processing jobs $\langle r-1 \rangle$ and $\langle r \rangle$ on stages $k-1$ and k of a large-scale flow shop for a case that the job order is completely identical in all stages.

(e) The makespan of the schedule is given by the lower half cell entry in cell (M, N).

--- **EXAMPLE 14.6** ---

Calculate the makespan for the optimal sequence obtained in Example 14.4. A table for calculating the makespan for this sequence is generated as indicated in Table 14.4. The makespan is 30 units of time, which is the same as the answer obtained by the method of Gantt charting represented in Figure 14.4. □

Three-stage Flow-shop Scheduling

Johnson's algorithm can be extended and used to solve a special three-stage (machine) flow-shop scheduling problem. The optimal job sequence for minimising the makespan can be obtained, if the maximum processing time of any job on machine 2 is less than or equal to the minimum processing time of any job on machine 1 or 3. Considering fictitious processing times such that:

$$\left. \begin{array}{l} T_{i1} = t_{i1} + t_{i2} \\ T_{i2} = t_{i2} + t_{i3} \end{array} \right\} \quad (i = 1, 2, ..., M), \tag{14.18}$$

Johnson's algorithm is applied to these fictitious processing times to obtain an optimal (minimum-makespan) solution.

Table 14.4 Calculating the makespan for an optimal schedule for a five-job, two-machine flow-shop scheduling given in Table 14.2. 30 units of time for the minimum makespan, which is indicated in the lower half entry in cell (5, 2) is equal to one obtained by the Gantt chart indicated in Figure 14.4.

Job $\langle i \rangle$	t_{i1} / $F_{\langle i \rangle 1}$	t_{i2} / $F_{\langle i \rangle}$
2	1 / 1	6 / 7
4	3 / 4	6 / 13
3	9 / 13	7 / 20
5	10 / 23	5 / 28
1	4 / 27	2 / 30

Large-scale Flow-shop Scheduling

Large-scale flow-shop scheduling, in which the number of stages or machines is over 3 ($N > 3$), is vital; in the following, two methods are explained:

- the branch-and-bound method for obtaining the exact solution, though a great deal of computational time is required.
- a heuristic algorithm—Petrov's method—for deriving a good schedule with much less computational effort.

Branch-and-bound Method for Large-scale Flow-shop Scheduling

The *branch-and-bound method* is an implicit enumeration algorithm for iteratively finding an optimal solution to discrete combinatorial problems by repeating branching and bounding procedures; it is usually more efficient than complete enumeration.

(I) The *branching procedure* partitions a solution set into several subsets. Thus *nodes*, each of which corresponds to a subset of the job sequence, are created. Node $\mathcal{N}_r$ is a node at which r jobs are chosen from among M jobs and sequenced: $\mathcal{N}_r = \{J_{(1)}, J_{(2)}, ..., J_{(r)}\}$. Then, if another job is chosen from among the remaining $(M - r)$ jobs and processed next, $\mathcal{N}_{r+1} = \{J_{(1)}, J_{(2)}, ..., J_{(r)}, J_{(r+1)}\}$. Hence, at node $\mathcal{N}_M$ a solution for the job sequence is indicated.

(II) The branch-and-bound method searches a best node $\mathcal{N}_M^*$ with the least iteration of evaluation. For this purpose the *bounding procedure* is executed such that the *lower bound* (LB) is computed for each node. This is an estimated value for measure of performance—the makespan under the current scheduling criteria, showing that this is the minimum makespan for the current node whatever the job sequence for the remaining jobs which have not yet been sequenced. Accordingly, such nodes with large values for LB can make no contribution to the solution and hence it is eliminated. Only a node with a least value for LB is considered and from this node the branching is conducted and new nodes with a higher level (larger value of r) than the current node are created. That is, from a node $\mathcal{N}_r$ which has a minimum LB value, $(M - r)$ nodes are newly created, which establishes $\mathcal{N}_{r+1}$ (branching procedure).

(III) *Calculating the lower bound.* The lower bound (LB) is expressed as follows for solving flow-shop scheduling problems minimising the makespan.

At node $\mathcal{N}_r$, r jobs have already been sequenced; from equation (14.15), the total completion time on M_k for them is:

$$\hat{F}'_{\mathcal{N}_r,k} = F_{(r)k} \quad (k = 1, 2, ..., N). \tag{14.19}$$

On the other hand, the total processing time for the remaining jobs which have not yet been sequenced can be evaluated as the sum of job processing times of the remaining jobs on M_k plus the minimum from among sums of processing times on the remaining stages ($M_{k+1}, M_{k+2}, ..., M_N$) for each of the remaining jobs:

$$\hat{F}''_{\mathcal{N}_r,k} = \sum_{\xi \in R_r} t_{\xi k} + \min_{\xi \in R_r} \sum_{\eta=k+1}^{N} t_{\xi \eta} \quad (k = 1, 2, ..., N) \tag{14.20}$$

where R_r is the set of the remaining jobs at $\mathcal{N}_r$. Hence, the lower bound (machine based) at $\mathcal{N}_r$ can be well evaluated as a maximum from among sums of equations

PRODUCTION SCHEDULING   249

(14.19) and (14.20) for all stages:

$$LB(\mathcal{N}_r) = \max_{1 \leq k \leq N} \left(\hat{F}_{(r)k} + \sum_{\xi \in R_r} t_{\xi k} + \min_{\xi \in R_r} \sum_{\eta=k+1}^{N} t_{\xi \eta} \right). \quad (14.21)$$

An optimal sequence for large-scale flow-shop scheduling problems minimising the makespan can be obtained by finding a terminal node N_M with a lower bound not exceeding that associated with any node created so far by the following algorithm.

Branch-and-bound algorithm for the minimum makespan flow-shop scheduling:

Step 1: Set $r \leftarrow 1$, and newly create nodes at $r = 1$.
Step 2: If $M - r = 1$, set $r \leftarrow M$.
Step 3: Calculate the lower bounds LB ($\mathcal{N}_r$) for the new nodes.
Step 4: Choose a node having the minimum lower bound from among nodes not yet branched. (In case of a tie, choose the node associated with the largest value of r. Break the tie arbitrarily for the same r.)
Step 5: If $r = M$, the job sequence associated with the current node is optimal and its lower bound is the minimum makespan. Stop. Otherwise, set $r \leftarrow r + 1$, create new nodes by branching at r, and return to Step 2. □

—— EXAMPLE 14.7 ——

(Flow-shop scheduling by the branch-and-bound method) Consider a four-job, five-machine flow-shop scheduling problem minimising the makespan, as shown in Table 14.5.

The tree structure for obtaining an optimal solution to this problem by the branch-and-bound method is represented in Figure 14.6; in this case, $M = 4$.

To begin with, four nodes at $r = 1$ are created (step 1): $\mathcal{N}_1 = \{J_1\}, \{J_2\}, \{J_3\}, \{J_4\}$. Proceeding to step 3, the lower bounds (LB) are calculated for all these nodes, and indicated in parentheses for each node. Calculation of LB for a node $\{J_2\}$ is illustrated in the following. In this case, only J_2 is specified for the job sequence, that is, $J_{(1)} = J_2$, the set of the remaining jobs being $R_r = \{1, 3, 4\}$. From (14.19),

$$\hat{F}_{N_1 1}(J_2) = F_{(1)1} = t_{(1)1} = t_{21} = 2,$$
$$\hat{F}_{N_1 2}(J_2) = F_{(1)2} = F_{(1)1} + t_{(1)2} = t_{21} + t_{22} = 2 + 5 = 7.$$

Similarly, $\hat{F}_{N_1 3}(J_2) = 10$, $\hat{F}_{N_1 4}(J_2) = 30$, $\hat{F}_{N_1 5}(J_2) = 34$.

Table 14.5 Basic data for a four-job, five-machine flow-shop scheduling problem minimising the makespan.

Job number	Processing time				
	Stage 1	Stage 2	Stage 3	Stage 4	Stage 5
1	5	8	20	4	13
2	2	5	3	20	4
3	30	4	5	3	21
4	6	30	6	14	8

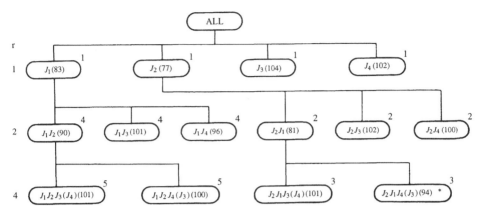

Figure 14.6 The branch-and-bound method for solving a four-job, five-machine flow-shop scheduling problem minimising the makespan (refer to Table 14.5). In each node the job sequence assigned is shown, and the lower bound is indicated in parentheses. Numbers written on the right shoulders of nodes designate the branching order indices of creating those nodes. After five cycles of the branching procedure including the backtracking process the optimal solution is found at the node with an asterisk; that is, the optimal sequence is: J_2–J_1–J_4–J_3 with the minimum makespan of 94.

From (14.20),

$$\hat{F}''_{\aleph,1}(J_2) = \sum_{\xi \in R_r} t_{\xi 1} + \min_{\xi \in R_r} \sum_{\eta=2}^{5} t_{\xi \eta}$$

$$= (t_{11} + t_{31} + t_{41}) + \min(t_{12} + t_{13} + t_{14} + t_{15},$$
$$t_{32} + t_{33} + t_{34} + t_{35}, t_{42} + t_{43} + t_{44} + t_{45})$$
$$= 41 + \min(45, 33, 58) = 74;$$

$$\hat{F}''_{\aleph,2}(J_2) = \sum_{\xi \in R_r} t_{\xi 2} + \min_{\xi \in R_r} \sum_{\eta=3}^{5} t_{\xi \eta}$$

$$= (t_{12} + t_{32} + t_{42}) + \min(t_{13} + t_{14} + t_{15},$$
$$t_{33} + t_{34} + t_{35}, t_{43} + t_{44} + t_{45})$$
$$= 70.$$

Similarly, $F''_{\aleph,13}(J_2) = 48$, $\hat{F}''_{\aleph,14}(J_2) = 29$, $\hat{F}''_{\aleph,15}(J_2) = 42$.
Hence from (14.21),

$$LB\{J_2\} = \max(2 + 74, 7 + 70, 10 + 48, 30 + 29, 34 + 42) = 77.$$

Then, proceeding to Step 4, we choose the node with the lowest valued LB; in this case, 77 corresponding to node $\{J_2\}$. In Step 5, we branch from this node and create new nodes at $r = 2$: $\mathcal{N}_2 = \{J_2J_1\}$, $\{J_2J_3\}$, $\{J_2J_4\}$. Returning to the previous steps, LBs are calculated for the new nodes. Then, a node $\{J_2J_1\}$ with the lowest LB is chosen. By branching from this node new nodes are created at $r = 3$; in this case, $M - r = 4 - 3 = 1$; hence, returning to Step 2, we set $r = M = 4$, and obtain complete job sequences such as $\mathcal{N}_4 = \{J_2J_1J_3J_4\}$, $\{J_2J_1J_4J_3\}$. For these new nodes LBs are calculated.

The node with the lowest LB is now node $\{J_1\}$ at $r = 1$. Accordingly, we go back to this node (this process is called *backtracking*), and repeat the process.

In summary, branching and creating new nodes as shown by the indices in Figure 14.6, we finally obtain an optimal solution at the node marked with an asterisk, where $r = M = 4$ and

PRODUCTION SCHEDULING   251

the minimum LB is the minimum makespan, as indicated in Step 5. The result shows that the optimal job sequence is $J_2-J_1-J_4-J_3$ and the minimum makespan is 94. The Gantt chart is depicted in Figure 14.7.   □

For the above particular example, we could obtain the optimal solution without backtracking, but this is not always the case. Without this process we usually obtain only a suboptimal solution.

The application of the branch-and-bound method to solving the large-scale scheduling problem assures the optimal solution; however, notice that this solution is for a case that the job order is identical in all stages, hence, this is suboptimal in the true sense.

— EXAMPLE 14.8

The exact optimal solution for the large-scale scheduling problem treated in the previous example (Table 14.5) is obtained by the trial-and-error or complete-enumeration method:

- $J_2-J_1-J_4-J_3$ on stages 1 and 2.
- $J_2-J_1-J_3-J_4$ on stages 3, 4, and 5.

The minimum makespan is 86.   □

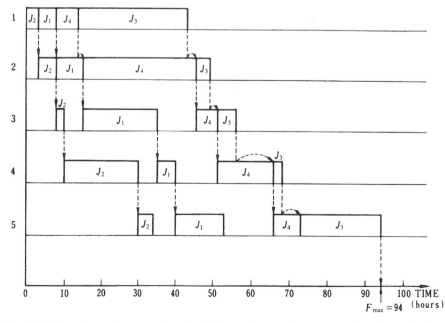

Figure 14.7 The Gantt chart associated with the optimal job sequence for a four-job, five-machine flow-shop scheduling problem illustrated in Table 14.5. This optimal solution was obtained by the branch-and-bound method represented in Figure 14.6.

Petrov's Method for Large-scale Flow-shop Scheduling

As mentioned previously, the branch-and-bound method needs great computational time to obtain the optimal solution for large-scale scheduling problems. Accordingly, methods of obtaining near-optimal solutions with less computational effort is necessary from the practical viewpoint. Petrov's method (Petrov, 1968) is one such method; it is actually an extension of Johnson's algorithm. This method does not ascertain optimality of the solution, but easily produces a fairly good job sequence through the following algorithm.

Petrov's algorithm for large-scale flow-shop scheduling:

Step 1: Divide N processing times for each of M jobs into two components:

$$\left. \begin{array}{l} T_{i1} = \sum_{k=1}^{h} t_{ik} \\ T_{i2} = \sum_{k=h'}^{N} t_{ik} \end{array} \right\} (i = 1, 2, ..., M) \qquad (14.22)$$

where $h = N/2$ and $h' = h + 1$ for even N and $h = h' = (N+1)/2$ for odd N.

Step 2: Apply Johnson's algorithm to M jobs, each of which has two fictitious processing times given in (14.22). □

─── **EXAMPLE 14.9** ───

(Petrov's method) Obtain a heuristic job sequence for the flow-shop scheduling problem already solved in Example 14.6. In this case, from Table 14.5 $N = 5$, which is odd, hence $h = h' = (5 + 1)/2 = 3$. T_{i1} and T_{i2} for $i = 1, 2, 3$, and 4 are calculated below (step 1).

Job	T_{i1}	T_{i2}
1	33	37
2	10	27
3	39	29
4	42	28

Applying Johnson's algorithm to these T_{i1}s and T_{i2}s (step 2), we obtain a job sequence J_2-J_1-J_3-J_4. The makespan for this sequence is 101. This figure is merely 7.5% larger than the optimal solution derived by the branch-and-bound method in Example 14.7, but 17.4% inferior to the exact solution obtained in Example 14.8. □

14.2.5 Job-shop Scheduling

Difficulty in Solving Job-shop Scheduling

General job-shop scheduling problems are complicated and hard to solve, since the work flow is not unidirectional. In this case the machine which the *j*th operation of job *i* requires is not identical as in the flow-shop case, but differs according to jobs to

PRODUCTION SCHEDULING

be processed. Hence, the data for a job-shop scheduling problem require:

- operation-machine assignments (technological order or routing) for jobs to be processed, and
- processing times for all associated operations.

The job-shop scheduling problem is then solved by constructing a schedule (Gantt chart) that is feasible, such that no two operations ever occupy the same machine simultaneously, and that, if possible, optimise a certain measure of performance.

--- EXAMPLE 14.10 ---

(Job-shop scheduling) The routings and operation processing times for a three-job, job-shop scheduling problem are shown in Table 14.6. A graphical representation of a feasible schedule is illustrated in a Gantt chart in Figure 14.8. The makespan for this schedule is 21 days. In this chart a series of operations marked with asterisks forms a critical path. □

General job-shop scheduling problems which are fairly easily solved are:

- two-job, job-shop scheduling, and
- two-machine, job-shop scheduling.

Two-job, Job-shop Scheduling

To this particular problem the following graphical solution procedure is effectively applied (Akers and Friedman, 1955):

Step 1: Mark on horizontal and vertical axes abscissae and ordinates with processing times in the technological order for J_1 and J_2, respectively. Let the product areas for each of the associated machines be denoted by prohibited regions.

Step 2: Proceed from the start point (0, 0) to the end point (its abscissa and ordinate are sums of processing times for J_1 and J_2, respectively), either taking horizontal direction (work on J_1 only), vertical direction (work on J_2 only), or diagonal (45°) direction (work on both jobs) without entering the prohibited regions (which would imply one machine working on both jobs simultaneously).

Table 14.6 An illustration of basic data for a job-shop scheduling problem. Each job has: (a) its own technological order of machines or production route, through which it is completed, and (b) processing times for operations contained in that job.

(a) Production route					(b) Processing time (days)				
	Operation (stage)					Operation (stage)			
Job	1	2	3	4	Job	1	2	3	4
J_1	M_1	M_2	M_3	M_4	J_1	2	5	3	2
J_2	M_4	M_2	M_1	M_3	J_2	1	4	4	5
J_3	M_1	M_2	M_3	–	J_3	4	1	6	–

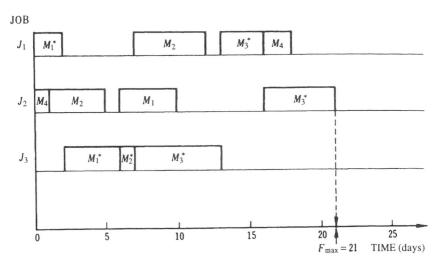

Figure 14.8 The Gantt chart for a feasible solution to a job-shop scheduling problem demonstrated in Table 14.6.

Step 3: Take the diagonal direction as much as possible, then the optimal solution would be obtained. □

--- **EXAMPLE 14.11** ---

Consider a two-job, job-shop scheduling problem minimising the makespan, as shown in Table 14.6 but excluding J_3. By marking the axes of abscissa and ordinate with processing times in the order: 2 on M_1, 5 on M_2, 3 on M_3, and 2 on M_4 for J_1 and 1 on M_4, 4 on M_2, 4 on M_1 and 5 on M_3 for J_2, respectively, the prohibited regions are indicated by shaded rectangular areas in Figure 14.9 (Step 1). Proceeding from the origin (0, 0) to the end point (12, 14) (Step 2) by taking the diagonal direction as much as possible (Step 3), we obtain an optimal solution as shown by a dashed line on the figure. The makespan for this schedule is 18, which is calculated by adding the sum of the lengths of vertical segments on that line to the sum of the processing times of J_1, or by Gantt charting. □

Two-machine, Job-shop Scheduling

A job-shop schedule for minimising F_{max} on two machines—A and B—is made by Jackson's method (Jackson, 1956), which applies the Johnson algorithm.

Step 1: Classify jobs to be performed into the following four groups:
(A) : jobs which are performed on A only;
(B) : jobs which are performed on B only;
(AB): jobs which are performed first on A, and then on B, and
(BA): jobs which are performed first on B, and then on A.

Step 2: Apply the Johnson algorithm to (AB) and (BA), and decide the sequences $(AB)^*$ and $(BA)^*$.

Step 3: The optimal sequences on A and B are:
- $(AB)^*(A)(BA)^*$ on A.
- $(BA)^*(B)(AB)^*$ on B. □

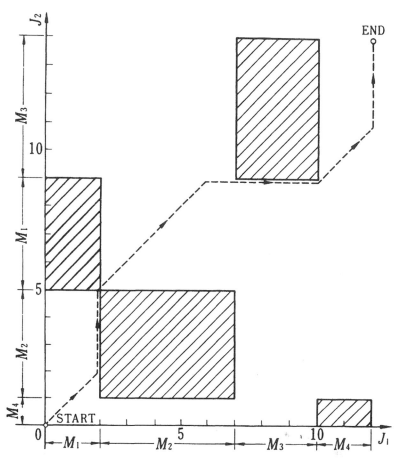

Figure 14.9 Graphical solution for a two-job job-shop scheduling problem (see Table 14.6 (excepting J_3)). By taking horizontal and vertical axes as J_1 and J_2, respectively, prohibited regions (shaded areas) are marked to prevent one machine working on both jobs simultaneously. An optimal schedule is obtained as a dashed line by taking the diagonal direction as much as possible from the start point to the end point.

--- EXAMPLE 14.12 ---

Eight jobs and their operation times run on two machines, A and B, are given by Table 14.7, which is already classified, based upon step 1. Following step 2,

- $(AB)^*: J_5-J_4-J_6$;
- $(BA)^*: J_8-J_7$.

Then the optimal sequences on A and B, which minimise the makespan are (step 3):

- $J_5-J_4-J_6-J_1-J_2-J_8-J_7$ on A;
- $J_8-J_7-J_3-J_5-J_4-J_6$ on B.

The minimum F_{max} is 30 (Question 14.19). □

Table 14.7 Basic data for a two-machine job-shop scheduling problem minimising the makespan.

Job group	Job No.	Processing time (days)		
		Machine A		Machine B
{A}	1	2		–
	2	5		–
{B}	3	–		3
{AB}	4	3	→	4
	5	1	→	2
	6	5	→	1
{BA}	7	1	←	2
	8	13	←	18

14.2.6 Scheduling Simulation

Why Job-shop Simulation?

In operations scheduling, optimality can be attained only in simple and idealised cases. In the dynamic job shop, where many varieties and numbers of production facilities are located in the plant so as to process various jobs, job processing times are of a stochastic nature, and unexpected work stoppages such as tool degradation and breakage, machine breakdown, absence of machine operators, etc. may occur. For such a dynamic case it is almost impossible to aim at optimality of scheduling analytically. An effective approach to this problem is a simulation experiment for the dynamic job-shop model, which is called *scheduling simulation* or *job-shop simulation* or, in some cases, *manufacturing simulation*.

—— EXAMPLE 14.13 ——

The 'dynamic scheduling simulator' (Hitomi, 1972) deals with simulation of a total manufacturing system including machining lines, sub-assembly lines, and assembly lines, and is designed for determining an optimal dispatching rule to minimise the total production time as well as determining an optimal lot size in dynamic job-shop environments, aiming at a complete synchronisation and balancing between the machining lines and the assembly lines, and the replacement of cutting tools and the recovery of machine tools which have failed. □

Dispatching Rules

When more than one job is in a queue in front of a machine, the order of processing such jobs must be determined to specify what the machine should do next at the completion of any job. Such scheduling decisions in a dynamic job-shop environment are usually determined by means of scheduling disciplines called *dispatching* (or *priority*) *rules*. Simulation studies and experiments are often made to determine a certain appropriate priority rule for a dynamic job shop.

Typical dispatching rules that are frequently used in practice are:[2]

- *Shortest-process-time (SPT) rule*. Jobs are processed in the increasing order of imminent operation times.
- *Least-work-remaining rule*. Jobs are processed in the increasing order of total processing times remaining to be done.
- *EDD rule*. Jobs are processed in the increasing order of due dates.
- *First-come, first-served (FCFS) rule*. Jobs are processed in the order of arrival at the machine.
- *Minimum-slack-time rule*. Jobs are processed in the order of non-decreasing slack times, where *slack time* signifies the remaining time obtained by subtracting the remaining processing times from the time between the present time and the due date.
- *Maximum-number-of-operations rule*. Jobs are processed in the order of decreasing number of remaining operations.
- *Minimum-number-of-operations rule*. Jobs are processed in the order of increasing number of remaining operations.
- *Penalty rule*. Jobs are processed in the order of decreasing delay cost.
- *COVERT* (cost-over-time) *rule*. Jobs are processed in the order of decreasing ratio of delay cost over processing time.

── **EXAMPLE 14.14** ──

The COVERT value is calculated by the following formula:

$$\text{COVERT} = \frac{\text{total remaining processing time}}{\text{processing time for operation being considered}}. \tag{14.23}$$

Then, the COVERT values for J_1 and J_2 on M_1 for basic data given in Table 14.6 are calculated as:

$\text{COVERT}(J_1, M_1) = (2 + 5 + 3 + 2)/2 = 6$
$\text{COVERT}(J_3, M_1) = (4 + 1 + 6)/4 = 2.75$

Hence, J_1 would proceed J_3 on M_1. □

One of the important objectives for scheduling simulation is to determine an appropriate dispatching rule as well as a detailed time schedule (Gantt chart) so as to achieve the production objective such as minimising the total flow time or cost or, on the other hand, maximising the machine utilisation or the total profit approximately.

Simulation Software

Usually computers are employed to conduct job-shop simulation. Several simulation languages for such manufacturing simulation have been developed as follows:

- DYNAMO (dynamic models) for dynamic information feedback systems described by a finite number of difference equations;

- GPSS (general purpose system simulator) for effectively describing complicated systems;
- SIMAN (simulation analysis language)/SLAM-FACTOR for effective design and analysis for manufacturing systems.

Other procedures effective to manufacturing systems analysis are:

- *Petri net* for analysis, design and control for discrete-event systems;
- OPT (optimised production technology) for obtaining near-optimal schedules in a combined procedure of production planning and production scheduling.

14.3 Project Scheduling—PERT and CPM

14.3.1 Scope of Project Scheduling

Techniques of Project Scheduling

A *project* is usually thought of as a one-off effort, and project scheduling is distinguished from operations scheduling in job shops by the non-repetitive nature of the work, such as the construction of a building, a ship, research and development (R & D) programs, and the like. The most popular techniques for this type of scheduling for large-scale projects are:

- PERT (program evaluation and review technique). Feasibility study of the project completion date is a major concern (PERT/time); in some cases, activity costs are taken into account (PERT/cost).
- CPM (critical path method). The optimal schedule is established by trading off required times and costs for performing activities.
- GERT (graphical evaluation and review technique). Network problems of a stochastic nature are solved by a procedure combining flow-graph theory, the semi-Markov process and PERT principles.

PERT was developed in 1959 as an R & D tool for the Polaris Missile project (Elsayed and Boucher, 1994) and it is now widely used for constructing a feasible schedule for large projects.

These techniques break the project into a collection of activities, represent the precedence relationships among the activities through a network (*arrow diagram*), and then determine a *critical path*, along which any delay in the activity causes a delay in the completion of the project, as pointed out in Example 14.5. Thus feasibility or optimality for project completion subject to time, money, and resource constraints is tested.

Structure of Arrow Diagram

An arrow diagram is made up of:

- *arrows* (or *arcs*) representing the activities or jobs of the project under consideration;
- *nodes* (or *events*) indicating the intersection (start or completion) of activities, and presenting the structure and the precedence relationships among activities.

Basic Rules for Constructing Arrow Diagrams

In construction of the network, information about all activities (times and costs required) included in the project and their precedence relationships is needed, and the following basic rules should be satisfied:

- Each activity is represented by one and only one arrow (see Figure 14.10).
- The length of the arrow has no meaning with respect to duration or cost of an activity relating to the arrow.
- The network should have a unique starting node (*origin* or *source*).
- The network should have a unique completion node (*terminal* or *sink*).
- The nodes should be numbered so that the number increases in the direction of arrows (*topological ordering* or *event numbering*).
- No activity should be represented by more than one arrow in the network.
- No more than one arrow should be drawn between two successive nodes. (A *dummy activity* represented by a broken line is used to avoid violation of this rule. This activity needs neither time nor cost.)
- Activities originating at a node can only begin after all activities terminating at the node have been completed.

── EXAMPLE 14.15 ──

(Use of dummy activities) A flow diagram of two parallel activities is represented in Figure 14.11(a). To avoid violating the rule that 'more than one arrow should not be drawn between two successive nodes', a dummy activity represented by a broken-line arrow is utilised to make an arrow diagram as represented in (b). □

14.3.2 PERT

Input Data for Constructing a PERT Network

A list of activities with their contents, precedence relationship and duration is required for analysis by PERT.

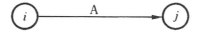

Figure 14.10 Basic representation of an activity A as an arrow with start node i and completion node j.

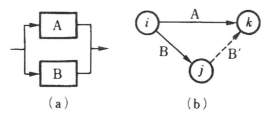

Figure 14.11 Representing parallel concurrent activities, A and B: (a) block (or flow) diagram, and (b) arrow diagram, utilising a dummy activity.

EXAMPLE 14.16

A project for constructing a new plant is indicated in Table 14.8. □

Drawing a PERT Network

Based upon basic data indicated in the list above, an arrow diagram is drawn following the rules described in the previous subsection.

EXAMPLE 14.17

(Arrow diagramming) Based upon the basic data indicated in Table 14.8, a flow diagram and an arrow diagram are drawn in Figure 14.12 for a new factory construction. □

Calculating Schedule Times

To determine the critical path and other scheduling and controlling information, the following node times are calculated by forward- and backward-path rules.

(1) *Earliest node* (or *schedule*) *time*. This is the earliest time at which activities originating at a node j can possibly begin, and is expressed:

$$\left. \begin{aligned} t_1^E &= 0 \\ t_j^E &= \max_{(i,j)\in Q} (t_i^E + \tau_{ij}) \quad (j = 2, 3, ..., S) \\ t_S^E &= \tau. \end{aligned} \right\} \quad (14.24)$$

where 1 and S stand for the source and sink of the network, respectively, (i, j) is an activity connecting node i and j, Q is a set of activities, τ_{ij} is the duration of activity (i, j), and τ is the completion time of the project.

Table 14.8 List of activities with their contents, precedence relationship and duration for analysis by PERT.

Activity code	Activity content	Preceding activities	Duration (weeks)
A	Demand forecast	–	8
B	Product development	–	9
C	Capital acquisition	–	15
D	Production planning	A	3
E	Product design	B	6
F	Plant planning	D, E	13
G	Profit planning	D, E	2
H	Materials acquisition	C, G	5
I	Operators acquisition	G	5
J	Production design	G	3
K	Facilities acquisition	C, J	10
L	Operators training	I	5
M	Plant layout	F, K	8
N	Manufacture	H, L, M	14

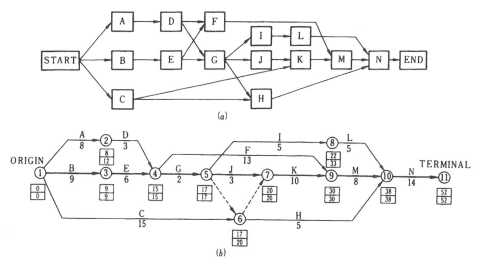

Figure 14.12 Construction of networks by flow diagramming (a) and arrow diagramming (b) based upon basic data given in Table 14.8. Letters in blocks or above arrows designate activities, and numbers below arrows are durations required to complete the associated activities. Topological ordering is indicated by numbers in circles (nodes). In the upper and lower boxes attached to nodes are shown the earliest and latest node times, respectively. The critical path is represented by bold arrows.

(2) *Latest node* (or *schedule*) *time*. This is the latest time at which the activity should be completed to avoid delay of the project beyond its due date, τ, and is expressed as follows:

$$\left.\begin{array}{l} t_S^L = \tau(=t_S^E) \\ t_j^L = \min_{(j,k) \in Q} (t_k^L - \tau_{jk}) \quad (j = S-1, ..., 1) \\ t_1^L = 0. \end{array}\right\} \quad (14.25)$$

Calculating Slacks

Slack or *float* is a measure of flexibility—allowable delay for an activity without causing delay of the project completion. For activity (i, j), with start node i and completion node j, the following four measures of slack are considered:

- node slack:
$$e_i = t_j^L - t_i^E \quad (14.26)$$
- total slack:
$$f_{ij}^T = t_j^L - t_i^E - \tau_{ij} \quad (14.27)$$
- free slack:
$$f_{ij}^F = t_j^E - t_i^E - \tau_{ij} \quad (14.28)$$
- independent slack:
$$f_{ij}^I = \max(t_j^E - t_i^L - \tau_{ij}, 0). \quad (14.29)$$

Among these slacks, there is a relationship:

$$f_{ij}^T \geq f_{ij}^F \geq f_{ij}^I \tag{14.30}$$

Determining the Critical Path

As explained in the previous section, the *critical path* is the longest path from the source to the sink. Along this path there is no room for any delay of activities. Consequently, this is the most important path for controlling the project. The critical path can be found by tracing nodes with no node slack from the source to the sink.

--- **EXAMPLE 14.18** ---

The earliest and latest node times for a PERT network represented in Figure 14.12 are calculated using equations (14.24) and (14.25), and are shown in the upper and lower boxes, respectively, adjacent to each node in the network of that figure. The critical path is represented by bold arrows. It is concluded that the project requires 52 weeks to be completed. □

Three-point Estimates for the Probabilistic Activity Duration

So far the activity duration has been estimated as only one deterministic value for each activity. In order to answer the question 'When will the project be complete?' with uncertain activity durations, PERT frequently provides the three-point estimates for the probabilistic activity duration (see Figure 14.13). For a given activity, these are:

- τ_o: an optimistic activity duration—the minimum value of the duration distribution;
- τ_m: a most likely activity duration—the mode of the duration distribution;
- τ_p: a pessimistic activity duration—the maximum value of the duration distribution.

Calculating the Mean and the Variance of the Activity Duration

The beta distribution is widely accepted as a reasonable model for the above duration distribution. In that case, the PERT method provides for the mean and the variance of

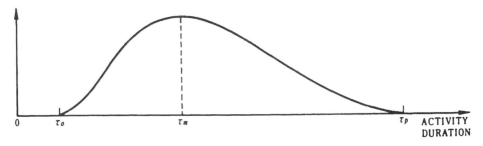

Figure 14.13 The density function of an activity duration expressed as a beta function. The minimum τ_o, the mode τ_m, and the maximum τ_p of activity duration are the optimistic, the most likely, and the pessimistic activity durations. The shape of this curve can be symmetrical, or skewed in either direction, depending upon the parameters τ_o, τ_m and τ_p.

the activity duration as follows:

$$\tau_e = \tfrac{1}{6}(\tau_o + 4\tau_m + \tau_p) \tag{14.31}$$

$$\sigma^2 = \left(\frac{\tau_p - \tau_o}{6}\right)^2. \tag{14.32}$$

Calculating the Mean and the Variance of Project Duration

Assuming that all the activities in the network are probabilistically independent and that the critical path contains a large number of activities, we can apply the central-limit theorem to analysing the length of the critical path, and, moreover, the mean of the length is equal to the sum of individual means of activities contained in the critical path, and the variance of the length is equal to the sum of individual variances of those activities.

— **THEOREM 14.6** —

(The central-limit theorem) The sum of a large number of independent random variables has a distribution that approaches a normal distribution as the number of components of the sum grows large, regardless of the shape of distribution of each component. □

Then the mean and the variance of the length of the critical path or project duration are obtained as follows:

$$t_{CP} = \sum_{l \in CP} \tau_{el} \tag{14.33}$$

$$\sigma^2_{CP} = \sum_{l \in CP} \sigma^2_l \tag{14.34}$$

where l denotes the activity contained in the critical path, and CP is a set of one chain of activities contained in the critical path. Furthermore, the distribution of project duration can be closely approximated by a normal distribution with mean t_{CP} and variance σ^2_{CP}.

Probability of the Project Completion

The probability that the project will be completed by a deadline t_d is then expressed:

$$P = \Phi\left(\frac{t_d - t_{CP}}{\sigma_{CP}}\right) = \frac{1}{\sqrt{(2\pi)}} \int_{-\infty}^{(t_d - t_{CP})/\sigma_{CP}} \exp(-x^2/2)\,dx \tag{14.35}$$

where $\Phi(\cdot)$ denotes the cumulative distribution function for a standard normal random variable, and we can easily obtain that value by the table of cumulative normal probability distribution (e.g. see Hald, 1952).

— **EXAMPLE 14.19** —

(Calculating the probability of the project completion) Consider the project described in Table 14.8 and Figure 14.12(b) in Example 14.17. Let the most likely duration of the activities contained in the critical path B–E–G–J–K–M–N be given as in Table 14.8, and the optimistic and the pessimistic durations of those activities are given below.

Activity	B	E	G	J	K	M	N
Optimistic duration	6	5	1	2	5	6	12
Pessimistic duration	11	8	3	4	12	11	15

With these data the mean and the variance of each of those activities are calculated from (14.31) and (14.32), and then the mean and the variance of the project duration are obtained from (14.33) and (14.34) as follows:

$$t_{CP} = 51.5 \text{ (weeks)}, \qquad \sigma_{CP}^2 = 3.47 \text{ ((week)}^2).$$

If the deadline of this project completion is given as $t_d = 51$ (weeks), the probability that the project will be completed by that deadline is, from (14.35):

$$P = \Phi\left(\frac{51 - 51.5}{1.86}\right) = 0.394.$$

For a deterministic case, it has taken 52 weeks to complete the project as described in Example 14.18. If the deadline of the project is determined as 51 weeks, then we are compelled to conclude that this project cannot be completed within the deadline. However, considering this problem as a probabilistic case, we can state that there is a probability of about 40% of completing this project. □

14.3.3 CPM

Trade-off Relationship Between Time and Cost

The operation time of an activity constituting a network can be reduced by increasing resources such as production facilities, human power, etc., which incur expenses. This means that there is a trade-off relationship between operation time and operation cost of an activity.

Denoting the maximum (standard or normal) operation time and operation cost by τ_{ij}^S and u_{ij}^S, and the minimum (crash) operation time and operation cost by τ_{ij}^C and u_{ij}^C, for activity (i, j), their relationship is depicted by a solid line in Figure 14.14. If this relationship is assumed a straight line—a dashed line in this figure, the relationship between operation time and operation cost of an activity (i, j) is formulated as:

$$u_{ij} = u_{ij}^o - \theta_{ij}\tau_{ij} \quad (\tau_{ij}^C \leq \tau_{ij} \leq \tau_{ij}^S) \tag{14.36}$$

where

$$u_{ij}^o = \frac{\tau_{ij}^S u_{ij}^C - \tau_{ij}^C u_{ij}^S}{\tau_{ij}^S - \tau_{ij}^C}, \qquad \theta_{ij} = \frac{u_{ij}^C - u_{ij}^S}{\tau_{ij}^S - \tau_{ij}^C}.$$

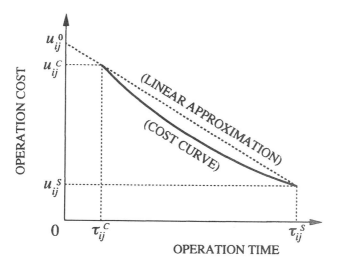

Figure 14.14 Activity cost curve, which represents the trade-off relationship between operation time and cost (solid line)—a broken line indicates linear approximation.

Aims of CPM

CPM takes the total cost into consideration, based upon the relationship between operation time and operation cost, in addition to the project completion time found by an arrow diagram—the same as in PERT analysis.

The project completion time is:

$$E(\tau) = t_N^E - t_1^E \equiv \tau \tag{14.37}$$

and the project completion cost is given by:

$$G(\tau) = \sum_{(i,j) \in Q} u_{ij} = \sum_{(i,j) \in Q} (u_{ij}^o - \theta_{ij}\tau_{ij}). \tag{14.38}$$

Hence CPM attempts to:

- minimise the project completion time within a specified fund available;
- minimise the project completion cost within a specified completion time.

CPM Procedure

CPM attempts to reduce the project completion time, whilst minimising the cost increase by the following algorithm.

Step 1: Find a critical path for normal operation times and costs by PERT analysis.
Step 2: If no activity exists that has the smallest cost-increase rate incurred by reduction of the activity time, stop (an optimal solution has been obtained). Otherwise, proceed.
Step 3: Reduce operation time of the activity until either
  (i) the other path(s) become(s) critical, or
  (ii) the current activity reaches its minimum operation time.
  Return to step 1. □

In Step 3 of the above algorithm, if state (i) occurs first, further decrease of that operation time makes decreasing τ more costly; hence it might cost less to reduce the operation time of another activity. If state (ii) occurs first, it is impossible to further reduce the operation time of the current activity.

Project Cost Curve

The project cost curve shows the relationship between the project completion time τ and project cost $G(\tau)$. This curve can be drawn from activity cost curves, based upon the CPM algorithm, as depicted in Figure 14.15, which is a convex, piecewise linear, decreasing diagram. In this diagram, U_S is the minimum cost for the project completion, the project completion time τ_S being the longest. U_C is the minimum cost needed to complete the project by the shortest time τ_C. In this case, all the activities on the critical path are performed in their crash operation times, but the activities on the slack path are not always performed in their crash time. If the entire activities are performed in their crash time, the project completion costs U'_C, but this is not recommended.

— **EXAMPLE 14.20** —

(Solution by the CPM algorithm) A project is given by Table 14.9. To begin with a network with the time scale is drawn as Figure 14.16(a), when all the activities are performed with their

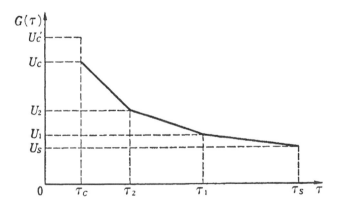

Figure 14.15 Project cost curve, which represents the relationship between the project completion time and cost—a convex, piecewise linear, decreasing diagram.

Table 14.9 List of activities with their precedence relationship, normal operation times and costs, crash operation time, and cost slopes for analysis by CPM.

Activity	Preceded activity	Normal operation time τ^S (days)	Crash operation time τ^C (days)	Normal operation cost u^S ($)	Cost slope α ($/day)
A	–	10	6	500 000	15 000
B	A	21	14	100 000	7 000
C	A	8	6	150 000	25 000
D	C	15	12	200 000	10 000

PRODUCTION SCHEDULING 267

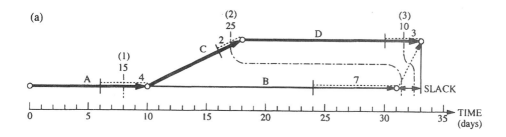

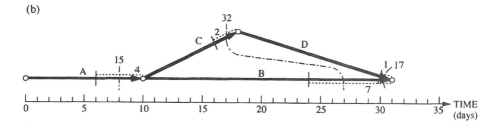

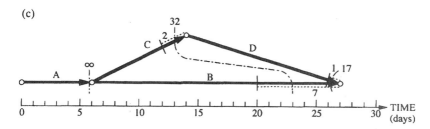

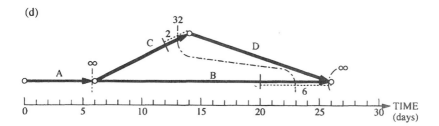

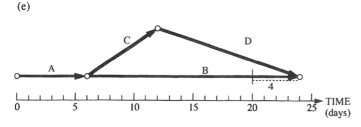

Figure 14.16 Solving by the CPM procedure—the optimal solution (e) still possesses the slack of 4 days on activity B, but this slack cannot be utilised for reduction of the total project time.

normal operation time. The project completion time is $\tau_S = 33$ days with the project completion cost $U_S = \$950\,000$. In this figure the critical path is indicated by a chain of bold arrows and the possible days which can be reduced are indicated by dashed lines.

In order to reduce the project completion time the possible reduction days are determined by (i) or (ii) in step 3 of the CPM algorithm for (an) activity or a combination of activities, with a chain line. This sort of chain line is called a *cut*. On each cut the cost gradient (thousand dollars per day)—i.e. the cost increase by one day's reduction—is indicated. This value is the cut capacity. A cut with the minimal value of this capacity is called the *minimum cut*.

Notice activity D with the minimum cut (3). This activity can be reduced by 3 days, but activity B which has been slack becomes critical when D is reduced by 2 days. This state is represented in Figure 14.16(b). The project completion time and cost for this state are: $\tau_1 = 31$ days and $U_1 = \$970\,000$.

In (b) activity A with the minimum cut is reduced as far as possible—that is, 4 days, obtaining (c). The project completion time and cost for this state are:

$\tau_2 = 27$ days and $U_2 = \$1\,030\,000$.

Now activity A can be reduced no more; hence this cut capacity is set at ∞. By choosing the minimum cut (B, D) and reducing by 1 day, activity D becomes crash (D is performed at utmost speed), obtaining (d). The project completion time and cost for this state are: $\tau_3 = 26$ days and $U_3 = \$1\,047\,000$.

Finally, taking note of cut (B, C), (e) is obtained after a 2-day reduction. In this state activity B can be reduced by 4 days more, but this does not bring reduction of the project completion time, merely causing a cost increase. (e) is the final optimum state; the minimum project completion time is: $\tau_C = 24$ days with the project cost of $U_C = \$1\,111\,000$. If activity B is reduced as much as possible, the total project cost is $U'_C = \$1\,139\,000$. □

Notes

1. These two cases—job waiting and machine idle—and the rare case—no occurrence of both patterns, hence neither job waiting nor machine idle—are illustrated in Figure 14.4.
2. A summary of over 100 dispatching rules is given in Panwalkar and Iskander (1977).

References

AKERS, S.B., Jr. and FRIEDMAN, J. (1955) A non-numerical approach to production scheduling problems, *Operations Research*, **3** (4).

BAKER, K.R. (1974) *Introduction to Sequencing and Scheduling* (New York: Wiley).

BEDWORTH, D.D. and BAILEY, J.E. (1982) *Integrated Production Control Systems* (New York: Wiley), p. 309.

CONWAY, R.W., MAXWELL, W.L. and MILLER, L.W. (1967) *Theory of Scheduling* (Reading, MA.: Addison-Wesley).

ELSAYED, E.A. and BOUCHER, T.O. (1994) *Analysis and Control of Production Systems* (2nd edn) (Englewood Cliffs, NJ: Prentice-Hall), p. 231.

FENCH, S. (1982) *Sequencing and Scheduling* (West Sussex: Ellis Horwood).

HALD, A. (1952) *Statistical Tables and Formulas* (New York: Wiley).

HITOMI, K. (1972) *Decision Making for Production* (in Japanese) (Tokyo: Chuo-keizai-sha), Sec. 8-2.

IGNIZIO, J.P. and GUPTA, J.N.D. (1975) *Operations Research in Decision Making* (New York: Crane, Russak & Company), pp. 287–288.
JACKSON, J.R. (1956) An extension of Johnson's results on job lot scheduling, *Naval Research Logistics Quarterly*, **3** (3).
PANWALKAR, S.S. and ISKANDER, W. (1977) A survey of scheduling rules, *Operations Research*, **25** (1).
PETROV, V.A. (1968) *Flowline Group Production Planning* (London: Business Publications).

Supplementary reading

BRUCKER, P. (1995) *Scheduling Algorithms* (Berlin: Springer).
CARRIE, A. (1988) *Simulation of Manufacturing Systems* (New York: Wiley).
ELMAGHRABY, S.E. (1979) *Activity Networks* (New York: Wiley).
MODER, J.J., PHILLIPS, C.R. and DAVIS, E.W. (1983) *Project Management with CPM, PERT, and Precedence Diagramming* (3rd edn) (New York: Van Nostrand Reinhold).
MORTEN, T.E. and PENTICO, D.W. (1993) *Heuristic Scheduling Systems* (New York: Wiley).

CHAPTER FIFTEEN

Inventory Management

15.1 Inventory Function in Manufacturing

Role of Inventory

As discussed in Chapter 6, 'inventory' plays the role of buffer—*stock*—in the material *flow* from the acquisition of raw materials to the finishing of products. Inventory management/control is made so as to:

- reduce inventory costs;
- stabilise the production level;
- increase the service level by preventing product shortages to customers.

The 'parts-oriented production [system]', which was mentioned in Section 4.3 and will be described in detail in Chapter 27, employs the parts made in advance of the order receipt positively for efficient multi-product, small-batch production, while the 'just-in-time (JIT) production [system]' expects to decrease item stocks as much as possible—even to zero, as was also described in Section 4.3 and will be explained in the next chapter.

Inventory in Logistics

As depicted in Figure 1.2, inventory is one of the basic functions concerning an integrated material flow. Other basic functions are:

- procurement;
- production;
- distribution;
- sales.

Stocks piled between procurement and production, in the production processes (stages) and between production and sales are called the [*raw*-] *material inventory*, the [*work*-]*in-process inventory* and the *product inventory*, respectively, as mentioned in Chapter 6.

15.2 Fundamentals of Inventory Analysis

Objectives and Policies of Inventory Analysis

In an inventory analysis, an attempt is usually made to minimise the total costs incurred such as:

- acquisition or production cost;
- inventory-carrying cost;
- penalty associated with shortages, etc.

Basic decision factors for attaining these aims of inventory analysis are (Buchan and Koenigsberg, 1963):

- reorder date (or point) P, and
- reorder quantity Q.

Inventory Models

The states of the above two factors produce various inventory models:

- the *fixed-order quantity system*, frequently referred to as the *Wilson model*: the most basic, classical model wherein the economic order quantity Q^* which minimises the total variable costs of managing inventory is optimally determined and this quantity is ordered whenever the inventory on hand drops to a particular level, referred to as the *reorder point*. Hence P varies.
- the *replenishment system*: P is fixed and Q varies according to the inventory level so as to fill the replenishment level.
- the (S, s) *system*: both P and Q vary.

Inventory in Practice

Practical inventory policies include the following:

- *Two-bin system*. The items are stored in two equal-sized bins. The demands are filled by one of the bins; when it becomes empty, the other bin is employed and parts are procured to fill the empty bin, e.g. inexpensive fasteners stocked in bulk in a two-bin system and on free issue to production.
- *ABC inventory classification* (or *analysis*). It is common that 5–20% of inventory items incur 50–60% of the inventory expense; these items are called 'A-items', and stock-leads are controlled accurately. 'B-items' typically account for 25% of the remaining inventory expense and represent 30–50% of the entire items. The remaining 'C-items' are of low cost, and control of their inventory is relaxed (Silver and Peterson, 1985; Vollmann *et al.*, 1988).

15.3 Inventory Systems

Economic Order Quantity for Fixed-order Quantity Inventory Systems

The *economic order quantity* (EOQ), which plays a fundamental role in the *fixed-order quantity* [*inventory*] *system*, minimises the cost of managing inventory. In determining this quantity the model assumes that the cost is made up solely of two components:

- ordering cost;
- carrying cost.

Hence the objective function is formulated:

$$U = c_o D/Q + c_h Q/2 \qquad (15.1)$$

where c_o is the cost of placing an order, c_h is the annual carrying cost per unit of inventory, and D is annual sales or demand in units (uniform distribution). c_h is often expressed:

$$c_h = c_p r \qquad (15.2)$$

where c_p is the unit purchase or production cost of an item and r is the carrying cost expressed as an annual percentage of this unit cost—annual inventory-carrying charge.

Then the EOQ is obtained by differentiating (15.1) and setting it equal to zero:

$$Q^* = \sqrt{2c_o D/c_h}. \qquad (15.3)$$

As in the case of the economic lot size mentioned in Chapter 13, it is usual that the inventory–cost curve is shallow near this EOQ; hence it is relatively insensitive to deviations from the EOQ.

In this model, the number of reorders a year is:

$$n = D/Q^* = \sqrt{c_h D/2c_o} \qquad (15.4)$$

and the minimum annual cost is:

$$U^* = \sqrt{2c_o c_h D}. \qquad (15.5)$$

--- **EXAMPLE 15.1** ---

(Calculating the EOQ) The annual demand for a product is 4000, the unit purchase cost is $100, and the ordering expense is $40. If the annual inventory-carrying charge is 20%, then from (15.3):

$$Q^* = \sqrt{\frac{2 \times 40 \times 4000}{100 \times 0.2}} = 126.$$

□

Reorder Point

The *reorder point* P is the point at which to reorder the above quantity Q^*. It is made up of the average expected sales during the lead time—the time lag between placing and fulfilment of a reorder—and the safety stock preventing a temporary out-of-stock condition which would lead to late deliveries and probably result in lost sales (see Figure 15.1), as follows:

$$P = Q_0 + \bar{d}L \qquad (15.6)$$

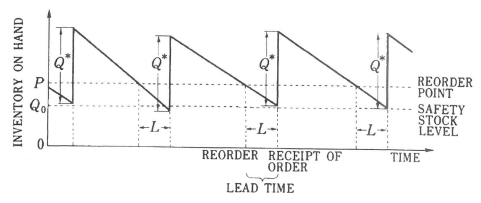

Figure 15.1 The fixed-order quantity inventory model. Wherever the inventory on hand drops to the reorder point P, the economic order quantity Q^* is ordered. L is lead time—the time length between reorder and receiving orders.

where $\bar{d}$ is the average daily sales in units, L is the lead time in days, and Q_o is safety stock in units, and is given by:

$$Q_0 = \theta_a \sigma_d \sqrt{L} \qquad (15.7)$$

where σ_d is the standard deviation for daily sales and θ_a is the reliability level for satisfying demands, e.g. $\theta_{0.95} = 1.65$ for satisfying 95% of demands.

The Replenishment Inventory System

In the *replenishment system* there is no fixed reorder quantity; instead, inventory is viewed at periodic intervals, and if there have been any sales since the last review, an order is placed. The order quantity is equal to the amount by which a fixed replenishment level exceeds the actual inventory level at the time of the review (see Figure 15.2). The *replenishment level* is determined by:

$$S = Q_0 + \bar{d}(L + R) \qquad (15.8)$$

where $Q_0 = \theta_{ad}\sqrt{L + R}$, and R is time between reviews in days.

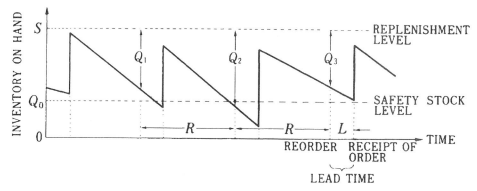

Figure 15.2 The replenishment inventory model. Reorder is made at periodic intervals R so as to fill the inventory up to the replenishment level S. L is lead time.

The reorder quantity is:

$$Q = \begin{cases} S - I, & \text{if } L \leq R \\ S - I - Q_r, & \text{if } L > R \end{cases} \qquad (15.9)$$

where I is inventory on hand at time of review in units, and Q_r is quantity on order in units.

The replenishment inventory system is effective in many real inventory situations, particularly when delivery lead times are long and shortage costs are extremely high or when stock counts are infrequent.

The (S, s) Inventory System

The (S, s) *system* is defined as a policy which generates an order when the inventory level falls to a prescribed value. In this model, by determining the reorder level corresponding to (15.6) as:

$$s = Q_0 + \bar{d}(L + R/2) \qquad (15.10)$$

along with the replenishment level S given by (15.8), at each review an order for

$$Q = S - I - Q_r \qquad (15.11)$$

units is placed only when $I + Q_r < s$. Thus the inventory on hand plus that on order never exceeds S.

The (S, s) policy responds more quickly to changes in demand values than other models.

—— **EXAMPLE 15.2** ——

Given the same demand distribution and the same review period, the behaviour of the three inventory policies mentioned above is demonstrated in Figure 15.3. □

Service Level

As seen in Figure 15.3, sometimes, in practice, all demands may not be met, since the amount of safety stock held depends on a certain measure of the maximum expected demand and the risk and associated cost of failure of the service. To prevent the shortage and express the measure of service of shipping the products to the customers, we often use the *service level*, which is defined as follows:

$$\text{service level} = \frac{\text{number of units shipped [without delay]}}{\text{number of units demanded}}$$

$$= \frac{\text{number of units demanded} - \text{number of units short}}{\text{number of units demanded}}. \qquad (15.12)$$

The service level is measured over a period which may be a week, a month, or a year.

As we raise the service level, the inventory cost necessary to obtain the indicated service increases. A typical relationship between service level and inventory cost is exhibited in Figure 15.4.

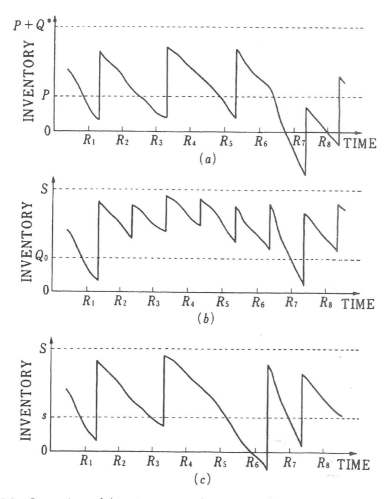

Figure 15.3 Comparison of three inventory policies, using the same demand data: (a) the fixed-order policy (P, Q*) in which the fixed economic order quantity Q* is ordered when the inventory I drops to the reorder point P at review time; (b) the replenishment policy (S) in which the quantity of the difference between the replenishment level S and the inventory I is ordered at each review time; and (c) the optimal replenishment policy (S, s) in which only when the inventory I is less than the reorder point s at the review time, the quantity of the difference between the replenishment level S and the inventory I is ordered. Review times are denoted by R_1, R_2, ..., and the length of lead time L is 3mm.

Inventory Model with Shortage Loss

If reorder (back order) is allowed to meet the shortage incurred by the excess demand, the optimum EOQ can be determined so as to minimise the annual variable cost consisting of:

- ordering cost,
- inventory-carrying cost, and
- shortage loss (see Figure 15.5).

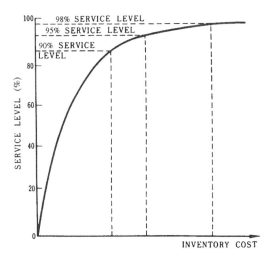

Figure 15.4 A typical relationship between service level and inventory cost. Inventory cost rises as service level increases.

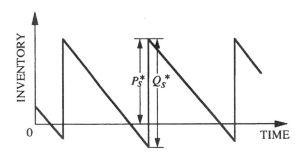

Figure 15.5 Inventory model with back orders—shortage loss occurs.

Hence the objective function to be minimised is:

$$U_s = c_o D/Q_s + c_s(Q_s - P_s)^2/2Q_s + c_h P_s^2/2Q_s, \qquad (15.13)$$

where c_s is the annual opportunity loss incurred by shortages of the unit product, Q_s is the order quantity, and P_s is the reorder level.

The inventory and the reorder level are obtained by partially differentiating (15.13) with respect to Q_s and P_s and setting them at 0:

$$Q_s^* = \sqrt{\frac{2c_o D}{c_h}} \sqrt{\frac{c_s + c_h}{c_s}} \qquad (15.14)$$

$$P_s^* = \frac{c_s Q_s^*}{c_s + c_h} = \sqrt{\frac{2c_o D}{c_h}} \sqrt{\frac{c_s}{c_s + c_h}}. \qquad (15.15)$$

In the inventory model with shortage loss, the average service level is given by:

$$s_1 = P_s^*/Q_s^* = c_s/(c_s + c_h). \qquad (15.16)$$

15.4 Multiple-product Inventory Management

Multiple-product Inventory Models

It is usual that a warehouse carries many different parts/products. A multiple-product inventory model is formulated so as to minimise the annual variable cost for M products in the case of uniform demands and no shortage allowed. The total cost is:

$$U_T = \sum_{j=1}^{M} \left(\frac{c_{oj} D_j}{Q_j} + \frac{c_{hj} Q_j}{2} \right). \tag{15.17}$$

If the warehouse, namely the maximum available inventory space, is large enough, the optimum (unconstrained) EOQs are obtained by simply differentiating the above equation with respect to the Q_js and setting them to 0:

$$Q_j^{**} = \sqrt{\frac{2 c_{oj} D_j}{c_{hj}}} \quad (j = 1, 2, \ldots, M). \tag{15.18}$$

Then the minimum unconstrained total cost is:

$$U_T^{**} = \sum_{j=1}^{M} (2 c_{oj} c_{hj} D_j)^{1/2}. \tag{15.19}$$

Restrictions on Multiple-product Inventory

Usually a few restrictions are imposed on multiple-product inventory as follows:

- *Space limitation.* Denoting the maximum space available by F (m²) and the area needed for a piece of item j by f_j (m²):

$$\sum_{j=1}^{M} f_j Q_j \leq F. \tag{15.20}$$

- *Fund restriction.* Since the inventory value of a piece of item j is c_{pj}, the fund restriction is expressed as:

$$\sum_{j=1}^{M} c_{pj} Q_j \leq C, \tag{15.21}$$

where C is the available fund.

Optimum Constrained EOQs for Multiple Products

The Kuhn–Tucker necessary condition, which was mentioned in Section 11.3, is applied to obtain the optimal solution for this multiple-product inventory analysis. That is, substituting (15.17), (15.20), and (15.21) into equations (11.108)–(11.110), we get:

$$-c_{oj}D_j/Q_j^{*2} + c_{hj}/2 + \lambda^* f_j + \mu^* c_{pj} = 0 \quad (j = 1, 2, \ldots, M) \tag{15.22}$$

$$\sum_{j=1}^{M} f_j Q_j^* - F \leq 0, \tag{15.23}$$

$$\sum_{j=1}^{M} c_{pj} Q_j^* - C \leq 0, \tag{15.24}$$

$$\lambda^* \left(\sum_{j=1}^{M} f_j Q_j^* - F \right) = 0, \tag{15.25}$$

$$\mu^* \left(\sum_{j=1}^{M} c_{pj} Q_j^* - C \right) = 0 \tag{15.26}$$

for positive Lagrange multipliers λ^* and μ^* (≥ 0) for space and fund restrictions.
Then the optimum EOQs are obtained from (15.22):

$$Q_j^* = \sqrt{\frac{2 c_{oj} D_j}{c_{hj} + 2(\lambda^* f_j + \mu^* c_{pj})}} \quad (j = 1, 2, \ldots, M). \tag{15.27}$$

λ^* and μ^* in the above equation are determined by simultaneously solving the following two equations:

$$\sum_{j=1}^{M} f_j \sqrt{\frac{2 c_{oj} D_j}{c_{hj} + 2(\lambda^* f_j + \mu^* c_{pj})}} = F, \tag{15.28}$$

$$\sum_{j=1}^{M} c_{pj} \sqrt{\frac{2 c_{oj} D_j}{c_{hj} + 2(\lambda^* f_j + \mu^* c_{pj})}} = C. \tag{15.29}$$

Then the constrained total cost is calculated by substituting (15.27) into (15.17).

— **EXAMPLE 15.3** —

(Optimum EOQs with space limitation) Annual demands (D_i) for three products, P_1, P_2, and P_3, are supposed to be 1000, 2000, and 4000. Their purchasing expenses (c_{pi}) are $4000, $2000, and $10 000. Each ordering ($c_o$) costs $4000 and the annual inventory-carrying charge (expense ratio) (r) is 20%.

Then from (15.18), the unconstrained EOQs are:

$$Q_1^{**} = \sqrt{\frac{2 \times 4000 \times 1000}{0.2 \times 4000}} = 100, \quad Q_2^{**} = 200, \quad Q_3^{**} = 126.$$

From (15.19), the unconstrained minimum cost is:

$U_T^{**} = \$412\,982.$

Let the inventory space (f_j) required by unit product be 4, 2, and 10 m² for P_1, P_2, and P_3, respectively. Then the total inventory space required (the left-hand side of (15.20)) for the unconstrained EOQs is:

$4 \times 100 + 2 \times 200 + 10 \times 126 = 2060$ (m²).

If the maximum available inventory space (F) is 3000 m², (15.20) holds with inequality. If, however, F is 1500, (15.20) no longer holds for the Q^*s.

Hence the constrained EOQs are calculated using (15.27) with the full use of the available inventory space; that is, from (15.29), neglecting μ^*:

$$4\sqrt{\frac{2 \times 4000 \times 1000}{4000 \times 0.2 + 2\lambda^* \times 4}} + 2\sqrt{\frac{2 \times 4000 \times 2000}{2000 \times 0.2 + 2\lambda^* \times 2}} + 10\sqrt{\frac{2 \times 4000 \times 4000}{10\,000 \times 0.2 + 2\lambda^* \times 10}} = 1500.$$

Then we get:

$$\lambda^* = 89.$$

Thus, the constrained EOQs are calculated using (15.27):

$$Q_1^* = \sqrt{\frac{2 \times 4000 \times 1000}{4000 \times 0.2 + 2 \times 89 \times 4}} = 73, \qquad Q_2^* = 145, \qquad Q_3^* = 92.$$

For this result the total inventory space needed is just equal to the available inventory space; that is, the entire inventory space available is exhausted. Substituting the above Q_i^*s into (15.17), we get the constrained minimum cost:

$$U_T^* = \$434\,080.$$

Hence the cost difference between the constrained and the unconstrained minimum costs is:

$$U = U_T^* - U_T^{**} = \$21\,098. \qquad \square$$

Optimising Algorithm for the Optimum EOQs for Multiple Products with Inventory Space Restriction

The above analysis leads to the following optimising algorithm for calculating the optimum EOQs and the minimum total cost for multiple products with inventory space restriction.

Step 1: Calculate the Q^{**}s using (15.18).
Step 2: Calculate the left-hand side of (15.20). If it is equal to or less than F, calculate U_T^{**} using (15.19) and stop (Q^{**}s and U_T^{**} are optimal). Otherwise, proceed.
Step 3: Obtain λ^* using (15.28) neglecting μ^*. Calculate the Q_i^*s using (15.27) and then U_T^* using (15.17). Stop (the Q_i^*s and U_T^* are optimal). $\square$

15.5 Probabilistic Inventory Models

Probabilistic Aspect of Inventory

In practice, the future demands are not deterministic, hence they are to be of a probabilistic (or stochastic) nature. It is assumed in this section that the future demand of an item, x, is continuous with density function $f(x)$.

Then a probabilistic inventory model determines the optimum value of inventory at the start of a period—'initial inventory' Q, so as to minimise the total cost.

Analysis of Probabilistic Inventory Models

If the initial inventory Q is less than the demand x, the entire quantity is sold out, and if $Q > x$, the amount of x is sold. Hence the sales volume is $\min(Q, x)$.

Denoting the unit purchase (or production) cost by c_p, inventory-carrying expense at the end of the period by c_h, and the unit shortage loss by c_s, the expected total cost to be minimised is:

$$u = c_p Q + c_h \int_0^Q (Q - x) f(x) \, dx + c_s \int_Q^\infty (x - Q) f(x) \, dx. \tag{15.30}$$

The optimum inventory at the start of the period is then obtained by differentiating the above equation with respect to Q and setting to 0 as follows:

$$0 = c_p + c_h \int_0^{Q^*} f(x) \, dx - c_s \int_{Q^*}^\infty f(x) \, dx = c_p + c_h F(Q^*) - c_s(1 - F(Q^*)) \tag{15.31}$$

where $F(x)$ is the distribution function of demand x; namely,

$$F(x) = \int_{0f}^x f(y) \, dy. \tag{15.32}$$

We obtain the optimum initial inventory:

$$F(Q^*) = \frac{c_s - c_p}{c_h + c_s}. \tag{15.33}$$

This is meaningful when $c_s > c_p$.

--- **EXAMPLE 15.4** ---

(Calculating the optimum initial inventory) Weekly demands of an article of food follow the following equation.

$$f(x) = 0.001 e^{-0.001x}, \quad x \geq 0.$$

This food is made on a weekly basis with a production cost of \$500/kg. The processing expense for unsold pieces is \$10/kg. If the demand exceeds the production, crash production costs \$800/kg.

Then the optimum production batch size for the initial inventory, Q^*, can be calculated by the following equation, which comes from (15.32) and (15.33):

$$\int_0^{Q^*} 0.001 e^{-0.001x} \, dx = 1 - e^{-0.001 Q^*} = \frac{800 - 500}{10 + 800} = 0.37.$$

Hence $Q^* = 462$ (kg). □

Optimum Order Policy

The above analysis of a probabilistic case of inventory leads to the following rule, if the initial inventory is I.

--- **POLICY 15.1** ---

(Optimum order quantity) Order $Q^* - I$ if $Q^* > I$, otherwise do not order. □

References

BUCHAN, J. and KOENIGSBERG, E. (1963) *Inventory Management* (Englewood Cliffs, NJ: Prentice-Hall), p. 285.
SILVER, E.A. and PETERSON, R. (1985) *Decision Systems for Inventory Management and Production Planning* (2nd edn) (New York: Wiley), Sec. 3.6.
VOLLMANN, T.E., BERRY, W.L. and WHYBARK, D.C. (1988) *Manufacturing Planning and Control Systems* (2nd edn) (Homewood, IL; Dow Jones-Irwin), pp. 744–745.

Supplementary reading

SHERBROOKE, C.C. (1992) *Optimal Inventory Modeling of Systems* (New York: Wiley).
WATERS, C.D. (1992) *Inventory Control and Management* (Chichester: Wiley).

CHAPTER SIXTEEN

Production Control

16.1 Scope and Problems of Production Control

Why Production Control?

Based upon information/instruction relative to the production schedule established by production planning and scheduling, production activities are implemented. This implementation stage of production is often not carried out in accordance with the schedule/instructions, even when the planning stage has been well established using exact and high-level analytical methods. There are several reasons, as follows:

(1) There is no exact way of production forecasting to meet the market demands in the future.
(2) Accordingly, aggregate production planning is made, based upon uncertain estimates for the future demand.
(3) Since data used for production planning and scheduling as setup and processing times are based upon standard data, deviation between the plan/schedule and the actual performance always occurs.
(4) Unforeseen conditions often occur at the implementation stage of production, such as breakdown of production facilities, absence of workers, delay in delivery of raw materials, etc.
(5) Defective products may be produced.

Meaning of Production Control

As a result of the above facts, modification or follow-up is more or less required during production implementation, in such a way that the desired production objectives for planned quantities of products in specified time periods are accomplished. This function is *production control*. The flexible utilisation of this function is particularly important in varied production to order.

Basic Functions of Production Control

The following two basic functions are contained in production control (Hitomi, 1975).

(I) *Control of logistics*. This controls the flow from raw materials to finished products, and includes:

(a) *production control* (in a narrow sense) or *process control* for controlling time (due date) and quantities to be produced;
(b) *quality control* for assuring the desired quality and reliability for finished products, as is discussed in the next chapter;
(c) *inventory control* for control for excess storages and shortages of raw materials, parts, and products, as was mentioned in the previous chapter.

(II) *Control of production resources*. This is a control function for resources of production, and includes:

(a) *productive maintenance* for preventing the breakdown of production facilities and repairing breakdowns, as will be discussed in Section 16.4;
(b) *cost control* for reducing/minimising the total production cost, as will be discussed in Chapter 19.

16.2 Process Control

Management by Exception

The quantity to produce and the time or due date identified at the planning stage are the production standards at the implementation stage. The actual performances of produced quantity and completion time are measured and compared with the planned standards. Where there is a large deviation between them, appropriate actions are taken to modify that deviation. Figure 16.1 shows a generalised feedback loop for production control. With this control loop a means of *management by exception* is provided by reporting the unusual cases to the management.

Production-control Procedure

Produced quantity is checked, and productive operations are assigned and adjusted in relation to the production capacity available, so that the desired quantity is acquired. Completion time for each product is measured and compared with the planned time

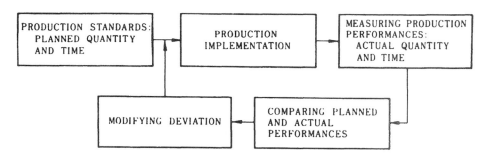

Figure 16.1 The production control system works as a generalised feedback loop: measuring production performances, comparing those with the planned standards, and modifying the deviation.

by means of, for example, the Gantt chart (refer to Chapter 14). When any job is delayed, we take action to expedite that job.

A production control system works more effectively if actual data of production performance are acquired as quickly and accurately as possible. For this purpose information processing on an on-line, real-time basis will be effective by installing terminals or factory computers in a workshop for gathering the actual production data and displaying the production instruction to the production stages (refer to Chapter 29: on-line production control systems).

Line of Balance

Production follow-up, that is, whether or not the production plan and schedule—production standards—established by production planning, process planning and production scheduling, are being timely and accurately implemented, can be investigated in aggregate by drawing a *production-progress chart*. This procedure is called *line of balance*.

—— EXAMPLE 16.1 ——————————————————————————

(Line of balance) Figure 16.2(a) shows the activities to be done some weeks in advance in order to ship (deliver) the finished product—event 5 at week 0. Figure 16.2(b) depicts the *production-progress chart*; the production performance is estimated at present (April) below the planned quantity = 6 indicated by a zigzag line (line of balance) shown in (c). The actual production quantity is determined by the present amount of event 5—namely 5.   □

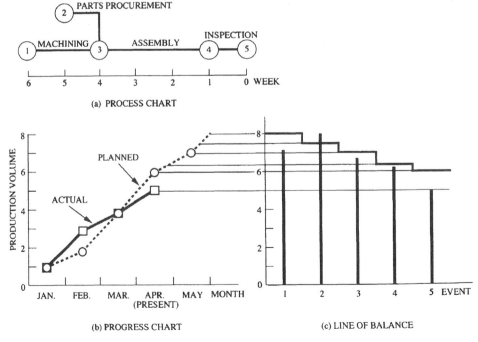

Figure 16.2 Production progress is controlled by the line-of-balance technique.

16.3 Just-in-time (JIT) Production

JIT and Kanban

As mentioned in Chapter 4, *just-in-time* (*JIT*) refers to the production and supply of the required number of parts at the right time, just as required, in order to minimise work-in-process inventory. This production system is the 'pull[-through] system' rather than the usual 'push system'. By inputting the production schedule only at the final stage, the demanded items are withdrawn from the previous stages with the circulation of *kanbans* (instruction cards), on which the items and their quantities needed are indicated (Ohno, 1978).

Preliminaries to JIT Production

In order to implement JIT production effectively, the following conditions should first be imposed:

(1) Standardise the individual operations.
(2) Provide U-shape layouts such that each operator can handle more than one machine (see Figure 16.3).

— EXAMPLE 16.2 —

(Effectiveness of U-shape layout) A straight-line flow shop with seven machines, each of which was run by an operator, produced a product with a cycle time of 9 s. This shop was restructured; the established U-shape layout contained seven machines as before, but with

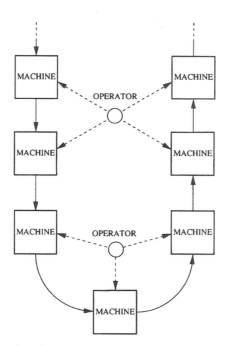

Figure 16.3 U-shape machine layout—each multi-functional operator handles three or four machines.

only three operators achieving the same cycle time. Each operator has been forced to handle three machines with on average 3 s at each. Of course such a system enables management to reduce the labour cost. □

---

(3) Multiple-job work by each operator. Job allotment to each individual operator changes frequently (e.g. every week); every operator is required to master multiple jobs. Again there are implications for job enrichment and satisfaction.
(4) Reducing setup times. A typical objective requires that setup is completed within ten minutes (called '*single*' *setup*). This may reduce batch sizes and work-in-progress inventory.
(5) Production smoothing. By this procedure stabilisation of a process for multiple-product, small-batch production is achieved. In this mode the cycle time—the required time per unit of production—is decided by dividing the available time by the production quantity, as formulated in equation (8.20a).

--- EXAMPLE 16.3 ---

(Production smoothing for multiple-product production) Suppose that a car-maker produces 10 000 of model A, 4000 of model B, and 6000 of model C in a month. There are 20 work days, the work conditions being 2 shifts and 8 hours a day. In this case, the three models are to be produced at the rate of 500 for A, 200 for B, and 300 for C a day. Then the average production time is calculated as follows:

$$\text{cycle time} = \frac{\text{available hours a day}}{\text{production volume a day}} = \frac{8 \times 2 \times 60 \times 60}{500 + 200 + 300} = 57.6 \text{ (s)}.$$

The production sequence for the three models is set: ABACABACAC. This is production smoothing (mixed production), and the production cycle is repeated 2000 times in a month. □

---

(6) '*Jidoka*' (self-actuation). When unusual events happen in a production line, the worker in charge stops the line and removes the cause of trouble.

JIT Production Process

As mentioned previously and as shown in Figure 16.4, the master production plan which has been decided through aggregate production planning is given only to the final stage of the whole production process. In this system, workers perform the required work on the material provided by the preceding workstation at the necessary time; it is hence called the *pull*[-*through*] *system*. For the purpose of providing production information, two kinds of *kanban* (instruction card) are used. These include 'withdrawal' and 'production ordering' information.

The withdrawal *kanban* indicates the items (parts) and their quantities to be brought from the previous stage to the current workplace. Once this *kanban* is issued, the items and their quantities indicated on the *kanban* are withdrawn and brought to the current workplace together with the *kanban*.

Then the production *kanban* is taken from the box which has stored the item parts withdrawn for the next stage. According to the instruction indicated by this production *kanban* production starts and continues until the quantities of the with-

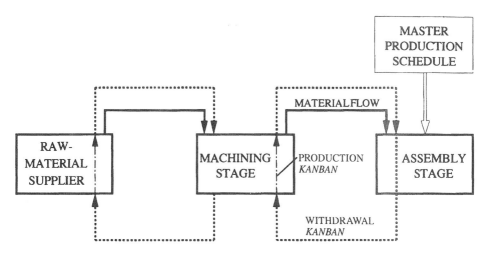

Figure 16.4 The pull-through type JIT production system receives the master production schedule at the final stage. Necessary parts are brought from the previous stage with the circulation of *kanbans*, which plays a role of information flow.

drawn parts have been completely replenished. This production aspect is represented in Figure 16.4.

Evaluation of JIT Production

As mentioned in Section 4.3, JIT production is a method of coping with inefficient multi-product, small-batch production. It is now receiving worldwide recognition as a Japanese-style manufacturing system with quality circle (QC) activities or TQC/TQM, which will be discussed in the next chapter. Effective application of JIT was largely through its development by T. Ohno at the Toyota Automobile Company, hence it is often referred to as the Toyota production system. It is acclaimed by many as the most significant development in production systems since Henry Ford introduced mass production flow-line systems for a single product—the Model T car.

However, it should be noted that the JIT principle was mentioned by L.P. Alford in 1928, and that the pull-through system procedure has been employed in supermarkets for many years. Instruction cards like *kanbans* had already been proposed by F.W. Taylor. J.M. Juran, who contributed to introducing quality control to Japan, developed a method, now called the *kanban* system, during World War II, and brought the technique to Japan after the war. It is rather difficult to identify the original source of the JIT and *kanban* ideas.

Issues of JIT Production

It is commonly believed that JIT production is very efficient; however, several issues or demerits are pointed out as follows (Hitomi, 1987):

(1) JIT production is effective only when the daily demands, hence daily production, are fairly stable. A small deviation can be met with the cooperation of the operators with multi-functional skills, but highly varied demands are difficult to cope with.

(2) The cycle time for assembling a piece of automobile is set by:

$$\text{cycle time} = \text{daily operating hours}/\text{daily production volume} \quad (16.1)$$

where

$$\text{daily production volume} = \frac{\text{monthly production volume required}}{\text{monthly work days}} \quad (16.2)$$

(see equation (8.20)(a)). Hence JIT production is effective if there is no variation of daily production and if production volume and the production capacity are well balanced.

(3) No consideration as to cycle times for varied models is given for mixed-model production. 'One-of-a-kind' production made in the JIT production mode as illustrated in Example 16.3 does not assure minimum cost, as pointed out in Note 2 in Chapter 13.

(4) No optimality is assured by JIT production in that none of the manufacturing criteria mentioned in Chapter 11 are used; only the concept of minimum or even zero, inventory is employed.

(5) The minimum or zero inventory can not be theoretically achieved; many medium/small companies supplying to large motor companies are required to hold stocks of parts in order to assure JIT delivery to them.

(6) No consideration is given to optimum dynamic scheduling; only the 'first-in, first-out' principle is applied by manufacturing items in the order of releasing *kanbans*.

(7) *Kanbans* and their circulation have roles to play as information medium and information flow. The use of factory computers/display terminals on the shop floor can be more beneficial for these purposes.

(8) Much needs to be done in connection with employer–employee cooperation, daily workstation rotation, training of operators for different kinds of jobs, and the system's adaptability to market fluctuation. The severe labour environment also lends itself to a variety of serious issues. Without the service of subcontract factories, this production system may not work at all.

(9) A particular, and perhaps not insignificant social issue arises from the frequent transportation by trucks of parts/products required to assure JIT delivery. It causes air pollution, noise, traffic problems, which are against the ideas of manufacturing excellence that will be discussed in the last chapter of this book.

16.4 Productive Maintenance

16.4.1 Failure and Productive Maintenance

Types of Failure

With age and use, physical productive facilities are susceptible to reduction or complete stoppage of their capability. This phenomenon is *deterioration* or *failure*. Failures can be classified under two headings, according to their nature.

(1) *Wear-out failure* arises as a result of a gradual change in the condition of the facility as a consequence of aging or wearing, and is often accompanied by loss of quality of the part being produced.

(2) *Catastrophic* (or *chance*) *failure*, a sudden cessation of functioning of the equipment due to abrupt change in the wear-out failure may occur. This is inevitably accompanied by an inability to continue manufacturing with the facility.

The process of wear-out failure of a cutting tool was illustrated in Figure 11.2. Such wear-out failure results in decrease of productive efficiency, deterioration of product accuracy and quality, rise in operating costs, and a loss of safety in the manufacturing environment, etc.

Effective action must be taken for both classes of failure:

- *maintenance* to reduce the rate of wear-out and protect equipment against complete breakdown;
- *repair* of defectives when equipment breaks down;
- *renewal* or *replacement* of parts with new ones prior to critical wear-out, to prevent catastrophic failure.

Productive Maintenance and Total Productive Maintenance (TPM)

Wear-out and eventual failure are unavoidable. However, to reduce the rate of their occurrence and to prolong the life of the equipment, i.e. the capacity for extended productive use of the equipment under the necessary technological functioning and servicing, an activity called *productive maintenance* can be performed. This predicts, discovers, and eliminates failures through periodical inspection, repair, and replacement.

For automated manufacturing systems, in which the role of human operative work is reduced and failure or breakdown of any portion of the system causes a major loss in productive efficiency, the function of productive maintenance takes on particular importance in the production control system.

Productive maintenance activity from a company-wide viewpoint is often called *total productive maintenance (TPM)*.

Types of Productive Maintenance

Productive maintenance is usually classified under the following three headings:

(1) *Breakdown* (or *emergency*) *maintenance*—often called *repair*. This is a maintenance operation performed after the facility has broken down or after it has deteriorated to a degree which renders proper productive operation impossible. Effective breakdown maintenance is aimed at reducing the repair time.
(2) *Preventive maintenance*—often abbreviated *PM*. This is a maintenance operation performed to prevent and eliminate the complete breakdown of the facility by taking appropriate actions according to predetermined maintenance schedules. Effective preventive maintenance reduces the number and often the severity of failures.
(3) *Corrective maintenance*. This is a maintenance operation performed to raise the capability of the facility by investigating causes of failures and improving it so as to protect against such failures.

Economics of Preventive Maintenance

Breakdown of a facility will be lessened as the amount of preventive maintenance increases; the associated maintenance cost increases and failure cost decreases.

Consequently, as shown in Figure 16.5, we have the optimal level of preventive maintenance from the minimum-cost viewpoint.

It is also self-evident that the service of preventive maintenance is useful to facilities which are associated with major losses in the event of their breakdown, while it is rather uneconomic for those which are associated with very little loss even at their failure. In the latter case breakdown maintenance is enough. In the former case an optimal service of preventive maintenance should be decided, together with optimal maintenance intervals and the optimal number of maintenance crews, for the production facilities under consideration.

16.4.2 Tools for Preventive Maintenance

Reliability Measures

Productive maintenance is closely related to reliability, since it is said that the concept of reliability is expressed by the set of three criteria:

- failure-free operation,
- life, and
- maintainability.

Facilities or products are desired to perform their prescribed function without incurring any deterioration or failure under given conditions for a specified time period. This is *reliability* of those facilities or products, and is expressed by various indices as follows:

- *Reliability* is the probability that a facility will perform satisfactorily without failure in its specified function for a specific time period under given working conditions.

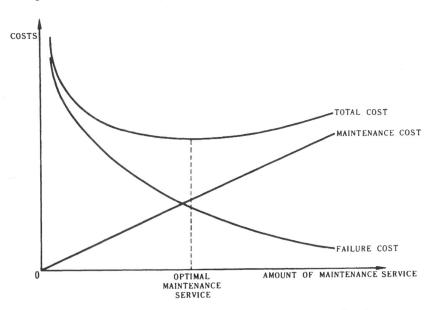

Figure 16.5 Economics of preventive maintenance. As the amount of maintenance service increases, the maintenance cost increases, while the failure cost that is generated due to stopped production decreases; hence, the total cost has a minimal point, which is the optimal amount of maintenance service.

- *Maintainability* is the probability that a facility will be repaired in a specific time period after its breakdown.
- *Availability* is the probability that a facility will be in service during a scheduled working period. In a steady state it is $T_u/(T_u + T_d)$, where T_u and T_d are up and down times, respectively.
- *Failure* (or *hazard*) *rate* is the probability that a facility which has survived up to a certain time will fail at the next time instant.
- *Mean life time* is the average time before a facility fails, such as *MTBF* (mean time between failures) for repairable (or renewable) facilities and *MTTF* (mean time to failure) for non-repairable facilities.
- *Useful life* or *longevity* is the operating time for which failure rate is less than a specified acceptable amount.
- *Up time* is the time during which a facility is capable of its function.
- *Down time* is the time during which a facility is in malfunction.

Reliability Function

'Failure rate' plays an important role in reliability analysis, and it is expressed by the probability density function of failure $f(t)$ and the reliability function $R(t)$, as follows (see Figure 16.6):

$$\lambda(t) = \lim_{\Delta t \to 0} \frac{1}{\Delta t} P\{t < \tau \leq t + \Delta t | t < \tau\} = \frac{f(t)}{R(t)}. \tag{16.3}$$

The *reliability function* is the probability that a facility will perform a failure-free operation during time t, and it is related to failure density function $f(t)$ by:

$$R(t) = \int_t^\infty f(x)\,dx. \tag{16.4}$$

This is a non-increasing function with time, and $R(0) = 1$ and $R(\infty) = 0$.
Then, 'mean life time' is given by:

$$\bar{\tau} = \int_0^\infty tf(t)\,dt = \int_0^\infty R(t)\,dt. \tag{16.5}$$

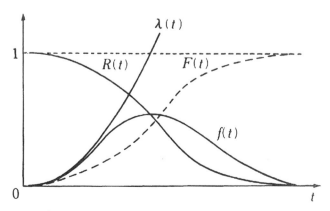

Figure 16.6 Various functions used in reliability analysis: the failure density function f(t); cumulative distribution function of failure or unreliability function F(t); reliability function R(t); and failure-rate function λ(t).

Phases of Failure

Failure types are distinguished by patterns of change of failure rate with time and are basically classified into the following three. These three phases of failure constitute the so-called 'bathtub' curve, as illustrated in Figure 16.7.

(1) *Initial (early* or *infant) failure.* This occurs during the initial phase of operation of a facility. The occurrence of failure is highest at the beginning, and failure rate decreases with increase in time—the *decreasing failure rate* phase. Initial failure is mainly based upon inherent assignable causes such as defects in raw materials, weak components, immature design and careless manufacture. It is important to find this type of failure and stabilise the operation of the facility. This process of reducing the failure rate is called *debugging* or *ageing.*

(2) *Chance (or random) failure.* There is *constant failure rate* in this phase, hence during this phase the age of an operating facility has no significance. Failure occurs at random due to such causes as sudden loading on the facility over its tolerable strength, unpredictable stress concentration, etc., which have not been eliminated. The duration of this phase is referred to as *useful life* or *longevity.*

(3) *Wear-out failure.* This is caused by mechanical wear and fatigue, chemical corrosion, change of property of materials associated with lapse of time, and so forth, as described previously. In this case, failure rate increases with time—the *increasing failure rate* phase.

Ways of Preventing the Failure Types

The above three failure types are closely connected with productive maintenance.

In the period of initial failure it is important to discover causes of failure by increasing the number of equipment inspections and to feed the information back to engineering departments to initiate equipment design modification or corrective maintenance. Sound quality-control procedures will help minimise initial failure.

In the period of chance failure daily maintenance such as cleaning, oiling, and readjusting will be carried out together with inspection, and efforts will be made to prolong the useful life by reducing a facility's failure rate.

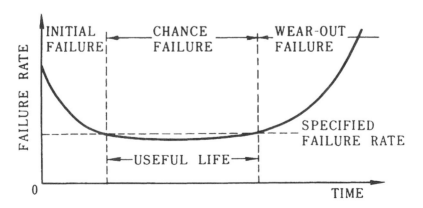

Figure 16.7 A typical failure pattern of the bathtub curve. With the lapse of time the failure rate decreases in the period of initial or early failure, it is constant in the period of chance or random failure, and it increases in the period of wear-out failure.

In the period of wear-out failure deterioration and wear-out of a facility may be postponed by improved maintenance, which reduces the failure rate. Proper cost-effective preventive or corrective maintenance will be justified from an economic standpoint.

Representative Reliability Functions

Two representative reliability functions are explained in the following examples.

EXAMPLE 16.4

(*Exponential distribution*) A facility has the failure density function of the exponential-distribution type, which is given as:

$$f_e(t) = \frac{1}{\theta} e^{-t/\theta}, \quad t \geq 0 \tag{16.6}$$

where θ is a constant.

In this case, the reliability function is:

$$R_e(t) = \int_t^\infty \frac{1}{\theta} e^{-x/\theta} dx = e^{-t/\theta}. \tag{16.7}$$

The failure-rate function is constant:

$$\lambda_e(t) = 1/\theta. \tag{16.8}$$

This phenomenon is very common for chance failures that occur at random as described before.

The mean lifetime (MTBF) is:

$$\bar{\tau}_e = \int_0^\infty e^{-t/\theta} dt = \theta. \tag{16.9}$$

Accordingly the failure rate is the reciprocal of MTBF for the exponential-distribution type. □

EXAMPLE 16.5

(*Weibull distribution*) A facility has the failure density function of the Weibull distribution type:

$$f_W(t) = \frac{\alpha}{\beta} t^{\alpha-1} e^{-t^\alpha/\beta}, \quad t \geq 0, \tag{16.10}$$

where α and β are constants, α being the shaping parameter and β the scale parameter. This is exponential in the case that $\alpha = 1$.

Then the reliability function is:

$$R_W(t) = \int_t^\infty \frac{\alpha}{\beta} x^{\alpha-1} e^{-x^\alpha/\beta} dx = e^{-t^\alpha/\beta}. \tag{16.11}$$

The failure-rate function is:

$$\lambda_W(t) = (\alpha/\beta) t^{\alpha-1}. \tag{16.12}$$

Hence with time the failure rate (see Figure 16.8)

- decreases if $\alpha < 1$ (initial failure)
- holds constant if $\alpha = 1$ (chance failure)
- increases if $\alpha > 1$ (wear-out failure).

Accordingly the Weibull distribution contains all three phases of failure.

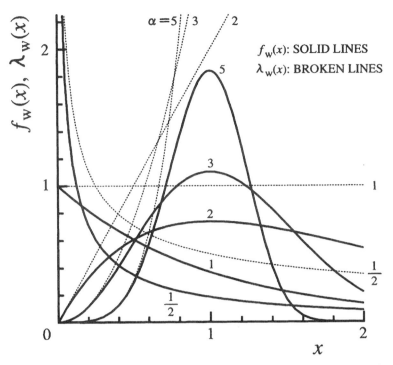

Figure 16.8 Weibull distribution varies with the shaping parameter α. It depicts the exponential distribution in the case that $\alpha = 1$.

By setting $t^\alpha/\beta = x$, the mean life time is:

$$\bar{t}_w = \int_0^\infty e^{-t^\alpha/\beta}\, dt = \int_0^\infty \frac{\beta^{1/\alpha}}{\alpha} x^{\frac{1}{\alpha}-1} e^{-x}\, dx = \frac{\beta^{1/\alpha}}{\alpha} \Gamma\left(\frac{1}{\alpha}\right) = \beta^{1/\alpha}\Gamma\left(\frac{1}{\alpha}+1\right), \quad (16.13)$$

where

$$\Gamma(z) = \int_0^\infty e^{-x} x^{z-1}\, dx \quad \text{(gamma function)}. \quad (16.14)$$

□

16.5 Replacement

16.5.1 Optimum Machine Replacement

Replacement Decision

As the production facility deteriorates during its use, the maintaining and operating expenses rise, resulting in decreased revenue. Accordingly, considering this fact and the investment needed for replacing the existing facility with a new one, the total profit obtained in the long term can be maximised by deciding on replacement at an appropriate time.

Optimum Replacement by Dynamic Programming

Assume that the revenue earned by a production facility, and its maintaining and operating expenses, depend upon the number of years in use, and denote their difference by $u(t)$. The replacement of this production facility with a new one costs c.

In an attempt to decide whether we should keep the existing facility after t years of use or replace it with a new one, both revenues:

- $u(t)$ in case of keeping the existing facility, and
- $u(0) - c$ if replaced for the coming year,

are compared. We decide so as to obtain the profit (Bellman and Dreyfus, 1962):

$$f_1(t) = \max \begin{cases} u(t)\text{: keep the existing facility;} \\ u(0) - c\text{: replace with a new facility.} \end{cases} \quad (16.15)$$

For H terms, these terms are divided into two: the initial one term and the remaining $(H-1)$ terms. For these two stages the profits obtained are:

- $u(t)$ and $f_{H-1}(t+1)$ in the case of keeping the existing facility, and
- $u(0) - c$ and $f_{H-1}(1)$ in the case of replacement.

Hence the replacement decision for obtaining the maximum profit is made by the following recursive functional equation, based upon dynamic programming:

$$f_H(t) = \max \begin{cases} u(t) + f_{H-1}(t+1)\text{: keep the existing facility;} \\ u(0) - c + f_{H-1}(1)\text{: replace with a new facility.} \end{cases} \quad (16.16)$$

The optimum periods for replacement of the facility are determined by deciding $f_1(t), f_2(t), \ldots$ sequentially by the above equation.

— **EXAMPLE 16.6** —

(Determining the optimum periods of replacement of the facility) A facility needs $10 million for replacement. The difference between the revenue earned and the maintaining and operating expenses of this facility in each year t is listed in Table 16.1 (a).

Table 16.1 Computing the optimum periods of facility replacement by the dynamic-programming approach.

	t	0	1	2	3	4	5	6	7
(a)	$u(t)$	1000	900	800	700	600	500	400	300
	$u(0) - c$	0	0	0	0	0	0	0	0
	$f_1(t)$	1000	900	800	700	600	500	400	300
	$f_2(t)$	1900	1700	1500	1300	1100	900(†)	900*	
(b)	$f_3(t)$	2700	2400	2100	1800	1700*			
	$f_4(t)$	3400	3000	2600	2400*	3000*			
	$f_5(t)$	4000	3500	3000†					

*Replace the existing facility with a new one.
† Either keep the existing facility or replace with a new facility.

For this case the optimum period of replacement of this facility is decided through Table 16.1(b). That is, no replacement is necessary for one year; considering two years, replacement is preferable after 5 years; etc.   □

16.5.2 Replacement Problems

Renewal Process

The *renewal process* is the process of renewing the function of a product/part by repair or replacement when it has failed.

Denoting the life (operation time) of the ith-replaced item by τ_i and the renewal time by τ'_i, the time periods at the failure and at the renewal are (see Figure 16.9):

$$t_i = \tau_1 + \tau'_1 + \tau_2 + \tau'_2 + \cdots + \tau_{i-1} + \tau'_{i-1} + \tau_i \qquad (i = 1, 2, \ldots) \qquad (16.17)$$

$$t'_i = \tau_1 + \tau'_1 + \tau_2 + \tau'_2 + \cdots + \tau_{i-1} + \tau'_{i-1} + \tau_i + \tau'_i \qquad (i = 1, 2, \ldots) \qquad (16.18)$$

It is usually assumed that τ_i and τ'_i ($i = 1, 2, \ldots$) are all independent and of a stochastic nature, denoting their density functions by $f(t)$ and $g(t)$, distribution functions by $F(t)$ and $G(t)$, the mean times by $\bar{\tau}$ and $\bar{\tau}'$, and the variances by σ^2 and σ'^2, respectively (Gnedenko et al., 1969).

Renewal Function

If the renewal time is small enough to be negligible, the average of $n(t)$—the number of failures that have occurred until time t—is called the *renewal function*. In order to express it, first, the probability that $n(t)$ is just k is:

$$P_k(t) \equiv P_k\{n(t) = k\} = F_k(t) - F_{k+1}(t), \qquad (16.19)$$

where $F_k(t)$ is the probability that $n(t) \geq k$:

$$P\{n(t) \geq k\} = P\left\{\sum_{i=1}^{k} \tau_i < t\right\} = F_k(t) = \int_0^t F_{k-1}(t-x) f(x)\, dx. \qquad (16.20)$$

Thus the renewal function is given by the following integral function:

$$H(t) = \sum_{k=1}^{\infty} k P_k(t) = \sum_{k=1}^{\infty} F_k(t) = F(t) + \int_0^t H(t-x)\, dF(x). \qquad (16.21)$$

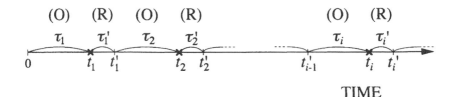

Figure 16.9 Cyclic operation (O), failure, and renewal (R) of a facility.

Average Number of Failures in a Unit Time

The average number of failures which occur in a unit time beginning at a certain time point is called the *intensity of failures* or *renewal density*. This is expressed as:

$$h(t) = \frac{dH(t)}{dt} = \sum_{k=1}^{\infty} f_k(t) \tag{16.22}$$

where

$$f_k(t) = \frac{dF_k(t)}{dt}.$$

--- **EXAMPLE 16.7** ---

(Determining the number of spare parts) Determining the minimum number of spare parts necessary for a machine to function during the time length [0, t] with the confidence level—$1 - \theta$. Denoting this figure by m,

$$P\{n(t) \leq m\} \geq 1 - \theta \tag{16.23}$$

hence,

$$P\{n(t) \geq m + 1\} < \theta \tag{16.24}$$

namely, the maximum number m which satisfies the following equation is what we should have for spare parts.

$$F_{m+1}(t) < \theta. \tag{16.25}$$

□

--- **EXAMPLE 16.8** ---

(Exponential and Poisson distributions) In order to obtain the discrete probability which gives the number of failures in [0, t], the time t is divided by a large number N, such that failure can occur once at the most in each infinitesimal range—t/N. The probability of failure occurrence is assumed to be $\lambda(t/N)$, where λ is a constant.

Then the discrete probability distribution to be obtained is:

$$P_k(t) = \lim_{N \to \infty} \frac{N!}{k!(N-k)!} \left(\frac{\lambda t}{N}\right)^k \left(1 - \frac{\lambda t}{N}\right)^{N-k}$$

$$= \lim_{N \to \infty} \frac{1}{k!}\left(1 - \frac{1}{N}\right)\left(1 - \frac{2}{N}\right)\cdots\left(1 - \frac{k-1}{N}\right)(\lambda t)^k \left(1 - \frac{\lambda t}{N}\right)^N \left(1 - \frac{\lambda t}{N}\right)^{-k} = \frac{(\lambda t)^k}{k!} e^{-\lambda t}, \tag{16.26}$$

where the following formula has been used:

$$\lim_{N \to \infty} \left(1 - \frac{\lambda t}{N}\right)^N = e^{-\lambda t}. \tag{16.27}$$

(16.26) shows the *Poisson distribution*.

For this distribution the probability that the time interval of two consecutive failures, i.e. the lifetime is t, is given by the product of the probability that no failure occurs until t:

$$P_0(t) = e^{-\lambda t} \tag{16.28}$$

and the probability that just one failure occurs during t and $t + \Delta t$: $\lambda \Delta t$.

Denoting that failure density function by $f(t)$, we get:

$$f(t)\,\Delta t = e^{-\lambda t} \lambda\, \Delta t,$$

that is,

$$f(t) = \lambda e^{-\lambda t}. \tag{16.29}$$

The above example leads to the following proposition. □

--- **PROPOSITION 16.1** ---

If the number of failures during a certain time length follows the Poisson distribution, then the MTBF follows the exponential distribution. □

--- **EXAMPLE 16.9** ---

(Renewal [density] function for the Poisson distribution) The renewal function for the Poisson distribution is obtained from (16.21):

$$H(t) = \sum_{k=1}^{\infty} k \frac{(\lambda t)^k}{k!} e^{-\lambda t} = \lambda t. \tag{16.30}$$

The renewal density function for the Poisson distribution is then:

$$h(t) = \lambda. \tag{16.31}$$

□

Calculating the Number of Spare Parts

The number of failures in a long time interval $[0, t]$ occurs within a range:

$$\frac{t}{\bar{\tau}} - \theta_{\alpha/2} \frac{\sigma\sqrt{t}}{\bar{\tau}^{3/2}} < n(t) < \frac{t}{\bar{\tau}} + \theta_{\alpha/2} \frac{\sigma\sqrt{t}}{\bar{\tau}^{3/2}} \tag{16.32}$$

with the confidence level $1 - \alpha$, where $(1/\sqrt{2\pi}) \int_{-\theta_{\alpha/2}}^{\theta_{\alpha/2}} e^{-x^2/2}\, dx = 1 - \alpha$.

--- **EXAMPLE 16.10** ---

(Calculating the number of spares) A facility has an average life of 100 h with a standard deviation of 10 h. Calculate the number of spares in order to operate this facility for 10 000 h with the confidence level of 0.99.

In this case, we merely estimate the upper limit; hence, $\theta_{0.005} = 2.56$ by the normal distribution table. From (16.32)

$$n(10\,000) < \frac{10\,000}{100} + 2.56 \frac{10\sqrt{10\,000}}{100^{3/2}} \approx 126.$$

□

Availability

If the renewal time can not be negligible, compared with the operating time, the probability that a certain facility performs its function at a certain instant is an important measure. It is called *availability*.

In the steady state (after a lapse of an infinite time) availability is given as:

$$A = \frac{\bar{\tau}}{\bar{\tau} + \bar{\tau}'}. \tag{16.33}$$

EXAMPLE 16.11

(Availability for a unit with the exponential distribution for operation and renewal times) If both operating and renewal times follow the exponential distribution:

$$F(t) = 1 - e^{-\lambda t}, \quad G(t) = 1 - e^{-\mu t}; \quad \lambda, \mu \text{ constants,}$$

then the availability in the steady state $(t \to \infty)$ is:

$$A = \mu/(\mu + \lambda). \tag{16.34}$$

Constants λ and μ are reciprocals of the mean life and the mean renewal time. The above equation coincides with (16.33). □

16.5.3 Optimum Preventive Maintenance

The Principle of Preventive Maintenance

In order to increase the availability given by (16.33), the principle of preventive maintenance is applied. As shown in Figure 16.10, if the renewal time for preventive maintenance, τ'', is smaller than the renewal time after failure, τ', then it is wise to replace this facility even though it is in operation at p (Morimura, 1966).

Formulating the Availability

The operating time of a facility includes:

- lifetime τ_i with renewal time τ'_i, and
- replacement time p with renewal time τ''_i $(<\tau'_i)$,

as shown in Figure 16.10. Distribution of the operating time is now expressed as:

$$F(t) = \begin{cases} F(t), & 0 \leq t < p \\ 1 - F(p) & t = p \\ 0 & t > p. \end{cases} \tag{16.35}$$

The mean operating time is calculated as:

$$\bar{\tau} = \int_0^p t \, dF(t) + p(1 - F(p)) = \int_0^p (1 - F(t)) \, dt \tag{16.36}$$

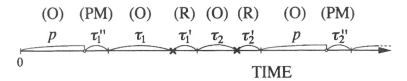

Figure 16.10 Process of operation (O), failure, and preventive maintenance (PM)/renewal (R) of a facility.

and the mean renewal time is: $F(p)\bar{\tau}' + (1 - F(p))\bar{\tau}''$, where $\bar{\tau}'$ and $\bar{\tau}''$ are the mean times for replacement after failure and for replacement by preventive maintenance, holding $\bar{\tau}' > \bar{\tau}''$.

Accordingly, from (16.33) the availability is formulated as:

$$A = \frac{\int_0^p (1 - F(t))\, dt}{\int_0^p (1 - F(t))\, dt + F(p)\bar{\tau}' + (1 - F(p))\bar{\tau}''}. \tag{16.37}$$

Maximising the Availability

The optimum replacement time p^* can be obtained by maximising the above equation. We try to minimise the reciprocal of this equation and obtain the following formula:

$$\lambda(p^*) \int_0^{p^*} (1 - F(t))\, dt - F(p^*) = \frac{\bar{\tau}''}{\bar{\tau}' - \bar{\tau}''}, \tag{16.38}$$

where $\lambda(t) = f(t)/(1 - F(t))$ is the failure rate.

--- **EXAMPLE 16.12** ---

(Optimum maintenance policy for a facility with the Weibull distribution) If the life of a facility follows the Weibull distribution, from (16.10),

$$F(t) = \int_0^t \frac{\alpha}{\beta} t^{\alpha-1} e^{-t^\alpha/\beta}\, dt = 1 - e^{-t^\alpha/\beta}. \tag{16.39}$$

As the failure-rate function is given by (16.12), (16.38) is:

$$2\alpha p^* \int_0^{p^*} e^{-\alpha t^2}\, dt - 1 + e^{-\alpha p^{*2}} = \frac{\bar{\tau}''}{\bar{\tau}' - \bar{\tau}''} \tag{16.40}$$

for $\alpha = 2$, setting $1/\beta = \alpha$. □

--- **EXAMPLE 16.13** ---

(Optimum maintenance policy for a facility with the exponential distribution) If the life of a facility follows the exponential distribution, from (16.6), setting $1/\theta = \lambda$,

$$F(t) = 1 - e^{-\lambda t}. \tag{16.41}$$

As the failure rate is a constant, λ, the left-hand side of (16.38) is:

$$\lambda \int_0^{p^*} e^{-\lambda t}\, dt - 1 + e^{-\lambda p^*} = [-e^{-\lambda t}]_0^{p^*} - 1 + e^{-\lambda p^*} = 0.$$

Hence, (16.38) has no root. This means that $p^* = \infty$. In other words, we need no preventive maintenance; that is, breakdown maintenance is enough. □

References

BELLMAN, R. and DREYFUS, S. (1962) *Applied Dynamic Programming* (Princeton, NJ: Princeton University Press).

GNEDENKO, B.V., BELYAYEV, Yu.K. and SOLOVYEV, A.D. (1969) *Mathematical Methods of Reliability Theory* (English translation from the Russian, edited by R.E. Barlow) (New York: Academic Press), Sec. 2.3.

HITOMI, K. (1975) *Theory of Planning for Production* (in Japanese) (Tokyo: Yuhikaku), pp. 50–51.

HITOMI, K. (1987) *Multi-product, Small-batch Production—A text* (in Japanese) (Tokyo: Business & Technology Newspaper Company), Chap. 10.

MORIMURA, H. (1966) *Probability and its Application* (in Japanese) (Tokyo: Diamond-sha), Chap. 4.

OHNO, T. (1978) *Toyota Production Systems* (in Japanese) (Tokyo: Diamond-sha).

Supplementary reading

NIEBEL, B.W. (1985) *Engineering Maintenance Management* (New York: Marcel Dekker).

CHAPTER SEVENTEEN

Quality Engineering

17.1 Quality Control (QC)

Background of Quality Control

Quality control (*QC*) is a management tool for producing goods with satisfactory quality characteristics by systematically establishing acceptable limits of variation in size, weight, finish, function, and so forth, for products to be produced and by maintaining the produced goods within these control limits. 'Statistical quality control' has been a main subject in this field since the pioneering work of Shewhart (1931). It applies statistical theory to solving problems of quality of the finished product.

TQC/TQM

Total quality control/management (*TQC/TQM*) has been developed (Feigenbaum, 1986) and used as company-wide management's major task for making appropriate decisions required for optimal operations in a system concerned with quality, particularly in Japan. These include:

- statistical decisions on product quality;
- feedback of the quality information to the product planning system, the process planning system, and appropriate technological action on product and process redesign;
- negotiation with suppliers of raw materials and components;
- feedback from users of products covering product failures and other troubles incurred during the use of the products;
- company-wide involvement from top management to the workers, organising *QC* (or *quality*) *circles* which are self-managed small groups, and so forth.

Control Charts

A powerful device often used in the workshop as the evaluation criterion of quality assurance is the *control chart*, in which some observed or calculated property of a

QUALITY ENGINEERING

product or process is plotted on a diagram with a *central line (CL)* and *upper-* and *lower-control limits (UCL* and *LCL)* in the order that it is produced as shown in Figure 17.1.

When the plotted points fall within the limits, the quality is stable and 'under control'. On the other hand, if a number of the plotted points fall outside the limits, or violate some other decision criteria, it means that a process is not in control;[1] in this case, after clarifying assignable causes of variation, corrective action is possibly needed.

How to Draw Control Charts

Among the various types of control charts $\bar{X}$- and R-charts are most commonly used to control the average and the general variability or range of the property of the product or process.

In a sample $(X_1, X_2, \ldots, X_k)$ the following are obtained (see Table 17.1):

- the average $\bar{X} = \sum_{i=1}^{k} X_i/k;$ (17.1)

- the range $R = X_{max} - X_{min}.$ (17.2)

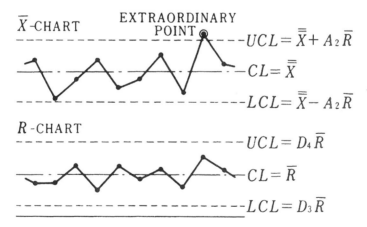

Figure 17.1 $\bar{X}$- and R-charts for controlling the average and the range of the property of the product or process. Where the plotted points for calculated property fall mostly within the upper- and lower-control limits, the process is 'under control'.

Table 17.1 Data sheet for plotting on $\bar{X}$- and R-charts.

Number	Data			Average	Range
1	X_{11}	X_{12} ...	X_{1k}	$\bar{X}_1$	R_1
2	X_{21}	X_{22} ...	X_{2k}	$\bar{X}_2$	R_2
⋮	⋮		⋮	⋮	⋮
h	X_{h1}	X_{h2} ...	X_{hk}	$\bar{X}_h$	R_h
Average				$\bar{\bar{X}}$	$\bar{R}$

These are plotted in the order of production.

The central line and the upper- and the lower- or 3σ-control limits are drawn using the following formulae:

- for the $\bar{X}$-chart:
$$CL = \bar{\bar{X}}, \ UCL = \bar{\bar{X}} + A\bar{R}, \ LCL = \bar{\bar{X}} - A\bar{R}; \quad (17.3)$$

- for the R-chart:[1]
$$CL = \bar{R}, \ UCL = D_1\bar{R}, \ LCL = D_2\bar{R}, \quad (17.4)$$

where $\bar{\bar{X}}$ = average of $\bar{X}$, $\bar{R}$ = average of R, as shown in Table 17.1, and A, D_1, and D_2 are constant factors depending upon the sample size, k, as shown in Table 17.2.

---- EXAMPLE 17.1 ----

Production of steel sheets of thickness 1.000 mm by a rolling mill resulted in the actual thickness for ten-days' production as represented in Figure 17.2. The CL, UCL, and LCL obtained are:

- 0.9999, 1.0030, and 0.9968 for the $\bar{X}$-chart.
- 0.0100, 0.0178, and 0.0022 for the R-chart. □

In addition to $\bar{X}$- and R-charts, which are for quantitative (variables) data, other representative control charts are p-charts for fraction defective and c-charts for number of defects per unit, which are for qualitative (attributes) data.

Inspection

To avoid shipment of defectives to customers or their transfer to a succeeding production stage, inspection of units manufactured may be required. Generally the more inspection done, the fewer the defectives that pass unnoticed. Ideally '100% inspection' is necessary to completely detect the defectives; however, this proves very costly and inspection itself is not necessarily 100% efficient. Hence, from an economic standpoint the optimal inspection level is decided as a compromise between the two costs—the cost of inspection and the subsequent cost of unnoticed defectives.

When the fraction defective is small, inspection time is long, cost of inspection is expensive, or inspection causes the destruction of the product, etc., a *sampling inspection* is beneficial.

Sequential Sampling

Sequential sampling is one of several sampling-inspection methods. As shown in Figure 17.3, samples are extracted one by one from the lot,[2] inspected and the

Table 17.2 Constant factors used in the construction of $\bar{X}$- and R-charts.

k	2	3	4	5	10
A	1.880	1.023	0.729	0.577	0.308
D_1	3.267	2.575	2.282	2.115	1.777
D_2	0	0	0	0	0.223

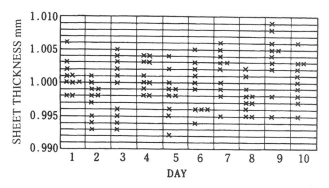

Figure 17.2 Observed values of thickness of sheet manufactured in 10 days.

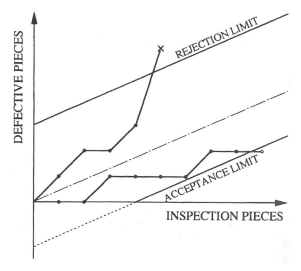

Figure 17.3 Sequential sampling procedure—acceptance or rejection is judged by the plotted point being below the acceptance limit or above the rejection limit.

number of defectives plotted on a graph. Whenever the number of defectives reaches either of the statistically determined acceptance or rejection limits, acceptance or rejection of the lot is decided.

Types of Error or Risk and OC Curve

With acceptance sampling there appear two kinds of risk:

(1) *producer's risk* or *type I error* rejects a good or acceptable lot;
(2) *consumer's risk* or *type II error* accepts a bad or unacceptable lot.

The probability of producer's risk, α, is closely related to the *acceptance quality level* (*AQL*) indicating the maximum percent defective (number of defects per hundred units) that can be considered satisfactory (usually set at about 5%). The probability of consumer's risk, β, is associated with *lot tolerance percentage defective* (*LTPD*) (often set at about 10%). Lots exceeding this level of defectives in the sample are unacceptable.

These stipulations define two points on an associated *operating characteristic (OC) curve* that indicates the percentage of lots which may be expected to be accepted under a specified sampling plan, as demonstrated in Figure 17.4.

A sampling plan is then established by indicating the number of product items from each lot to be inspected (the sample size), and the criteria for determining the acceptability of the lot (acceptance and rejection numbers).

Zero Defects Concept

Quality control in an organisation should cover not only quality inspection in actual manufacturing but also quality consciousness in the whole manufacturing system. The quality culture must include such functions as:

- quality in design;
- planning for quality;
- purchasing of raw materials with high quality;
- control of quality at every production stage in a workshop;
- inspection of products for quality control.

It is important to take appropriate and permanent corrective actions whenever non-conformance to requirements is found during consumption of the finished products. Such action might be:

- proper modification of the product design;
- feedback of the information to design, purchasing, and manufacturing functions;

and so forth.

The *zero-defect (ZD)* concept is one such system where problems are continually identified and permanently solved with the objective of eliminating the occurrence of defectives.

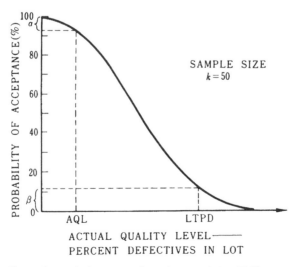

Figure 17.4 An illustration of the operating characteristic (OC) curve. Producer's and consumer's risks, α and β, are associated with the desired acceptance quality level (AQL) and lot tolerance percent defective (LTPD), respectively, on the OC curve.

Tools of Quality Control

Beside control charts, sampling inspection, OC curves, and zero defects, several other useful tools are employed to design, manufacture and assure high-quality products. Those are:

- *cause-and-effect (fishbone) diagrams* for gathering information regarding the causes of a problem guiding a creative problem-solving activity (developed in 1943 by K. Ishikawa);
- *histograms* (credited to the French statistician A.M. Guerry in 1833) used to provide clear presentation and understanding of data;
- *Pareto charts* for identifying the few problems that cause the greatest loss of profit (defined by J. Juran in 1950 after V. Pareto in the 19th century);
- *scatter charts* or *correlation diagrams* for indicating the mutual relationship between two kinds of data;
- *check sheets* for recording events, such as the occurrence of specific defects, and to assist in developing appropriate corrective actions;
- *failure mode and effects analysis (FMEA)* to analyse qualitatively, usually at the design stage, the effect of potential defects on product quality, prioritising the problems, aiding corrective action and indicating the effect of such action.

International quality assurance is at present made through ISO (International Standardisation Organisation) 9000.

17.2 Quality Function Deployment (QFD)

Scope of Quality Function Deployment

The method of *quality function deployment (QFD)* is a hierachical design methodology involving converting the market/consumers' needs into the surrogate characteristics and determining the design quality of the product. This design quality of the product is broken down into the quality characteristics for every part/process/function. In a narrow sense, this method develops the detailed steps of tasks/functions for the whole activity of assuring product quality.

QFD Procedure

The procedure for quality function deployment is depicted in Figure 17.5 (Ohfuji *et al.*, 1990), the so-called 'house of quality'.

Step 1: *Gathering raw data.* Market/consumers' needs are collected through questionnairing, interviewing, claim and complaint information, etc.

Step 2: *Establishing required qualities.* From the collected market data are established the required qualities for the commodity. These quality elements or customer attributes should be expressed in a simple unambiguous way. In some cases, several detailed elements are grouped. In this stage engineering quality characteristics and policies are not included.

Step 3: *Determining engineering characteristics.* The measure for evaluating every quality element is called the *engineering characteristic*. Extraction of these characteristics, such as physical, functional, economic etc. is required to transform market needs into technical specifications.

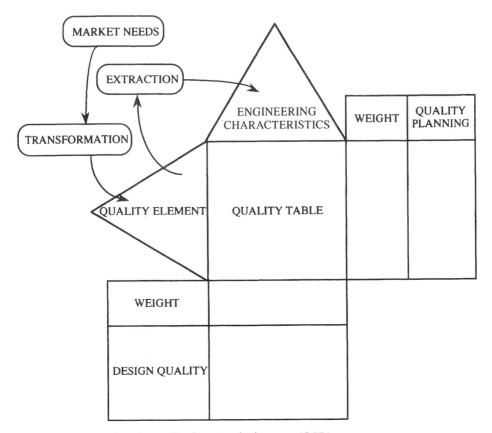

Figure 17.5 Procedure for quality function deployment (QFD).

Step 4: *Making a quality table.* By correlating required elements deployed in Step 2 and engineering characteristics deployed in Step 3, a *quality table* (or *chart*) is made indicating the positive or negative extent by which engineering characteristics influence the various quality elements.

Step 5: *Calculating the required quality weights.* A relative importance weighting is given to every required quality element in a similar manner as was discussed in Section 9.2. In some cases, the analytic hierarchy process (AHP) is employed.

Step 6: *Setting planned quality.* Planned qualities are decided with typically five priority levels to indicate the key factors or achievement levels for the success of new product development. This is benchmarked by comparison to that in the competitive companies. Comparative analysis is made between a newly developing product and its competitive products, and the potential sales advantages of the new product are stressed.

Step 7: *Correlation of engineering characteristics.* The degree of importance among engineering characteristics and their interrelationships is expressed for each characteristic, or relatively against all other characteristics.

Step 8: *Setting design qualities.* The *design quality* is the planned target, objective specification etc. for the quality of a product intended to be designed and manufactured. Design qualities for the product are established from the planned qualities and the weights given to engineering characteristics which have been set in Steps 6 and 7. Particular attention is given to those with high priorities, and then concrete design values for parts having these characteristics are reasonably determined. Failure mode and effects analysis (FMEA) and other techniques are useful to apply in this stage. □

17.3 Quality Engineering

Quality Defined

In quality engineering or the Taguchi method, *quality (loss)* is defined as the losses imposed on society from the time a product is released for shipment or as the losses incurred due to deviations of product characteristics from their target values (Taguchi *et al.*, 1989).

Procedure of Quality Engineering

The purpose of *quality engineering* or the *Taguchi method* is to produce a product that is robust with respect to all variability or 'noise' factors, and we follow three phases:

(1) *System design* involves the product phase during its life.
(2) *Parameter design* selects the values of controllable design parameters to minimise the effect of uncontrollable 'noise' factors on the functional characteristics of the product.
(3) *Tolerance design* applies if the reduction in variation of the functional characteristics achieved in the above phase is not sufficient.

Loss Function

The *loss function* is used to evaluate the effect of quality improvement.
As depicted in Figure 17.6, the quadratic loss function is expressed as:

$$L(y) = k(y-m)^2 \tag{17.5}$$

where m is the target value, y is the functional characteristic value, and k is a proportionality constant. k can be determined by:

$$k = \frac{\text{cost of a defective product}}{\text{tolerance}^2}. \tag{17.6}$$

If the loss or cost of adjustment is $\hat{L}$, the functional characteristic value is:

$$y = m + \sqrt{\hat{L}/k}. \tag{17.7}$$

── **EXAMPLE 17.2** ──

A unit produced has a tolerance of 5 and the cost of a defective product is $10. Then, $k = 10/5^2 = 0.4$.

If the deviation $(y-m)$ of this product's functional characteristic value is 2, the loss caused by deviation in the production of this unit is:

$L = 0.4 \times 2^2 = \$1.6/\text{unit}$. □

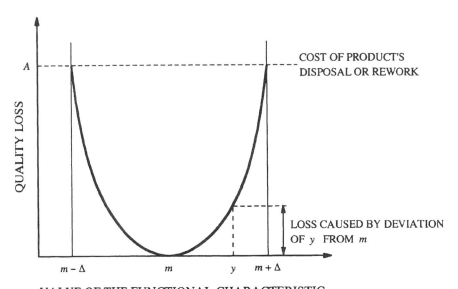

Figure 17.6 Quality loss function in terms of the value of functional characteristics—quality loss increases with deviation from the target value, m (Taguchi et al., 1989).

S/N Ratio

The key technology of quality engineering is to measure the functional robustness of the functions of products, processes, or technologies by calculating appropriate *signal-to-noise* (S/N) ratios (Taguchi, 1993). The output y is ideally expressed as:

$$y = \beta M, \tag{17.8}$$

where M is the signal and β is a constant. Practically, however, noise occurs due to controllable variables $x_1, x_2, \ldots, x_m$, and error factors $z_1, z_2, \ldots, z_n$:

$$y = f(M, x_1, x_2, \ldots, x_m, z_1, z_2, \ldots, z_n). \tag{17.9}$$

Hence,

$$y = \beta M + \{f(M, x_1, x_2, \ldots, x_m, z_1, z_2, \ldots, z_n) - \beta M\}. \tag{17.10}$$

The first term of the right-hand side represents the ideal state (*useful part*), and the second part, the *harmful* part, deviation from the ideal state, meaning the robustness of y. An improved state is specified by increasing the former and decreasing the latter.

A measure which represents this improved state is the S/N ratio—the ratio between the useful part and the harmful part—expressed as:

$$\eta = \left(\frac{\partial f}{\partial M}\right)^2 \bigg/ \sum_{i=1}^{m} \left(\frac{\partial f}{\partial x_i}\right)^2 \sigma_{x_i}^2, \tag{17.11}$$

where σ_{x_i} is the observed standard deviation for x_i.

If the harmful part is expressed by a single error, and its standard deviation be σ, then,

$$\eta = \beta^2/\sigma^2. \tag{17.12}$$

This S/N ratio is a *measure of robustness*; the higher the ratio, the less harm variations cause to the product. Hence the 'robust' conditions are explored at the design stage by increasing the effect of the S/N ratio.

Notes

1. It should be noted that if points fall outside the lower control limit for the range, this is indicative of a significant improvement in variability, the cause of which might well be exploited in a continuous improvement programme.
2. The term 'lot' here is 'inspection lot', a collection of product items from which an inspection sample is to be drawn.

References

FEIGENBAUM, A.V. (1986) *Total Quality Control* (New York: McGraw-Hill).
OHFUJI, T., ONO, M. and AKAO, Y. (1990) *Method of Quality Function Deployment* (in Japanese) (Tokyo: Nikkagiren Publishers).
SHEWHART, W.A. (1931) *Economic Control of Quality of Manufactured Product* (New York: Van Nostrand).
TAGUCHI, G. (1993) *Taguchi on Robust Technology Development* (New York: ASME Press), pp. 91–92.
TAGUCHI, G., ELSAYED, E.A. and HSIANG, T.C. (1989) *Quality Engineering in Production Systems* (New York: McGraw-Hill), p. 3 and p. 9.

Supplementary reading

BOUNDS, G., YORKS, L., ADAMS, M. and RANNEY, G. (1994) *Beyond Total Quality Management* (New York: McGraw-Hill).
FUTAMI, R. (1988) *Introduction to Quality Control Techniques* (in Japanese) (Tokyo: Nikkagiren).
KOGURE, M. (1988) *TQC in Japan* (in Japanese) (Tokyo: Njkkagiren).
ROSS, P.J. (1996) *Taguchi Techniques for Quality Engineering* (2nd edn) (New York: McGraw-Hill).
SHORES, A.R. (1990) *A TQM Approach to Achieving Manufacturing Excellence* (Milwaukee, Wis.: Quality Press).

PART FOUR

Value Systems for Manufacturing

This part describes basic principles of production economy or economical production—the importance of 'flow of value/costs' is pointed out against the 'flow of materials' and the 'flow of information', which are mainly concerned with technical production. The concept of cost is mentioned and the time-series value of money is formulated (Chapter 18).

Classifying the costs from morphological and economical viewpoints, the product cost structure is constructed, and price-setting methods are introduced. Typical cost accounting and control procedures are also mentioned (Chapter 19).

Appropriate profits are obtained through profit planning. The rate of return on investment and equity and the break-even point are formulated. The optimum plant size and production scale are discussed by the marginal principle (Chapter 20).

Effective capital investment is mentioned for manufacturing, and typical evaluation methods for investment are introduced with sample examples (Chapter 21).

CHAPTER EIGHTEEN

Value and Cost Flows in Manufacturing Systems

18.1 Value/Cost Flow in Manufacturing Systems

Technical Production

As was mentioned in Chapter 1, production/manufacturing is to convert raw materials into products. This is *technical production*, and constitutes the 'flow of materials'. In order to perform this flow efficiently, effective management—planning and control—is needed; namely, the 'flow of information'. The importance of these two flows was pointed out by Church (1913).

Economical Production

The above technical production is not enough in effective manufacture; creation of value added is inevitable. This is concerned with 'C' (cost), one of the important evaluation criteria for manufacturing as mentioned in Chapter 1. To produce products at possible low cost and sell them at acceptable prices is a role of production management. This mode is *economical production*. The importance of value added, g, created through procurement, production, and distribution/sales processes was emphasised in the capital circulation process as depicted in Figure 1.5.

In passing through a multiple-stage manufacturing system costs are accumulated by production activities at individual stages as well as by transfer/material handling between the stages and by stock/in-process inventories when needed. Hence economical production is really concerned with the 'flow of costs'. Since creation of value added is most important in manufacturing, this flow is also called the 'profit stream'. From a social standpoint value/utility is output through production, creating possession utility. In this respect this flow is also called the 'flow of value'. This term—'Wertfluß' was coined by Nicklish (1922).

Three Fundamental Flows—Material, Information, and Cost—in Manufacturing

Manufacturing consists of three basic flows: the *flow of material[s]*, the *flow of information*, and the *flow of cost[s]*, as was pointed out in Section 1.2 and in

Figure 1.1. In practical factories, reduction of costs is especially important; this activity is continually processed through industrial engineering (IE) and value engineering (VE) techniques. Cost engineering, engineering economics or cost management greatly assists this activity.

Cash Flow

A term 'cash flow' is also used to express the cash stream inside/outside a company. Cash is provided as the funds or capital for operating a manufacturing firm either by its own assets obtained through sale of stocks, or by borrowing from banks. The funds are used to install production facilities and procure raw materials, and also used as wages paid to the employees, to manage the company, and to sell the products.

After selling the products revenue is obtained to recover cash; cash is also obtained through the sales credit. Cash thus obtained is used to pay for the stock dividend to the stock holders and the interests to the banks.

The above aspect of capital circulation or the 'flow of money' is depicted in Figure 18.1.

Historical Review of Economic and Management Aspects in Manufacturing

Historically recognition of economic and management aspects in manufacture[ing] was pointed out by C. Babbage, H.R. Towne, and others in the 19th century. As

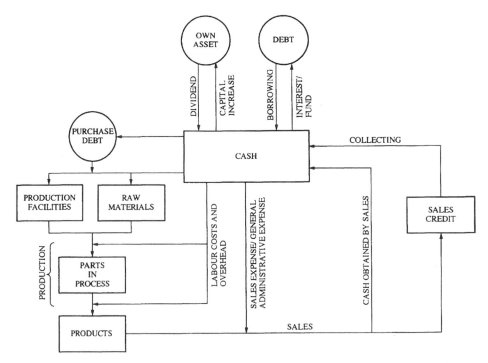

Figure 18.1 Circulation of cash for a manufacturing firm to execute production activities—funds are acquired from the stocks and banks and employed to purchase the resources of production. Revenue obtained through production and sales is used to pay dividends and interests (after Wakasugi, 1988).

early as 1832 Babbage mentioned the economy of manufacturing (Babbage, 1832), as described in Chapter 1. In 1886 Towne pointed out the importance of management recognition in his famous article, 'The engineer as an economist' (Towne, 1886). This became connected to factory management developed by F.W. Taylor and his associates early in the 20th century (Taylor, 1911). Prior to Towne, a mechanical engineer, C.H. Metcalfe discussed the necessity of product-cost and factory accounting (Metcalfe, 1885). It is important to notice that the prototype of the present accounting methods was developed by around 1925 by mechanical engineers.

18.2 Concepts of Cost and Time-series Value of Money

What is Cost/Expense?

The general concept of *cost* is: the amount of money sacrificed economically to achieve a particular objective, whereas *expense* is conversion of product cost when the product is sold—expired cost contributed to profit acquisition. The amount which has vanished without contributing to profit acquisition is *loss*. In this book, however, both cost and expense are not strictly distinguished; the words are used synonymously.

Time-series Value of Money

Cost can be materialised by monetary units. Money/capital changes with time. Time-series monetary values have been investigated in the fields of cost engineering, engineering economy, and cost accounting. The basic computing methods are as follows (see Figure 18.2).

(1) Worth after h periods:

$$S = P(1 + r)^h \qquad (18.1)$$

where P is the present worth—capital at time 0—and r is the rate of interest per period. $(1+r)^h$ is called the *compound-interest factor* or *final-worth factor*.

If the rate of interest is not constant with years, that is, $r_1, r_2, ..., r_h$, then $(1 + r)^h$ is replaced by $(1 + r_1)(1 + r_2) ... (1 + r_h)$. In this case, the average rate of interest is:

$$r = \sqrt[h]{(1 + r_1)(1 + r_2) ... (1 + r_h)} - 1. \qquad (18.2)$$

(2) Funds required at time 0 to obtain S after h periods. From (18.1)

$$P = S/(1 + r)^h. \qquad (18.3)$$

$1/(1+r)^h$ is called the *present-worth factor*.

(3) Uniform recovery M at the end of each period from the present worth P:

$$M = Pr/[1 - (1 + r)^{-h}]. \qquad (18.4)$$

$r/[1 - (1 + r)^{-h}]$ is called the *capital-recovery factor*; it converges to r after an infinite number of periods have passed (Question 18.6). The reciprocal of this factor is the *uniform series present-worth factor*.

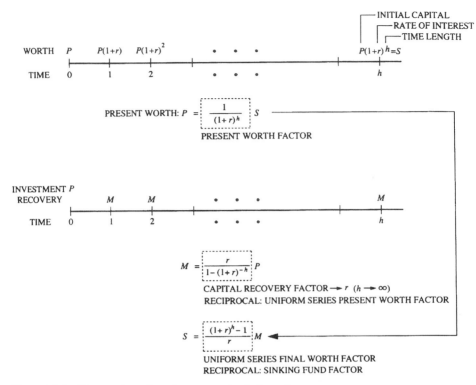

Figure 18.2 Time-series value of money—basic formulae.

(4) From (18.1) and (18.4)

$$S = M[(1 + r)^h - 1]/r. \tag{18.5}$$

$[(1 + r)^h - 1]/r$ is called the *uniform series final-worth factor*. The reciprocal of this factor is the *sinking-fund factor*.

The above factors are tabulated as numerical values (e.g. Barish and Kaplan, 1978).

References

BABBAGE, C. (1832) *On the Economy of Machinery and Manufactures* (4th edn enlarged 1835) (London: Charles Knight).
BARISH, N.N. and KAPLAN, S. (1978) *Economic Analysis for Engineering and Managerial Decision Making* (2nd edn) (New York: McGraw-Hill).
CHURCH, A.H. (1913) Practical principles of rational management, *Engineering Magazine*, **44** (6).
METCALFE, C.H. (1885) *The Cost of Manufactures and the Administration of Work Shops, Public and Private* (New York: John Wiley).
NICKLISH, H. (1922) *Wirtschaftliche Betriebslehre* (C.E. Poeschel), S. 173.
TAYLOR, F.W. (1911) *The Principles of Scientific Management* (New York: Harper & Brothers).
TOWNE, H.R. (1886) The engineer as an economist, *Trans. ASME*, 7.
WAKASUGI, K. (1988) *Corporate Finance* (in Japanese) (Tokyo: University of Tokyo Press), pp. 231–232.

CHAPTER NINETEEN

Manufacturing Cost and Product Cost Structure

19.1 Classification of Costs

Costs Classified

Manufacturing costs are generally classified from the two standpoints.
  I. Morphological classification:

(1) *Material cost*—occurs by consuming materials.
(2) *Labour cost*—occurs by utilising human labourforce.
(3) *Overhead*—occurs by consuming cost elements other than the above two.

  II. Economical classification:

(1) *Direct cost*—incurred directly for producing a piece of product.
(2) *Indirect cost*—not directly associated with a particular product.

19.2 Product Cost Structure

Manufacturing Cost and Total Cost

By combining the above two classifications the *manufacturing* (or *factory*) *cost* for producing a piece of product is established, as indicated in Figure 19.1. Further addition of *non-manufacturing* (or *commercial*) *cost* which consists of general administrative expense and selling cost establishes the *total cost*.

Figures indicated in Figure 19.1 are ratios of cost elements against the manufacturing cost in brackets and against the total cost in parentheses for Japan's small-medium-sized manufacturing firms (Agency of Small- and Medium-sized Companies, 1995).

Cost Accounting

Cost elements are accumulated in the process of completing a piece of product, which is the 'flow of costs'. To trace, record, and calculate the cost elements is *cost*

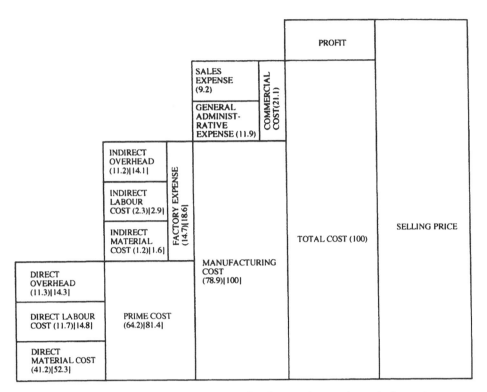

Figure 19.1 Product cost structure. Figures in brackets and parentheses indicate the average percentage of the cost components against the manufacturing cost and against the total cost, respectively, in the case of Japan's small and medium manufacturing firms in 1993.

accounting. This consists of three phases:

(1) calculation by cost elements for goods and services consumed for production and sales;
(2) calculation by departments or places where costs occur;
(3) calculation by products—accumulates cost elements for a piece of product.

19.3 Selling Price

19.3.1 Full-cost Pricing

Calculating Selling Prices

Adding a suitable profit to the total cost comprises the selling price:

$$\text{selling price} = \text{total cost} + \text{profit} = \text{total cost} \times (1 + \text{markup rate}), \quad (19.1)$$

where the *markup rate* means the profit against the total cost.
On the other hand the *margin rate* is the profit against the selling price:

$$\text{margin rate} = \text{profit}/\text{selling price}. \quad (19.2)$$

Hence the selling price is also given as:

selling price = total cost/(1 − margin rate). (19.3)

The above pricing method is called the *full-cost* (or *markup*) *pricing*. This procedure is common in manufacturing industries for establishing the product price.

Value Added and Depreciation

From the manufacturing firm's standpoint selling price becomes the gross revenue after the sale of the product. Subtracting the outside expenses such as material cost, subcontracting cost, etc. from the gross revenue is the *gross value added*. Further subtraction of depreciation cost comprises the *net value added*, as indicated in equations (1.9)–(1.11) in Section 1.5.

Here *depreciation* is a decrease in value through such reasons as wear, deterioration, obsolescence, style changes, customers' whims, etc. It is necessary to consider it to provide for the recovery of capital which has been invested in physical facilities, including property, and to enable the expense of depreciation to be charged to the cost of producing products that are turned out by the facilities (de Garmo, 1960). The straight-line depreciation is one of the most common methods in use, in that the same depreciation charge is made each year; hence, the yearly depreciation is determined by dividing the first invested capital minus the salvage value by the estimated life.

19.3.2 Walras-type Pricing

Equilibrium Price and Social Production Volume

In contrast to microscopic pricing described above, a socially appropriate price can be decided such that the social demand and the social supply be well matched from the macroscopic standpoint.

Under perfect competition the point where the market demand and the supply from the firms are coincident is an *equilibrium price*, as represented in Figure 19.2.

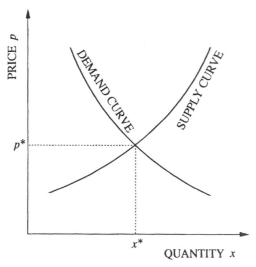

Figure 19.2 Coincidence of the demand and the supply—decides the social production x^* and the social price p^*.

As this figure shows, as the product price rises, the demand reduces while the supply increases; hence excess supply produces unsold products. This decreases the price. On the other hand low price incurs excess demand; hence the price rises.

Thus the price falls to the equibrium price p^* under the Walras-type price adjustment, which is a stability condition. Once p^* has been determined, a social[ly appropriate] production volume x^* can be decided.

19.4 Computing the Manufacturing Cost

Cost Estimation

Accurate cost estimation for producing a piece of product plays a role for capital budgeting.

If the product is standardised, or nearly so, the product cost can be easily calculated based upon the standard cost. Simply it is the sum of the costs of the parts which comprise the product. Multiplying this sum by production volume generates the total cost for the product.

In the process of producing the whole volume of the product, if the production lead time exceeds the due date, production is to be expedited, which raises the production cost.

If the product is individually produced by order, check the cost of the similar product which has been made before. If this procedure is impossible, the times for processing the parts and for assembling these parts into a product are estimated. Then the total production cost is estimated by multiplying this obtained production time by the labour cost and overhead and further adding the material cost.

Calculating Cost Components

The details of cost components for the total production cost are as follows:

(1) *Material cost*—value of the materials necessary for producing a piece of product. This consists of:
  (a) direct material cost for raw materials, parts, subassemblies, etc.;
  (b) supplementary material cost for paints, solders, etc.;
  (c) indirect material cost for consumable articles, such as cutting tools, grinding wheels, cutting fluids, oils, electricity, wastes, etc.

  By establishing a bill of materials for a product (see Section 7.3), the material cost for the product can be calculated by multiplying the unit material cost for each component by the number of that component and summing up for all the components.

(2) *Direct labour cost*—wage for direct labour to produce a unit of product. By determining the standard times for direct labour with industrial engineering (IE) techniques, the direct labour cost can be calculated by multiplying the obtained standard time by the wage rate (hourly wage) and summing over all the processes.

(3) *Direct overhead*—overhead cost necessary to manufacture directly, such as the subcontract expenses, royalties, etc.

(4) *Indirect overhead* (excepting indirect material cost: (1))—expenses supplementing direct manufacturing operations for a piece of product, such as

indirect labour cost, depreciation expense, insurance, maintenance cost, warehousing charge, electricity, water, and gas charges, etc. It is not easy to standardise these expenses, but simply summing up all the indirect manufacturing costs in a certain fixed period, and then obtaining the indirect manufacturing costs per hour, the indirect manufacturing costs for a piece of product are estimated by multiplying the obtained hourly indirect manufacturing costs by the production time needed. This expense is in some cases allocated to each work centre (production stage).

---- **EXAMPLE 19.1** ----

(Individual cost accounting) A product is made up of parts: a, b, and c. The manufacturing cost for this product is calculated as Table 19.1. ☐

Cost Accounting Techniques

In cases where several products are manufactured with common production equipment, the allocation of the fixed cost to these products is difficult; *direct cost accounting* does not allocate the fixed cost.

---- **EXAMPLE 19.2** ----

When two kinds of products—A and B—are manufactured, the fixed costs are treated in bulk and not allocated to individual products, as illustrated in Table 19.2. ☐

Table 19.1 Calculating the product cost by individual cost accounting.

Part	Quantity	Material cost ($)	Work centre	Operation time (h/pc)	Direct labour cost ($) hourly	Direct labour cost ($) total	Indirect cost ($) hourly	Indirect cost ($) total	Manufacturing cost ($)
a	1	30	30						
			101	1.0	4	4	4	4	
			101	0.5	6	3	2	1	42
						7		5	
b	1	50	50						
			101	0.25	4	1	4	1	
			103	1.0	5	5	3	3	70
			104	2.0	3	6	2	4	
						12		8	
c	2	10	20						
			105	3.0	3	18	1	6	44
						18		6	
				Total manufacturing cost ($/pc)					156

Table 19.2 Calculating the total manufacturing cost by direct cost accounting.

Cost component	Product A ($)	Product B ($)	Total ($)
Direct material	10	20	
Direct labour	15	10	
Indirect (variable)	+20	+10	
Manufacturing (variable)	45	40	
Production quantity	×10	×20	
Total manufacturing (fixed)	450	800	1250
Indirect (fixed)			1000
Total manufacturing			2250

If actual costs are known upon calculating the fixed cost, *actual cost accounting* can be done *post factum*. However, *standard cost accounting* is implemented beforehand, based on the standard costs which have been predetermined. Variances in material cost, material use, total operation time, etc. appear between the standard and the actual costs.

Cost Control

Cost control is a function to establish the above variances quantitatively, investigate the causes, and take action in reducing the variances, thereby resulting in reduction of the manufacturing cost.

References

Agency of Small- and Medium-sized Companies (ed.) (1995) *Cost Indices for Small- and Medium-sized Companies* (in Japanese) (Tokyo: Association of Diagnosis for Medium Sized Companies), p. 41.
DE GARMO, E.P. (1960) *Engineering Economy* (New York: Macmillan), p. 93.

Supplementary reading

COOPER, R. and KAPLAN, R.S. (1991) *The Design of Cost Management Systems* (Englewood Cliffs, NJ: Prentice-Hall).
FABRYCKY, W.J. and BLANCHARD, B.S. (1991) *Life-cycle Cost and Economic Analysis* (Englewood Cliffs, NJ: Prentice-Hall).
JELEN, F.C. and BLACK, J.H. (1983) *Cost and Optimization Engineering* (2nd edn) (Tokyo: McGraw-Hill International).

CHAPTER TWENTY

Profit Planning and Break-even Analysis

20.1 Profit Planning

20.1.1 Profit Planning as Business Planning

What is Profit Planning?

One of the most important objectives for a firm dealing in manufacturing is to obtain and accumulate profits, other company objectives being an increase of market share, return on capital, growth rate, social contribution, etc.

An individual company establishes the management policy and determines the long-range business planning as strategic decision-making activities. In this stage *profit planning* is done by estimating and calculating the sales volume and the cost, product by product, department by department, thereby attaining the profits for the future.

A firm is justified in obtaining profits, because of

- reward for taking business risk;
- result of imperfect competition;
- challenge to innovation.

Profit Obtained and Target Costing

An amount of profit is calculated *post factum* as:

actual sales − actual cost = actual profit. (20.1)

Hence, reducing the actual cost ensures larger profit. This calculation is useless for future business activities. Accordingly the target profit, hence the allowable total production cost is preset as

anticipated sales − target profit = allowable cost (20.2)

before the actual production implementation in order to ensure the desirable profit amount. This is called *target costing* or *managerial accounting*.

20.1.2 Calculating Profits

Firm's Profits

Profits obtained by a firm are calculated as follows (Nishizawa, 1978):

$$\text{operating profit} = \text{sales} - \text{production cost} - (\text{sales expense} + \text{general administrative expense}); \quad (20.3)$$

$$\text{current profit} = \text{operating profit} + \text{non-operating profit} - \text{non-operating expense}; \quad (20.4)$$

$$\text{departmental profit} = \text{departmental sales} - \text{departmental expense}; \quad (20.5)$$

$$\text{product marginal profit} = \text{sales by product} - \text{variable cost by product}; \quad (20.6)$$

$$\text{profit by product} = \text{product marginal profit} - \text{fixed cost by product}; \quad (20.7)$$

$$\text{operating profit by product} = \text{profit by product} - \text{common expense allotted to the product}. \quad (20.8)$$

20.1.3 Return on Capital

Rate of Return on Investment (ROI)

The above profit is obtained by the use of capital invested. *Rate of return on investment* (or *invested capital*) or *ROI* means the efficiency of invested capital to obtain the return—the amount of profit gained. It is expressed:

$$\begin{aligned} \text{ROI} &= \frac{\text{profit}}{\text{capital}} \\ &= \frac{\text{profit}}{\text{sales}} \times \frac{\text{sales}}{\text{capital}} \\ &= \text{profit on sales} \times \text{capital turnover}. \end{aligned} \quad (20.9)$$

The detailed structure of ROI is shown in Figure 20.1, which is called the *DuPont chart*.

EXAMPLE 20.1

Top companies in the world in 1993 were (according to *Fortune*, 25 June 1994):

- General Motors for sales: $133 622 million;
- Exxon for profit: $5280 million;
- LTV for return on sales: 103.6%/second—Intel: 26.1%. □

EXAMPLE 20.2

In 1990 the Japanese manufacturing industry showed for (20.9):

$$5.8 = 4.8 \times 1.21.$$

ROI, the profit on sales, and the capital turnover, have worsened year by year; in 1994:

$$3.1 = 2.9 \times 1.05 \quad □$$

PROFIT PLANNING AND BREAK-EVEN ANALYSIS 327

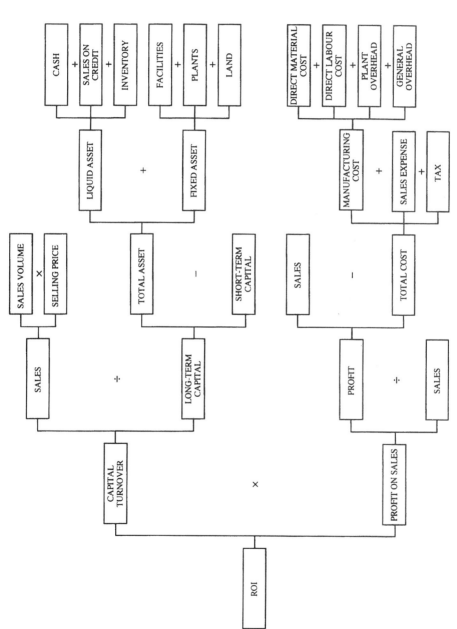

Figure 20.1 Structure of return on investment (ROI)—DuPont chart.

EXAMPLE 20.3

In 1984 Japanese chemical industries showed $8.3 = 9.2 \times 0.90$ for (20.9); these financial indices became poor in 1993, that is, $0.8 = 1.3 \times 0.63$, because of a huge amount of capital investment toward modern production facilities during this decade. □

EXAMPLE 20.4

In 1992 ROI produced by foreign direct investment in the United States was 0.6% on average, while Japan's was -1.9% (a loss!) □

Return on Equity (ROE)

If the share (stock) capital—stockholders' equity—is taken into account, profit over this is called *return on equity (ROE)*:

$$\text{ROE} = \frac{\text{profit}}{\text{equity}}. \qquad (20.10)$$

EXAMPLE 20.5

In 1994 ROE for the world's top thousand companies was 21.2% for Britain, 20.8% for the United States, 10.4% for Germany, while Japan's was only 4.9% and had decreased every year from 8.9% in 1990 to 7.3% in 1991 and to 4.8% in 1992/93 (*Business Week*, 11 July 1995). □

EXAMPLE 20.6

The world's top three home-appliance companies in 1994 with respect to sales were: Japan's Hitachi ($76 431 million) and Matsushita Electric (69 947), and US's General Electric (GE) (64 687) (*Fortune*, 7 August 1995). Their profit on sales and ROE (%) were:

- 1.5, 3.3 for Hitachi;
- 1.3, 2.4 for Matsushita;
- 7.3, 17.9 for GE.

Setting GE's profit and profit per employee at 100, these figures are:

- 24, 16 for Hitachi, and
- 19, 16 for Matsushita. □

20.2 Break-even Analysis

20.2.1 Linear Break-even Chart

What is a Break-even Analysis/Chart?

After the profit is obtained *post factum* or the anticipated (or target) profit is preset, the relationship among the profit, the sales and the cost is to be clarified in order to ensure the target is achieved. *Cost–volume–profit (CVP) analysis* or *break-even analysis* is effectively used for this purpose.

In this analysis the cost is classified into two:

- fixed cost—invariable during a certain period, and
- variable cost—variable through an increase/decrease in utilisation of production facilities, production volume, and sales.

For the case in which the variable cost varies proportionally with production volume, the [linear] *break-even chart* (originated by H. Hess in 1903) is drawn as depicted in Figure 20.2.

The break-even analysis and chart are very often utilised to investigate the gain which would be generated by a specified plan or to decide the best alternatives (see Figure 8.7).

Formulating the Break-even Point

Denoting the fixed cost by a, the unit variable cost by b, and production or sales volume by x, the cost function is expressed in the same way as equation (4.1):

$$f(x) = a + bx. \qquad (20.11)$$

Denoting the unit revenue or selling price by c, the revenue function is:

$$g(x) = cx. \qquad (20.12)$$

Gross revenue is the difference between (20.12) and (20.11):

$$p(x) = g(x) - f(x) = (c - b)x - a. \qquad (20.13)$$

Hence:

$$\text{profit} = \text{marginal profit} - \text{fixed cost}. \qquad (20.14)$$

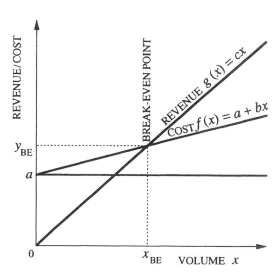

Figure 20.2 Linear break-even chart. The intersection of cost and revenue curves is the break-even point x_{BE}. The production and sales volume above this point generates profits.

If $p(x)$ is positive, a profit is obtained; if negative, a loss is incurred. A point where the gross revenue is 0 is called the *break-even point*; it is:

$$x_{BE} = a/(c - b). \qquad (20.15)$$

Ways to Increase Profits

A profit is obtained by producing and selling a volume greater than x_{BE}. The break-even point depends upon three parameters, a, b, and c, and their sizes as follows:

(1) An increase of a raises the break-even point; hence, in order to obtain the profit, production and sales volumes should be raised.
(2) An increase of b raises the break-even point, and reduces the profit.
(3) An increase of c lowers the break-even point, and raises the profit.

Accordingly, in order to obtain a large amount of profit, a and b should be small and c large. Decreases of a and b mean reduction of cost in manufacturing, and an increase of c means pricing up of the product.

Sales at the Break-even Point

An amount of sales or revenue at the break-even point is given by:

$$y_{BE} = a/(1 - b/c) \qquad (20.16)$$

where b/c is the rate of variable cost, and the denominator is the rate of marginal profit.

--- **EXAMPLE 20.7** ---

In 1994 Japanese small- and medium-sized manufacturers used $6.543 million for the fixed cost (a) and $9.695 million for the variable costs (bx), and obtained $16.889 million for sales (cx) on average. Hence,

- profit obtained ($cx - bx - a$): $661 thousand;
- marginal profit ($cx - bx$): $7.194 million;
- sales at the break-even point ($y_{BE} = a/(1 - bx/cx)$): $15.362 million/91%;
- rate of variable cost (b/c): 57.4%;
- rate of marginal profit ($1 - b/c$): 42.6%. □

--- **EXAMPLE 20.8** ---

In 1991 the Japanese manufacturers indicated:

- fixed cost ratio (a/cx): 30.9%;
- variable cost ratio (bx/cx): 63.8%;
- profit on sales (($cx - bx - a$)/cx): 5.3%;
- break-even point ($a/(cx - bx)$): 85.4%—this increased to 90.9% in 1992, and further to 93.7% in 1993; however, it reduced to 91.1% in 1994. □

Sharing Profits

The profit (gross revenue) minus tax is the *net revenue* which is allotted to stock dividend, labour share, and internal reserve.

EXAMPLE 20.9

The effectiveness of the manufacturing industry in Japan, the United States, and Germany for 1990 is listed in Table 20.1. In Japan the ROI and stock dividend ratio are lowest, and labour shares against sales and against value added are lower than in Germany; hence the internal reserve is high. □

20.2.2 Nonlinear Break-even Chart

Increasing/Decreasing Returns

In the case that linearity does not hold, nonlinear break-even analysis is performed. It is often assumed that the cost curve $f(x)$ shows *increasing returns*,[1] while the revenue curve $g(x)$ obeys the law of *diminishing returns*.[1] This means the unit revenue decreases with the volume of production, as depicted in Figure 20.3.

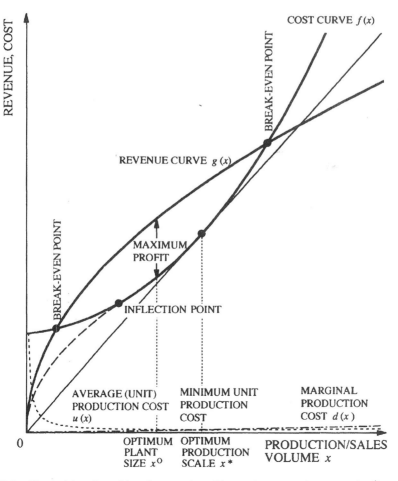

Figure 20.3 General (nonlinear) break-even chart. The optimum production scale x^* minimises the unit production cost, whilst the optimum plant size x^o generates the maximum profit.

Table 20.1 Manufacturing efficiency: International comparison among Japan, USA, and Germany (1990).

Country	ROI	Profit on sales	Capital turnover	Tax rate	Stock dividend ratio	Internal reserve ratio	Labour share on sales	Labour share on value added
Japan	5.2	5.3	0.87	44.2	16.3	40.5	10.3	53.1
USA	6.1	5.7	1.06	30.1	39.0	31.0	—	—
Germany	5.8	4.4	1.32	45.7	27.9	26.4	20.3	82.4

(*Data source*: *Comparative Economic and Financial Statistics*, 1993, Bank of Japan, pp. 119–120).

Optimum Plant Size

The *optimum plant size* which ensures the maximum total profit, that is, the maximum-profit production volume x^o can be derived by differentiating $p(x) = g(x) - f(x)$ with respect to x and setting it to 0 as follows:

$$\left.\frac{df(x)}{dx}\right|_{x=x^o} = \left.\frac{dg(x)}{dx}\right|_{x=x^o}. \tag{20.17}$$

The left-hand side of this equation is the *marginal cost*,[2] and the right-hand, the *marginal revenue*.[2]

Hence, we get (Baumol, 1977):

—— **PROPOSITION 20.1** ——

The marginal principle. The marginal cost is equal to the marginal revenue at the point of maximum profit. □

Optimum Production Scale

In the case that the cost function shows increasing returns as above, as depicted by a broken line in Figure 20.3, the average (or unit) production cost:

$$u(x) = f(x)/x \tag{20.18}$$

is minimal at a point where a straight line through the origin is tangent to the cost curve. The production volume x^* at this point is the *optimum production scale*, as was mentioned in Section 4.2. The tangent to the cost curve at this point is coincident with the slope of the straight line; accordingly the slope of the tangent to the cost curve, namely, the marginal cost:

$$d(x) = df(x)/dx \tag{20.19}$$

is an increasing function, as shown by a chain line in Figure 20.3, and provides the following proposition.

--- **PROPOSITION 20.2** ---

The marginal cost curve lies under the average production cost curve below the optimum production scale x^*, intersects with it at x^* ($d(x^*) = u(x^*)$), and lies above it for x exceeding x^*.

□

This important fact also holds in the case that the cost curve starts from the origin, increases with diminishing returns until a point of inflection, as shown by a dash line in Figure 20.3, and then increases with the increasing returns.[3]

Notes

1. A function $F(x)$ is subject to the law of
   - diminishing returns if $d^2F(x)/dx^2 < 0$;
   - constant returns if $d^2F(x)/dx^2 = 0$;
   - increasing returns if $d^2F(x)/dx^2 > 0$.
2. The *marginal cost/revenue* is an increase of cost/revenue incurred by a slight (one unit) increase of production volume.
3. This sort of cost curve is often discussed in 'production economics' or 'theory (or system) of production'.

References

BAUMOL, W.J. (1977) *Economic Theory and Operations Analysis* (4th edn) (Englewood Cliffs, NJ: Prentice-Hall), p. 68.

NISHIZAWA, O. (1978) *Introduction to Profit Planning* (in Japanese) (Tokyo: Tax-Management Association), pp. 67–69.

Supplementary reading

CASSIMATIS, P. (1996) *Introduction to Managerial Economics* (London: Routledge).
WOOD, A. (1975) *A Theory of Profits* (Cambridge: CUP).

CHAPTER TWENTY-ONE

Capital Investment for Manufacturing

21.1 Investment for Manufacturing Automation

Need for Capital Investment in Manufacturing

New factory construction and industrial automation—particularly computer-integrated manufacturing (CIM), renewal of the existing production facilities, etc.—requires huge investment.

The rationale of the capital investment for production facilities is judged from the standpoint of engineering economy and from the strategic viewpoint of management.

21.2 Evaluation Methods of Capital Investment

Basic Formula for Capital Investment

If a machine tool for which capital I was invested produces profits P_j ($j = 1, 2, \ldots, H$) over the economic life H in which the facility is capable of operation, the total sum of the present worth of profits is expressed as:

$$P = \sum_{j=1}^{H} P_j/(1+r)^j, \tag{21.1}$$

where r is the average rate of interest or discount.

Using the above basic formula, several methods of evaluation for capital investment are described below (Kobayashi, 1968).

Payoff (or Pay-back) Period Method

The anticipated annual savings in direct costs, resulting from the introduction of the proposed investment, are first determined. The *payoff* (or *pay-back*) *period* is calculated as the number of years in which the initial cost of investment, after allowing for resale of the old equipment, is all repaid by those savings.

CAPITAL INVESTMENT FOR MANUFACTURING

Then an alternative associated with the shortest payoff period or one having the payoff period shorter than the predetermined reasonable period (h_0) is considered acceptable. That is, an alternative with the smallest h such that $P = I$ or any alternative such that $h \leq h_0$ is acceptable.

--- **EXAMPLE 21.1** ---

In Japan the payoff method is widely utilised for investment decisions for factory automation, and h_0 is usually set at 2–6 years. ☐

Rate-of-return (on Investment) (or simply Return on Capital) Method

This is the most commonly used method of investigating the effectiveness of a capital investment for production facilities by calculating anticipated annual net profit (after allowing for depreciation) expressed as a percentage of the capital invested—$(P/h)/I$.

Where there exist several alternatives for investment, such as in general- and special-purpose machine tools, automatic machine tools, etc., we will choose an alternative with the largest rate of return on investment.

Rate-of-interest Method

Calculating the discount rate r such that $P = I$, if r is greater than the satisficing level r_0, that is, $r \geq r_0$, then that alternative is chosen. Or an alternative associated with the largest r is chosen.

A variation of this method is one known as the new MAPI method.[1]

Minimum-cost Method

Unit production cost is estimated for each proposal for investment, and an alternative associated with the least cost is chosen. If an alternative with larger cost is chosen, the difference from the least cost is called *opportunity cost* (or *loss*).

The old MAPI[1] method is a variation of this method.

Present-worth (or Present-value) Method

This is a type of discounted cash flow method, and the total sum of the present value of discounted cash earnings[2] (before depreciation and after tax) over the economic life of a facility is calculated as in (21.1). If this value exceeds the capital invested, then the proposal is evaluated as acceptable.

Selection of the Evaluation Method for Investment

Which method should be used for capital investment depends basically upon the size of the capital invested. When it is comparatively small, the minimum-cost method is sufficient. In the case of large investment planning rate-of-return on capital invested is evaluated for each alternative, and by also referring to the analytical results from other methods a best plan is chosen.

EXAMPLE 21.2

(Cost comparison) The existing production facility was purchased for $25 000 four years ago. Its present value is $17 000. The anticipated remaining life of this machine is estimated as six years, where the resale or salvage value is estimated as $5000. This machine can produce 3 pieces per hour of a certain article. Since monthly demand of this article is 480 pieces, 8 hours-per-day and 20 days-per-month are worked and the machine requires an hourly operating cost of $15, together with a direct labour cost of $7.5 per hour.

A new automatic machine proposed is of high efficiency, producing 5 pieces per hour with an hourly operating cost of $10; hence, it is only necessary to operate this machine for 96 hours per month to satisfy the monthly demand. Owing to the automatic capability of this machine we can employ an unskilled worker for $6-per-hour direct labour cost. The purchase cost is $60 000, and the salvage cost is estimated as $10 000, with an estimated life of 10 years.

A piece of raw material costs $10. Assuming that depreciation is calculated on a straight-line basis and the rate of interest is 7% of cost of capital, an annual cost comparison is made for the existing general-purpose machine and the new automatic machine proposed, resulting in Table 21.1. It is concluded from this table that the existing machine requiring an annual cost of $103 570 should be replaced by the new machine which will require $82 887 annual cost. □

EXAMPLE 21.3

(Present-worth approach) A new automated production facility costs $100 000, and its salvage value after an estimated life of 5 years will be 10% of the initial cost. A straight-line depreciation and the discount rate of 6% are assumed to be adequate. If the annual sales amount and operating cost for the coming 5 years are given as in columns (1) and (2) of Table 21.2, then the total of present worth of discounted cash earnings is calculated as in that table, and we have $S = \$129\ 832$. Since this is greater than the invested capital, this investment is justified. □

EXAMPLE 21.4

(Rate-of-return evaluation) Suppose an investment plan for a new automation plant requires $I = \$1\ 000\ 000$ and the existing plant equipment may be sold at $R_0 = \$200\ 000$. Hence, the net investment is $I - R_0 = \$800\ 000$.

Table 21.1 Cost comparison of the existing general-purpose machine tool and a new automatic machine.

	New automatic machine ($)			Existing machine tool ($)		
Direct labour cost	$6 \times 96 \times 12$		$= 6912$	$7.5 \times 8 \times 20 \times 12$		$= 14\ 400$
Material cost	$10 \times 480 \times 12$		$= 57\ 600$	$10 \times 480 \times 12$		$= 57\ 600$
Operating cost	$10 \times 96 \times 12$		$= 11\ 520$	$15 \times 8 \times 20 \times 12$		$= 28\ 800$
Depreciation cost	$\dfrac{60\ 000 - 10\ 000}{10}$		$= 5000$	$\dfrac{25\ 000 - 5000}{4 + 6}$		$= 2000$
Interest cost[*]	$\dfrac{60\ 000 - 17\ 000 + 10\ 000}{2}$	$\times 0.07$	$= 1855$	$\dfrac{17\ 000 + 5000}{2}$	$\times 0.07$	$= 770$
Total			$\$82\ 887$			$\$103\ 570$

[*]Simply obtained by multiplying the rate of interest by the average investment which is given by half of the sum of the actual investment and the salvage cost.

CAPITAL INVESTMENT FOR MANUFACTURING

Table 21.2 Calculation of discounted cash earnings for a new investment.

Year	(1) Sales amount ($)	(2) Cash expense ($)	(3) Depreciation ($)	(4) Assessable income ($)	(5) Tax (50%) ($)	(6) Cash earning ($)	(7) Discount rate	(8) Discount cash earnings ($)
1	80 000	30 000	18 000	32 000	16 000	34 000	0.9434	32 075.6
2	80 000	30 000	18 000	32 000	16 000	34 000	0.8900	30 260.0
3	75 000	35 000	18 000	22 000	11 000	29 000	0.8396	24 348.4
4	75 000	35 000	18 000	22 000	11 000	29 000	0.7921	22 970.9
5	71 000	35 000	18 000	18 000	9 000	27 000	0.7473	20 177.1
Total								129 832.0

⟨Computational procedures⟩ (1), (2): given

$$(3) = \frac{I - 0.1I}{5} = \frac{100\,000 - 0.1 \times 100\,000}{5} = 18\,000$$

(4) = (1) − (2) − (3)
(5) = (4) × 0.50
(6) = (1) − (2) − (5)

$$(7) = \frac{1}{(1+r)^n}, \; r = 0.06, \; n = 1, 2, 3, 4, 5$$

(8) = (6) × (7).

The next year operating advantage from this project is estimated as $S = \$300\,000$, and the use value of the new equipment is estimated to reduce to $V = \$800\,000$ in a year. The capital consumption incurred is $I - V = \$200\,000$.

If the existing plant continues to be used until the following year, the salvage value will be reduced by an amount of $U = \$50\,000$. Hence, the [relative] rate of return is evaluated as:

$$P = \frac{S + U - (I - V)}{I - R_0} \times 100 = 18.8\%.$$

This is comparatively large; hence, it may be concluded that the existing plant should be replaced by the proposed new automation plant. □

Notes

1. The old and new Machinery and Allied Products Institute (MAPI) formulae and systems, which deal with the economic replacement of equipment, were developed by Terborgh (1944).
2. *Cash earnings* are calculated by subtracting the sum of increases in cash expense including material cost, wage cost, and other overheads and in tax from the increase in sales.

References

KOBAYASHI, Y. (1968) *Management of Capital Investment* (in Japanese) (Tokyo: Hakuto-shobou), Chap. 5.

TERBORGH, G. (1944) *Dynamic Equipment Policy* (New York: McGraw-Hill).

Supplementary reading

PRIMROSE, P.L. (1991) *Investment in Manufacturing Technology* (London: Chapman & Hall).

SHARPE, W.F. and ALEXANDER, G.J. (1990) *Investments* (4th edn) (London: Prentice-Hall).

PART FIVE

Automation Systems for Manufacturing

This part describes basic principles and hardware of manufacturing automation for automated 'flows of materials and technological information'. To begin with, three steps towards automation—tool, mechanisation, and automation—meanings and kinds of automation—factory automation (FA) and office automation (OA)/transfer automation, flexible automation, and computer-controlled automation—are introduced (Chapter 22).

Today's most advanced manufacturing hardware and software—computer-integrated manufacturing (CIM) is defined as a system integration of computer aids to design, production and management functions (Chapter 23).

Computer aids to design and production are discussed—computer-aided design (CAD: Chapter 24) and computer-aided manufacturing (CAM: Chapter 25). Computer-aided design for the automated 'flow of technological information' includes computer-aided design and drafting (CADD), computer-automated process planning (CAPP), autoprogramming systems, and computerised layout planning.

With regard to the automated 'flow of materials', the following are explained: automated machine tools for mass production, numerically controlled (NC) machine tools, flexible manufacturing systems (FMS), automated assembly including flexible assembly system (FAS), and automatic materials handling including industrial robots and the automated warehouse, and automatic inspection/testing (CAI/CAT).

The possibility of fully automated manufacturing or an unmanned factory is discussed with a real-life example.

CHAPTER TWENTY-TWO

Industrial Automation

22.1 Towards Automation

Automation

'Automation' is automatic operation; it is basically concerned with automatic 'goods' production or manufacturing.

Three Steps Towards Automation

Steps towards automation constitute a history of increasing manufacturing efficiency and labour productivity; three steps have been followed (Hitomi, 1994):

- (I) introduction of tools;
- (II) mechanisation; and
- (III) automation.

Tool[isation]

It was around four million years ago when human beings appeared on the earth.[1] Since then they have made 'tools' (so they are called *tool makers* or *tool-making animals* (B. Franklin, 1778)). A *tool* (or an *instrument*) is an extention of a hand which performs effective actions. Human beings have utilised tools for productive activities, thus differing from the other animals.[2] In this sense the human being can be called *Homo faber* as well as *Homo sapiens*. Thus the first step in history of increasing manufacturing efficiency is the birth and utilisation of tools.

Mechanisation

The second step is the replacement of human physical labour by machines. A machine consists of three components:

- *prime mover* converting a variety of energy to mechanical energy;
- *transmitter* transmitting a driving force to the workplaces;
- *machining unit* conducting mechanical work.

'Machine tools' play an important role in increasing manufacturing efficiency for the production of industrial products—tangible goods; they are machines which can in turn produce machines, hence often called 'mother machines'. Tools, which played the basic role at the first stage of increasing manufacturing efficiency, are now attached to the machine tools and implement machining operations. With the Industrial Revolution, two hundred years ago in England, metal manufactured machine tools appeared. The mass-production mode, enhanced by 'standardisation' and the 'principle of interchangeability', appeared around the middle of the 19th century in the United States; hence this production mode is often called the 'American system of manufacture', as mentioned in Section 4.2. The development of precision machine tools has contributed to modernisation of manufacturing—the conversion of 'manufacture' to 'big industry' (the highest level of capitalism, as will be explained in Section 31.2). This is a phase of 'mechanisation'.

Automation

The third step was the replacement of human mental labour by machines. The setup, operation and control of machine tools came to be operated automatically rather than by skilled workers as in the second stage. This is the movement toward 'automatisation'. Its first realisation was the 'automatic lathe' which was invented in 1873. Developments in electronics and control engineering advanced the tendency toward automatisation. This word changed to 'automation' after World War II, though this term was originated in 1936 by D.S. Harder (*Morris Dictionary of Word and Phrase Origins*, 1988).

Automation has developed greatly since the war, being accelerated by the invention of numerical control (NC) in 1952. At that time this tendency was recognised as a revolution and a new philosophy of manufacturing. Automation thus had a great impact on industry, together with 'cybernetics'[3] which was developed by Wiener (1961).

Mechatronics

'Mechatronics', which is a 'Japanese–English' word meaning a unification of '*mecha*nics' with regard to physical labour and 'elec*tronics*' with regard to mental labour, now plays an important role in automation.

22.2 Meanings of Automation

Three Theories on Automation

In the beginning, when D.S. Harder coined the term 'automation' in 1936 whilst working for the General Motors Corporation, he meant it as the transfer of workparts between the machines in a production process, without human operation. In 1946 he established the 'Automation' Department when he was a vice-president of the Ford Motor Company.

After the war, a management consultant, J. Diebold, wrote two books on automation. In his book published in 1952 he defined automation as automatic operation or a process of automatically making tangible goods. In 1955 he mentioned that automation had two meanings (Veillette, 1959):

(1) automatic regulation by feedback;
(2) integration of a plural number of machines.

J.B. Bright presented the development stages of mechanisation and automation, and D.F. Drucker recognised automation as a conceptual system beyond technology. These three theories are typical in understanding the concept and meanings of automation (Munekata, 1989).

Origin of Automation

'Automation' may be considered as the abbreviation of 'automatisation' or 'automatic operation'; alternatively, automation is a combination of Greek *automatos* (meaning self-acting) and Latin '*-ion*' (meaning a state).

Meaning of Mechanisation/Automation

'Mechanisation' is the replacement of human physical labour by machines, but the control of this machine operation is effected by human operators. However 'automation' also replaces this control action by machines as mentioned in the previous section; that is, 'automation' means the replacement of both human physical and mental activities by machines.

22.3 Kinds of Automation

Automaton vs. automation

As has been suggested, 'automation' means automatic operation. However, such items as the 'south-indicating statue' fitted to a wheel in ancient China, and automata, such as shrines with automatically opening doors, automatically adjusting lamps, etc., invented by Heron in 1BC, self-moving machines, clocks, automatically performing birds or dolls made in the medieval age, *Karakuri* (Japanese automaton made in the pre-modern age) cannot be identified with automation.

Automation implies processes adopting automatic *production* methods or full automatic *production* in factories (Einzig, 1956). Hence this term has been concerned with, and confined to, production activities.

Three Kinds of Automation

Three directions have pointed the way toward automation (Department of Scientific and Industrial Research, 1956):

(1) automatic flow-type production in manufacturing industries;
(2) automatic control of continuous production in process industries;
(3) increase in business efficiency by computers.

These tendencies created the following three types of automation (Inoue, 1978):

(I) *mechanical* (or *Detroit*[4]) *automation* for manufacturing industries;
(II) *process automation* for process and chemical industries;
(III) *office* (or *business*) *automation* for office work.

Factory Automation (FA)

Mechanical automation and process automation are concerned with direct production processes which convert raw materials into products; i.e. the 'flow of materials'. This type of automation is now called *factory automation (FA)*. This terminology appeared in 1961.

Office Automation (OA)

On the other hand, *office automation* (OA) is concerned with management/control of productive activities, namely, the 'flow of information'. Once FA and OA are integrated, 'corporate (or enterprise) automation' emerges (Wakuta and Hitomi, 1983).

Low-cost Automation (LCA)

In general, high-level automation needs a vast amount of capital investment. *Low-cost automation (LCA)* emerged in 1965 at the Pennsylvania State University, aiming at automation with a small amount of capital by employing standardised automatic equipment such as actuators, sensors, etc. This type of automation is particularly useful for small businesses, because the cost-effectiveness for automatic [mass] production of a single product is great.

Extented Automation

The word 'automation' is attached to other words, such as:

- *design automation* for speedy automatic design and drawing of parts and products,
- *laboratory automation* for automatic measurement, collection and analysis of test data,
- *store* (or *sales*) *automation* for sales management by computers using POS (point of sale) techniques, and others.

Even 'home automation' is suggested for a pleasant life in the home.

22.4 Development of Automatic Manufacturing

22.4.1 Development of Automation

Early Automation

In the latter half of the 19th century, conveyors, which were first employed in an automatic flour-producing plant by O. Evans in 1787, were widely used for transfer of goods. Around 1873 C. Spencer invented an automatic screw lathe.

Transfer Automation

A simple type of automation, which has been widely used until now, is the 'transfer machine' for continuous production, combining in one unit the functions of several separate machining units unified with attendant loading and transferring mechanisms.

The first transfer machine was built by Morris Motors of England in 1924 for making cylinder blocks, but it was unreliable because of its reliance upon mechanical controls. The development of transfer machines reached its climax in the mid-1950s, when the main car factories installed this integrated type of transfer line. This is called *transfer automation*, only oriented towards the mass-production industries.

Flexible Automation

For automation of non-mass (variety of product, small-batch) production, a versatile machine, capable of automating this production mode and equipped with a control device that enables it to read the instructions for specified machining operations and operate automatically according to those commands under the control of an operator, has been developed. Typical of this 'program-controlled automation' or *flexible automation*, which was mentioned in Section 4.3, is numerical control (NC). In NC the machining instructions are given to the control device by means of punched or magnetic tape or disk.

The idea of a punched-card control system, however, was first introduced in 1725 by B. Bouchon for a loom to weave correctly the desired pattern, and extended to culminate in the perfection of the Jacquard loom in around 1804. It is still being used today. In the manufacturing area the control of a machine tool by means of a punched tape was first carried out in the USA in 1952 when a numerically controlled milling machine was developed for the US Air Force at the Massachusetts Institute of Technology. This type of program-controlled machine tool has been used in industrial workshops since about 1956.

22.4.2 Development of Computers

Invention of Computers

The principles of automatic calculation and the 'difference engine', which was realised and developed in about 1833 by the English mathematician, C. Babbage, were extended in H. Hollerith's tabulating and calculating machines, which were used in the United States census of 1890 by processing punched cards, and further in H.H. Aiken's Automatic Sequence-Controlled Calculator, using electro-magnetic relays, in 1944.

The general-purpose electronic computer having a fixed stored program and the most basic principles for automatic calculation was proposed by J.V. Atanasoff in 1939, and the Atanasoff–Berry Computer (ABC) was made at the University of Iowa. The Electronic Numerical Integrator And Computer (ENIAC) based upon this idea was completed at the University of Pennsylvania in 1946 for calculating trajectories of artillery shells for the US Army Ordinance Department. A similar electronic computer, COLOSSUS, was developed in 1943 in England for decoding cryptograph codes of the German Army.

Modern Computers

The first computers as we know them today, based upon variable stored-program principles, appeared in 1949 as the Electronic Discrete Variable Automatic Computer (EDSAC). It was built by a team led by M.V. Wilkes in England. Since then, computers have improved rapidly through the use of more refined electronic and technological techniques, such as large-scale integrated circuit and memory chips. Such improvements and developments have led to today's high-speed calculating and data-processing machines with large-capacity memories.

22.4.3 Application of Computers to Manufacturing

Computer Use in Production Management

Computers have not only been utilised in many fields of science and technology, but have also played an essential role in handling a vast amount of the clerical work involved in modern business administration. Since 1953, when J. Lyons and Company of London first carried payroll calculations on a computer, computers have been employed in production planning and control to reduce human clerical work significantly and improve the efficiency of these operations.

This trend has been extended further to the information-oriented system for the firm's total production management by computers, which is described in Chapter 30.

Computer-controlled Automation

The progress of automatic production beyond transfer automation and program-controlled automation, so far described, for manufacturing single components, has been quickened by the utilisation of computers, and 'computer-controlled automation' has become reality. With this, an individual job is performed by automated machinery connected to a computer. Each operation and entire production processes can be supervised by computers, by continually receiving progress reports from the workstations, analysing them and transmitting timely operation orders back to each of the workstations, on an on-line, real-time basis, thereby creating an efficient total factory management system.

Although generally today the situation is still slightly removed from a fully computer-controlled automated (unmanned) factory, early versions for producing machined parts in different shapes, even for jobbing and variety production, have been realised by computer control of a group of machines with flexibility of program changes.

Automated Factories

Automated factories were first introduced in 1961 in a North Carolina factory of the Western Electric Company for making carbon resistors (Lilly, 1965). This was followed by a 'flexible manufacturing system (FMS)', which consisted of machining/turning centres, conveyors, automated guided vehicles (AGVs), automated warehouses and robots all under the control of computers. Hence FMS was called the 'computerised manufacturing system' (CMS) until 1979. Since then FMS, which was a trademark of the Kearney & Trecker Corporation, has become the common terminology. In an FMS the flow of materials and the flow of information are integrated in an on-line, real-time mode, as will be discussed in Chapter 25.

An early mode of FMS was 'System 24' developed by the Molins Company in England in 1967. The next year the Sundstrand Company in the United States developed a similar direct numerical control (DNC), the 'Omnicontrol System', and the Japanese Railway developed 'group-control systems' used to make a variety of repaired parts for railway cars.

Social Effect of Automation

The most obvious effect of automation is that it increases the labour productivity and reduces the production cost. The introduction of automatic inspection devices improves quality. These clearly raised the standard of living—higher wages with shorter working hours, cheaper products of good quality, and timely delivery of products.

On the other hand, automation needs an enormous capital expenditure. In order to pay for this investment in a capitalist economy, market expansion is required; otherwise, automation may increase unemployment. So it is desirable that automation be planned and installed slowly and cautiously for the maximum benefit to people and society.

There has been a tendency in today's society to increase the wage rate rapidly due to lack of skilled labour. Associated with this is a change of feelings towards the value of labour saving, also scorn for physical labour. To cope with this situation it has been necessary to enhance labour saving by converting industry from a labour-intensive to a capital-intensive structure, thereby promoting automation of production. Since it requires a vast investment, the assessment of the profitability of that investment is economically very important.

Accordingly, it is often necessary not to introduce the highest degree of automation from the beginning, but to enhance the degree of automation gradually, by:

- first, automating individual machines;
- then, automating machining or assembly lines;
- lastly, introducing computer-integrated automation for the whole plant, aiming at the unmanned factory.

The last, and highest, level of factory automation is presently recognised as and implemented by 'computer-integrated manufacturing (CIM)' systems. CIM integrates various automated functions such as design, production, management, and others, as is discussed in the next chapter.

History of Automation

The historical progress of mechanisation and automation for manufacturing and management is traced and summarised in Table 22.1.

Table 22.1 Landmarks in the rise of mechanisation and automation.

Year	Event (manufacturing technology)	Year	Event (management technology)
4 million ago	Tools/instruments appeared (*Ramapithecus*)		
2 million ago	Tool-making (*Australopithecus*)		
532BC	Lathe (Greece)	c.2600BC	Abacus (China)
c.6BC	Mass production (Phoenician)	2000BC	Management (Babylonian)
c.6–2BC	Machine tools appeared	394BC	Xenophon: Division of labour (Greece)
c.1BC	Heron: Automatics (Greece)	c.4BC	Mengzi: Division of labour (China)
c.AD8–9	Mass production (Japan)		
1306	Pin-producing machine (Germany)	AD1202	Pisano: Abacus-book (Italy)
1613	Serra: Principle of mass production (Italy)	1494	Pacioli: Double-entry bookkeeping (Italy)
		1605	Profit calculation (Holland)
		1642	Pascal: Mechanical calculating machine (France)
1713	Gun-boring machine (Switzerland)		
1725	Bouchon: Punch-card system		
c.1760	Industrial Revolution (UK)	1751	Principle of division of labour (France)
1765	Principle of interchangeability (France)		
1775	Wilkinson: Boring machine (UK)	1776	Smith: *Wealth of Nations* (UK)
1778	Evans: Conveyor-driven flour mill (USA)		
1796	Watt: Modern factory (UK)		
1797	Maudsley: Screw lathe (UK)		
1800	Blanchard: 'Gun-stocking' lathe (USA)		
1804	Jacquard loom with punch-cards (France)		
1808	North: Milling machine (USA)		
1811/2	Luddite movement (UK)	1815	Owen: Factory system (UK)
1818	Blanchard: Mechanical copying machine (USA)		
1818	Whitney: Milling machine (USA)	1823	Babbage: Automatic calculating machine (UK)
1820	Planer (UK)		
1835	Whitworth: Automatics (UK)	1826	Tunen: Factory location (Germany)
1840s	Turret lathe (USA)	1832	Babbage: Manufacturing economy (UK)
1843	Turret machine (France)		

1855	Braun: Universal milling machine (USA)
1860	Principle of assembly line (USA)
1861/2	Universal milling machine (USA)
1864	Cylindrical grinding machine (USA)
1867	Marx: Capitalist production mode (Germany)
c.1873	Spencer: Automatic screw machine (USA)
1875	Metcalfe: Materials management (USA)
1877	Wellington: Engineering economy (USA)
1881	Taylor: Time study (USA)
1885	Gilbreth: Motion study (USA)
1887	Hollerith: Punch-card tabulating and calculating machines (USA)
1880s	Special-purpose machine/Hobbing machine (USA)
1895	Multiple-spindle lathe (USA)
1898	White–Taylor: High-speed steel tool (USA)
1900	Hydraulic/electric copying machines (Italy)
1901	Gunn: Terminology—IE (USA)
1903/11	Taylor: Shop/scientific management (USA)
1908	Kimball: Lecture 'IE' (USA)
1912	Emerson: Twelve principles of efficiency (USA)
1913	Harris: Economic lot sizing (USA)
1913	Ford: Conveyor-driven, flow-type car assembly (USA)
1914	Gantt: Gantt charting (USA)
1914	Harris: Inventory model (USA)
1916	Fayol: General principles of administration (France)
1920	Čapek: Terminology 'Robot' (Czechoslovakia)
1923	Gilbreths: Principles of motion economy (USA)
1923	Keller: Copying shaper (USA)
1924	Transfer machine (Morris Motor Car Co., UK)
1924–32	Mayo: Hawthorne experiments/Human relations (USA)
1927/8	Shewhart: Quality control (USA)
1924–6	Carbide tool (Krupp, Germany)
1928	Ermanski: Production rationalisation (Russia)
1928	Cobb–Douglas: Production theory (USA)
1930	Patent of numerical control (USA)
1932	Mogensen: Work simplification (USA)
1934	Tippett: Work sampling (USA)
1934	Quick: Work-factor (USA)
1936	Harder: terminology 'Automation' (USA)
1936	Wright: Learning effect (USA)
1939	Roethlisberger: Human relations (USA)
1939	Kantorovich: LP production planning (Russia)
1939	Atanasoff–Berry: ABC(first digital computer) (USA)
1940	Maynard: MTM (USA)
1940	William: Relay-calculator (USA)
1941	Leontief: Input–output analysis (USA)
1941	Hitchcock: Transportation-type LP

(continued)

Table 22.1 (Continued)

Year	Event (manufacturing technology)	Year	Event (management technology)
		1941–5	Operational research (OR) (UK)
		1943	Flowers: COLOSSUS computer (UK)
		1944	Aiken: Automatic sequence-controlled calculator (USA)
1945	NC milling machine (Kearney & Trecker, USA)	1945	Terborgh: MAPI (USA)
		1946	Eckert-Mauchly: ENIAC (USA)
		1946	Mitrofanov: Group technology (Russia)
1947	Harder: Detroit (mechanical) automation (Ford Motor Co., USA)	1947	Dantzig: Simplex method for linear programming (USA)
1947	Remote manipulator (USA)	1947	Wiener: Cybernetics (USA)
1950	Fully automated press machine (Ford Motor Co., USA)	1949	Wilkes: EDSAC (stored-program type computer) (UK)
1950	Fully automated piston production (Russia)	1950s	Office automation
1950s	Process automation (USA)	1951	Commercial digital computer (UNIVAC, USA)
1952	Parsons: Three-axis NC milling machine (MIT, USA)	1953	Computerised production management (Lyons & Co., UK)
1954	Devol: Patent of industrial robot (playback manipulator) (USA)	1954	Johnson: Flow-shop scheduling (USA)
		1954	Salveson: Line balancing (USA)
1955	Auto-programming system—APT (USA)	1955	Muther: Systematic layout planning (USA)
1958	Machining centre (Kearney & Trecker, USA)	1957	Bellman: Dynamic programming (USA)
		1957	CPM (Du Pont, USA)
c.1958	Automatic drawing machine (Gaber Co, USA)	1958	PERT (US Army, USA)
1959	Polar-coordinate robot (Unimation Co., USA)	1959	Silberston: Optimum production scale (USA)
1960	Adaptive-controlled milling machine (Bendix Co., USA)	1960s	MIS (USA)
		1961	Feigenbaum: TQC (USA)
1960	Terminology 'FMS' (Kearney & Trecker, USA)	1961	Production information system MOS (IBM, USA)
1961	Computer-assisted resistor machining line (Western Electric Co., USA)	1962	On-line, real-time production management—TOPS (Westinghouse, USA)
1962	Coordinate robot (AMF Co., USA)	1962	QC (Quality) circle (Japan)
1962	Two-dimensional CAD (USA)	1963	On-line, real-time computer system—SABRE (American Air Lines, USA)
1963	Sutherland: Computer-graphical CAD 'Sketchpad' (MIT, USA)	1964	IBM 360 computer (USA)
c.1965	Low-cost automation (Pennsylvania State University, USA)		
c.1965	Direct digital control for process automation (USA)		

1966	Automatic programming—EXAPT (Germany)
1967	DNC 'System 24' (Molins Co., UK)
1968	PICS (Production Information and Control System) (IBM, USA)
1967	CAD/CAM software—CADAM (Lockheed, USA)
1969	CAM (USA)
1970	IMS—Welding line by robots (General Motors Co., USA)
1970	Patent of FMS (Molins Co., UK)
1970	Orlicky: MRP (USA)
1972	COPICS (Communications-Oriented Production Information and Control System) (IBM, USA)
1973	Harrington: terminology 'CIM' (USA)
1973	CAD—three-dimensional solid models (UK, Japan)
1977	Small-group assembly without conveyor (Volvo Co., Sweden)
1980	Manufacturing automation protocol—MAP (General Motors, USA)
1980	CAE (SDRC, USA)
1989	Patent of CIM 'Product realisation method' (USA)
1978	Ohno: JIT principle (Toyota Automobile, Japan)
1983	Office automation protocol—TOP (Boeing, USA)
1985	Wiseman: Strategic information system (SIS) (USA)
1987	CALS (USA)
1987	ISO 9000: Quality systems (EU)
1988	Taguchi: Quality engineering (Japan)
1988	Production information system—BAMCS (UNYSIS, USA)
1988	Karmarkar: LP software patent (AT&T, USA)
late 1980s	TQM (USA)
1991	IMS (Intelligent Manufacturing System)-project (Japan/USA/EC)

Notes

1. Human beings and chimpanzees were separated 4.9(±0.2) million years ago; the oldest pithecanthrope which has been discovered is *Australopithecus lamidas* (4.4 million years ago).
2. Some apes, e.g. chimpanzees, use pieces of wood, stones, etc., but they do not create/fabricate tools/machines at all.
3. A technique seeking an optimum state by adequately controlling the controllable variables, based upon the past data of the uncontrollable parameters.
4. Named after the place where Harder established the Automation Department in the Ford Motor Company.

References

Department of Scientific and Industrial Research (1956) *Automation* (London: Stationary Office).
EINZIG, P. (1956) *Economic Consequences of Automation* (London: Martin Secker & Warburg).
HITOMI, K. (1994) Automation—its concept and a short history, *Technovation*, **14** (2).
INOUE, K. (1978) *Management and Theory of Industrial Production* (in Japanese) (Kyoto: Minelva), pp. 181–182.
LILLY, S. (1965) *Men, Machines and History* (London: Lawrence & Wishart), p. 260.
Morris Dictionary of Word and Phrase Origins (2nd edn) (1988) (New York: Harper & Row), p. 31.
MUNEKATA, M. (1989) *Theory of Technology* (in Japanese) (Tokyo: Dobunkan), p. 36.
VEILLETTE, P.T. (1959) The rise of the concept of automation, in Jacobson, H.B. and Roucek, J.S. (eds), *Automation and Society* (New York: Philosophical Library), Chap. 1.
WAKUTA, H. and HITOMI, K. (1983) *Factory Automation and Office Automation—Towards Corporate Automation* (in Japanese) (Tokyo: Business & Technology Newspaper Company).
WIENER, N. (1961) *Cybernetics* (2nd edn) (Cambridge, MA: MIT Press).

Supplementary reading

DIEBOLD, J. (1952) *Automation* (New York: Van Nostrand).
HALL, G.M. (1995) *The Age of Automation* (Praeger).
NOBLE, D.F. (1986) *Forces of Production—A Social History of Industrial Automation* (Oxford: Oxford University Press).

CHAPTER TWENTY-THREE

Principles of Computer-integrated Manufacturing (CIM)

23.1 Essentials of Computer-integrated Manufacturing Systems

Three Major Functions in Manufacturing

As mentioned in Section 5.3, an integrated manufacturing system (IMS) integrates the following three major functions:

(1) *production function* for converting the resources of production (especially raw materials) into products—the 'flow of materials';
(2) *design function* for transformation of customer specifications into designs and detailed drawings, and the design of process plants, prior to production—the 'flow of technological information';
(3) *management function* for planning and controlling production activities—the 'flow of managerial information'.

Three Computer Aids in CIM

Computer-integrated manufacturing (*CIM*) is a computerised system which integrates computer aids to the above three different functions via a common database[1] (see Figure 23.1), namely, it incorporates the following (also see Figure 5.3) (Hitomi, 1993):

(I) *computer aid to the production function* (automated flow of materials: procurement–production–quality control–process control–cost control–distribution/sales): computer-aided manufacturing (CAM);
(II) *computer aid to the design function* (automated flow of technological information: research and development–product design–process design–planning process (CAPP)–layout design): computer-aided design (CAD).
(III) *computer aid to the management function* (automated flow of managerial information: sales planning–production planning–operation scheduling): computer-aided production planning (CAP^2).

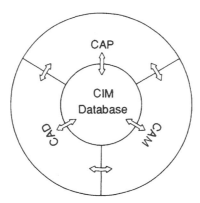

Figure 23.1 The basic framework of computer-integrated manufacturing (CIM)—system integration of computer aids to three different functions: design (CAD), production (CAM), and management (CAP) by use of a common database.

Concept and Origin of CIM

The integration of the 'automated flow of materials' and the 'automated flow of information' is a concept of CIM. CIM is also a realised mode of 'system integration' mentioned in Section 2.1, for the above three functions.

Historically computer-aided manufacturing (CAM) was advocated in the United States in around 1969. It extended to integrated computer-aided manufacturing (ICAM). In 1973 J. Harrington coined the term 'computer-integrated manufacturing (CIM)' including computer-aided design (CAD) (Harrington, 1973).

23.2 Definition of CIM

CIM in Practice

In application, CIM is perceived as the integration of production and sales, and further includes technology. It is intended to ensure reduction of the lead time and flexible adaptation to highly varied, small-batch production through the computerised processing of the entire activities from order receipt to product shipment, possibly in a just-in-time (JIT) mode, for an integrated manufacturing system (IMS) described in Section 3.5. In Figure 3.6, which shows the IMS procedure, the computer aids are also expressed; an integrated chain of these computer aids constitutes the CIM.

CIM Defined

A common definition of CIM has not yet been well established, but it can be understood as follows: CIM is a 'flexible market-adaptive strategic manufacturing system which integrates three different functions and systems—design, production, and management—through the information network with computers'.

It should be noted that the M of CIM means 'manufacturing' in a broad sense as described in Section 1.1; that is, manufacturing is a chain of business functions: planning–acquisition–production–distribution–inventory–marketing and sales–management, and includes 'management' and 'marketing' activities.

23.3 Effectiveness of CIM

Major Merits of CIM

Merits of installing the CIM [system] are summarised as follows.

(1) *Systems integration of design, production, and management.* As mentioned in Section 23.1, CIM integrates three basic computerised functions in a manufacturing firm; namely, design, production and management, holding individual autonomy in the CIM system. Production and sales are also closely connected together, so as to meet market/consumer needs in a just-in-time mode.

(2) *Common database.* Effective utilisation of computer aids for the above three functions—CAD, CAM, and CAP—is done by constructing and using a common database, which every function can access. The database should be always maintained to store up-to-date data/information.

(3) *Flexible automation to multi-product, small-quantity production.* Automation for single-item production is easily done by conventional automatic machines, special-purpose machine tools, and transfer machines, as mentioned in the previous chapter. However, multi-product, small-batch production, which constitutes 75% of manufacturing in the United States and 85% in Japan, can only be automated by a CIM which contains flexible manufacturing systems (FMSs) for producing a variety of items.

(4) *High quality, low cost, and timely delivery (quick response).* Conventional multi-product, small-batch production tends to be low in productivity/efficiency; however, CIM attempts to achieve high productive fabrication of high-quality products with low cost for timely (just-in-time/quick respondent) delivery.

(5) *Towards manufacturing strategy.* A manufacturing firm can be oriented toward manufacturing strategy. This vital aspect will be discussed in Chapter 32.

Sample of CIM

An ideal CIM system is not easily realised; however, installing the CIM system is vital for future production from the strategic and competitive viewpoints.

—— **EXAMPLE 23.1** ——————————————————————

A CIM system is demonstrated in Figure 23.2, in which three major computerised functions—CAD, CAM, and CAP—are represented as the 'development and design system', the 'factory management system', and the 'production management system'. In addition the 'sales system' is stressed in an attempt to connect both activities of production and sales directly, which is an important aspect of CIM. The 'management information system' deals with office automation (OA) of the firm's management. □

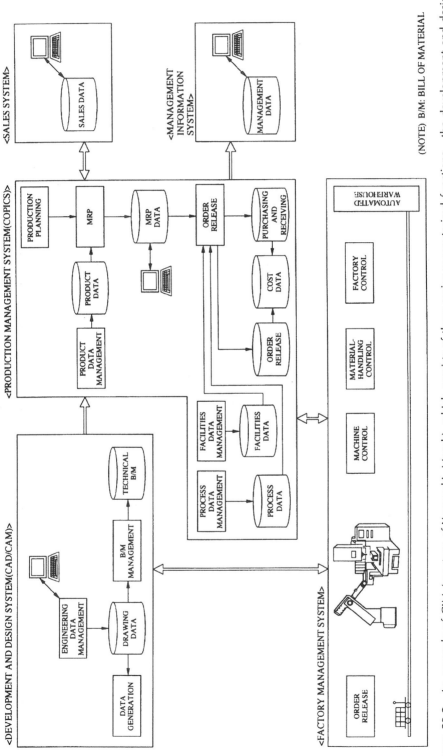

Figure 23.2 An example of CIM (courtesy of Yamazaki–Mazak), which consists of three major computerised functions—the development and design system, the factory management system, and the production management system, together with the sales system and the management information system.

(NOTE) B/M: BILL OF MATERIAL

Advantages/Disadvantages of CIM Installation

To demonstrate the effectiveness of CIM implementation, the British Society of Production Management published (data by UNYSIS, Japan):

- reduction of design cost: 15–30%;
- increased efficiency of design activity: 4–20 times;
- decrease of lead times: 30–70%;
- improved quality for products: 2–5 times;
- reduction of inventory: 30–70%;
- total productivity increase: 30–60%;
- reduction of labour costs: 5–25%.

On the other hand, it should be noted that implementation of CIM requires large capital investment; economy of investment is to be carefully investigated (refer to Chapter 21).

Notes

1. Group technology (GT) is also emphasised in CIM systems.
2. This is a three-letter acronym coined by the author. In Germany it has a different meaning, namely computer-aided process planning, rather than computer-aided production management. Theoretically, since 'manufacturing' covers a wide range of functions in a manufacturing firm and 'production' means a narrow function of fabrication, as mentioned in Section 1.1, the usual CAM should be 'CAP' (computer-aided production) and computer-aided [production] management 'CAM'.

References

HARRINGTON, J., Jr. (1973) *Computer Integrated Manufacturing* (New York: Industrial Press).
HITOMI, K. (ed.) (1993) *Principles of Computer-Integrated Manufacturing* (in Japanese) (Tokyo: Kyoritsu Publishing), Chap. 1.

Supplementary reading

SINGH, N. (1996) *Systems Approach to Computer-integrated Design and Manufacturing* (New York: John Wiley).
TEICHOLZ, E. and ORR, J.N. (1987) *Computer-Integrated Manufacturing Handbook* (New York: McGraw-Hill).

CHAPTER TWENTY-FOUR

Computer-aided Design (CAD)

24.1 Brief History of Computer-aided Design (CAD)

Computer Use in Three Fields of Design/Drafting

Design and drafting by the use of computers have developed separately in the following three fields (Hitomi, 1993):

(1) interactive graphics for modelling, drafting;
(2) process and operations planning for automatic machining;
(3) engineering analysis and evaluation.

Computer Graphics and CAD

Interactive computer graphics first appeared in 1963 as *Sketch Pad* by I.E. Sutherland at the Massachusetts Institute of Technology. This software used a cathode ray tube (CRT) with a light pen to input and output figures and forms, thereby designing interactively on an on-line basis through both human and computer actions. S.A. Coons proposed a CAD concept integrating conceptual, detail, and production designs and developed a 'patch' for free surfaces.

CAPP

Automatically programmed tools (APT), an autoprogramming system for numerical control (NC) machining, was constructed in 1952, also at the Massachusetts Institute of Technology. It performs mathematical calculations required to determine the geometrical contours of the tool path and the sequence of operations from the part geometry, which is geometrical processing. Extended APT (EXAPT) which was developed in Germany, contains an element of process planning—technological processing.

The first computer program for process planning was developed as automated manufacturing planning (AMP) by the International Business Machines Corporation. Later, Computer Aided Manufacturing-International (CAM-I) constructed one of the best-known systems called computer-automated process planning (CAPP).

CAE

Optimum engineering design by computer is called computer-aided engineering (CAE), which was proposed by the Structural Dynamics Research Corporation in 1980.

24.2 Computer-aided Design/Drawing (CADD)

24.2.1 Preliminaries to CAD[D]

CAD Defined

Computer-aided design (CAD) utilises computers, graphic display, automatic drawing machines, and other peripheral devices for interactive or automatic design and drawing. The interactive method of computer graphics (CG) deals with the generation of shapes and images, change of location, drawing projection, etc.

CAD Facilities

The facilities used for CAD are depicted in Figure 24.1. The computers employed are mainframe type, stand-alone type, engineering work station (EWS) type, or small micro- (or personal) computers.

Functions Involved in CAD

Functions treated in CAD are:

- geometric modelling (pattern/parametric design);
- optimum design;
- automatic drawing.

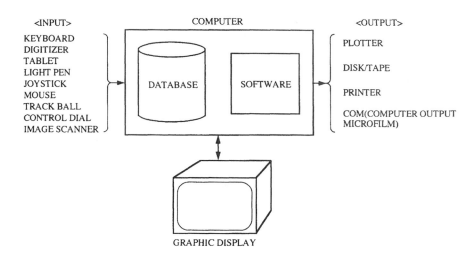

Figure 24.1 The CAD configuration consists of a computer with graphic display, input and output devices.

24.2.2 Geometric Modelling

Geometric Models

Geometric modelling or pattern design is a major task of CAD, constructing geometric patterns of parts/products on the display. For this purpose the following models are used:

- regular two-dimensional (2-D) models;
- wireframe models, three-dimensional (3-D) models;
- surface models;
- solid models as 3-D shapes.

—— EXAMPLE 24.1 ——

A 2-D model is illustrated in Figure 24.2. □

—— EXAMPLE 24.2 ——

A 3-D surface model for a crank shaft is demonstrated in Figure 24.3. □

Solid Models

The two most visual solid models are:

- *Constructive solid geometry (CSG)*. Primitives such as rectangle, cylinder, triangle, pyramid, cube, etc., are assembled by set calculation, as represented in Figure 24.4.
- *Boundary representation (B-rep)*. Boundaries are combined together, as shown in Figure 24.5.

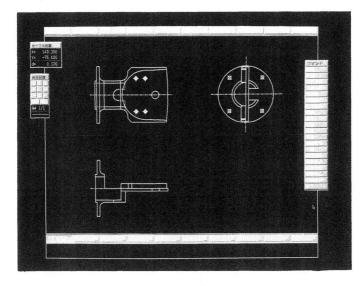

Figure 24.2 Two-dimensional model—an example (courtesy of Unix).

COMPUTER-AIDED DESIGN (CAD) 361

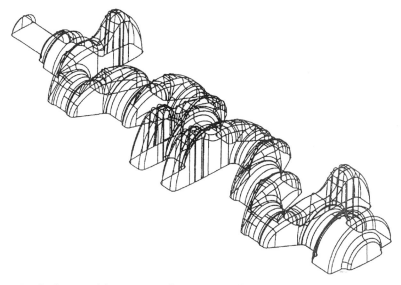

Figure 24.3 Surface model—an example (courtesy of Sumitomo Metal Industries).

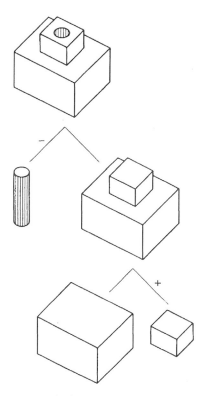

Figure 24.4 CSG procedure—set calculation of primitives.

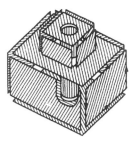

Figure 24.5 B-rep combines boundaries of a solid together.

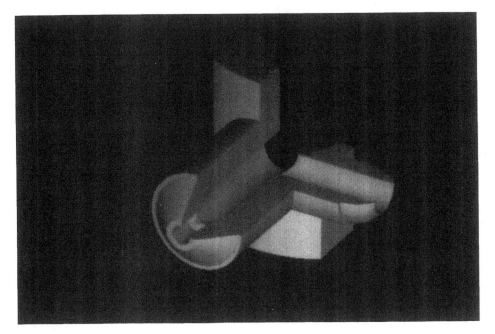

Figure 24.6 Three-dimensional solid model—an example of CSG (courtesy of Dr N. Okino).

— **EXAMPLE 24.3** —

A three-dimensional solid model using CSG is demonstrated in Figure 24.6. ☐

CSG is simple in data structure, easy for inputting data and converting the system into B-rep, and requires less memory for form information. On the other hand, CSG requires the operation of setting the domain of the shape whenever analysis and processing are conducted. The characteristics of B-rep are the reverse of those for CSG.

Free Form

Complicated free-form (or sculptured) surfaces, which cannot be expressed by basic primitives or mathematical expressions, such as car wings, moulds, etc., are demonstrated by assemblage of surface segments satisfying certain continuity conditions. Examples are:

- the Coons curve;
- the Bézier curve;
- the non-uniform rational B-spline (NURBS).

Characteristics of CAD and Design

The above geometric models are configured to automate drawing by automatic drafting apparatus; then, design and drawing lead times and manpower are significantly reduced. Meanwhile designers can concentrate on creative design.

'Design' should be directed towards new products; it requires a great deal of creativity and professionalism. However, today's CAD still plays a large role in engineering drawing. For real design activities, the application of artificial intelligence or knowledge engineering may be required. In this process expert systems based upon production rules—'if (conditions), then (execution)'—are effectively used. These systems comprise (Figure 24.7):

- knowledge acquisition from professional experts or published data, etc.;
- expression of 'if–then' rules;
- implementation of an inference mechanism.

24.2.3 Optimum Design and CAE

Methods of Optimum Design

Another major application of computer aid to design assists product development by attaining optimality, which may be difficult by manual calculation. This is *optimum design*, often called computer-aided engineering (CAE). CAE and CAD often proceed concurrently, resulting in CAE/CAD. Effective CAE methods are:

- structural analysis by finite element methods (FEM);
- characteristics analysis using vibration, fluid and heat transfer;
- optimisation analysis using systems theory, mathematical programming techniques, and expert systems based upon production rules using 'if' and 'then' reasoning.

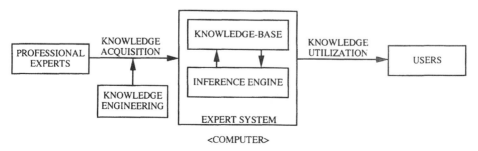

Figure 24.7 An expert system based upon production rules acquires production knowledge from professional experts, implements the inference mechanism, and generates suitable information for utilisation.

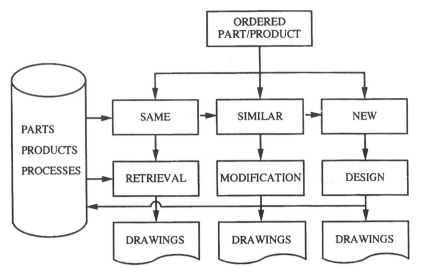

Figure 24.8 Design process based upon group technology retrieves the same pattern from the file, or executes parametric design on a similar pattern; only a new part/product is designed and the created pattern is stored in the file for future use.

Parametric Design

CAD can reduce design lead time by using group technology (GT) principles. As shown in Figure 24.8, if a required product/part has already been designed and details are stored in the design file/database, it may be rapidly retrieved and a hard copy drawing produced via an automatic drafting machine. If an ordered product/part is similar to one stored in the design file/data, *parametric design* modifies the shapes and dimensions of the stored (standard) design to achieve the required features. Design/drawing is restricted to completely new products/parts.

In this sense, configuration, construction, utilisation, management and maintenance of a database for design information is particularly important for effective design/drawing. A logical extension provides for the flow of design information to other functions such as process planning, production, cost estimating via computer networks to remote workstations or other companies, which constitute the 'virtual corporation', and cooperate in design activities for new products.

24.3 Computer-automated Process Planning (CAPP)

24.3.1 Scope of CAPP

Automated Process Planning

Production process planning or routing design, dealing with the conversion process of a raw material into the finished product, is based upon the shape and machinability of the raw material, the shape and accuracy of the finished product, and the available production facilities, jigs and tools, etc.

This requires a vast amount of knowledge and experience in manufacturing methods and technology. Accordingly, in order to automate process planning, referred to as *computer-aided process planning (CAPP)*, it is necessary to find technical laws concerning the sequence of operations associated with each given shape to be machined, based upon past experiences of process design, and to make a process file.

Two Methods for CAPP

The following two major methods have been presented for CAPP.

- The *variant* (or *retrieval*) *method*. Appropriate sequences of operations for producing a given shape are retrieved from the process file. Since generally a number of feasible alternative process routes are available to complete a finished product from a raw material, the best one is chosen.
- The *generative method*. An optimal sequence of operations is determined by logical laws of production technology and by minimising the total production time or cost, together with the proper selection of the production facilities to be utilised.

At present no fully satisfactory algorithm to automate process planning is available. Several of the procedures for CAPP will be mentioned in the following, describing their configurations and functions.

24.3.2 Automated Manufacturing Planning

Scope of AMP

'Automated manufacturing planning (AMP)' is one of the first computer programs for automated process planning developed by the International Business Machines Corporation. It is a basic, simple system for automatically converting engineering specifications and design data into manufacturing instructions. These are made up of a sequence of operations, manufacturing methods, and operation time standards, together with production cost estimates through the use of decision rules on the documented 'planning logic' of routing, operation methods, and time standards, based upon a vast amount of past experience of manufacturing technology.

Functions Involved in AMP

Figure 24.9 represents a flow chart of AMP. Its major functions are briefly explained below.

(1) *Reading input data*. The final shape, dimension, finish, and lot size of parts to be manufactured, properties and shape of raw materials, routing logic, and other data are read into the computer. Based upon these data, necessary operations for producing the finished part are generated.
(2) *Routing analysis*. A representative part is selected, and a routing pattern or a sequence of operations for this part is established.
(3) *Operations method analysis*. Machine tools and tooling that are suitable for each operation analysed above are searched and determined by means of

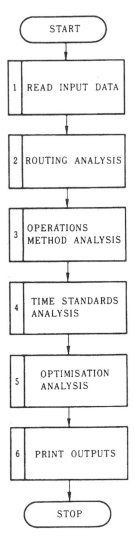

Figure 24.9 General flow chart for the automated manufacturing planning (AMP) system for automatically converting engineering specifications and design data into manufacturing instructions.

operations method logic tables by taking into account a complete list of machine tools, jigs and tools as well as measurable values of their capabilities.

(4) *Time standards analysis.* Standard times for setup and actual machining required for a series of operations are determined by standard data sheets that have been established by means of manufacturing technology and method and time study. Production costs required for those operations are also calculated.

(5) *Optimisation analysis.* With the criterion of the minimum time or cost, the optimal sequence of operations and their associated machine tools and tooling are selected.

(6) *Printing the outputs.* Manufacturing instructions, which include the process route (the sequence of operations), machine tools, jigs and tools to be utilised, machining conditions, time and cost standards, etc., are printed out.

24.3.3 Computer-aided Routing

Scope of CAR

'Computer-aided routing (CAR)' was developed in Japan in collaboration with the Kansai Institute of Information Systems. It determines a best process route (the sequence of operations) automatically and, in addition, supplementary information such as dimensions, accuracy, additional machining, a standard production time for a part to be manufactured, etc. This is done by sequentially generating and combining 'unit patterns' to produce the final shape of the part from the original shape of the raw material registered in the memory of the computer, and projecting this on a graphic display in an interactive mode.

Unit patterns are basic elements of geometrical shapes to be removed from the new material by machining operations, such as internal cylinder, taper, hole, groove, key, etc. These are prepared beforehand and registered in the memory of the computer, and retrieved to be sequentially combined so as to complete the final shape. Thus the proper portions of raw material to be eliminated by machining can be decided.

CAR System Flow

The CAR system flow consists of the following four stages, as shown in Figure 24.10.

(1) *Processing geometrical shape*: In a human–computer interactive mode, the original shape of raw material is first registered and is retrievable whenever needed by means of the graphic display. Unit patterns that are suitable for producing the final shape of the part against the raw material shape are retrieved with additional information such as dimensions, accuracy, etc. on the display. All the necessary processing items are given as a menu, and so the operator only selects those required and puts them in with specified data, whilst the technological feasibility of the operations specified is checked.

(2) *Deciding the process route*: After all the necessary unit patterns for producing the part in question are registered in the above program, the optimum sequence of operations is automatically determined. This is done, in connection with machine tools to be utilised, by arranging the unit patterns, with the primary routing program, for the principal machining of constitutional connected shape elements, and with the secondary routing program for supplementary machining of technological surface elements such as precision machining, drilling, grooving, screw cutting, etc. As a result a sequence of operations for processing the part, as well as portions to be machined and their size, are indicated on the graphical display.

(3) *Calculating production times*: Additional information as to machining conditions and supplementary operations such as size measurements and inspections of the machined surfaces, if needed, are registered into the computer. Unit production times required for each of the operations registered by the previous program are calculated, based upon standard time data, by taking into account alteration of machine tools, jigs and tools as well as setups of workpieces.

(4) *Printing the outputs*: The results obtained in the above procedures are printed out as the recommended sequence of operations and other necessary

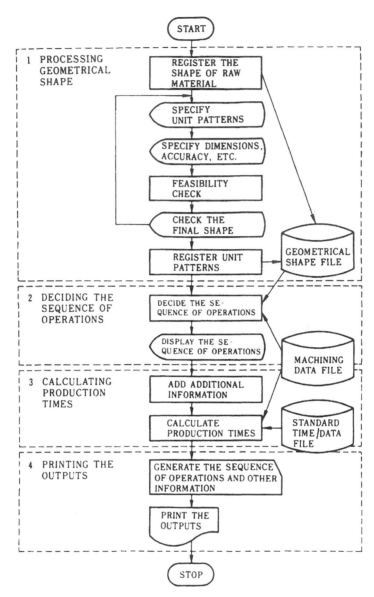

Figure 24.10 General flow chart for the computer-aided routing (CAR) system. This is made up of geometrical shape processing by means of the 'unit pattern' method and process route decision in a human–computer interactive mode, together with production time calculation and print-out of the outputs.

information such as dimensions, accuracy, machining conditions, production times, etc. for the part to be manufactured.

— EXAMPLE 24.4

The final shape of a product converted from a cylindrical shape of a raw material through the CAR procedure is illustrated in Figure 24.11. □

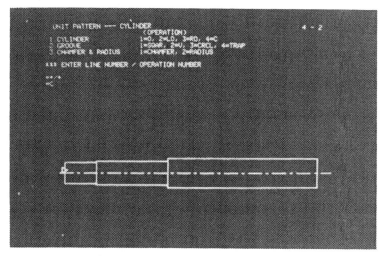

Figure 24.11 Illustration of a part configuration represented on the face of the graphic CRT display by the computer-aided routing (CAR) program (courtesy of the Kansai Institute of Information Systems). The original shape of the workpiece–cylinder–is converted into the shape as illustrated by means of the 'unit pattern' method on a human–computer interactive basis.

24.3.4 Comments on CAPP

Incompleteness of CAPP

As pointed out previously, few theoretical methods have yet been developed to create optimal process plans for the conversion of raw material into the product. An exception is the design theory of Suh (1990), in which a few corollaries are presented, such as minimisation of the number of required functions, integration of required functions in a part, use of standardised and interchangeable parts, simplification of product structure, etc.

Optimisation theory is mostly directed to select a best process route from among alternatives established on the past experiences of design/manufacturing engineers. Consequently, computer programs for automated process planning, which have been presented up to now, do not guarantee optimality in a true sense. They are based upon the method of dividing the conversion process into basic operations such as work elements or complex processes or unit patterns, then building them up on the basis of past experiences and determining the sequence of those operations by assigning a proper machine tool for performing each of such operations. This results in the construction of sequential arrangements of machine tools, irrespective of a true optimality.

24.3.5 CAD/CAM System

Integrating CAD and CAPP

Combining CAD (concerning decisions on the shape of parts) and CAPP (the machining process associated with NC machining), the drawing of parts can be

CAD/CAM by Concurrent Engineering

The 'CAD/CAM system' thus integrates design and production using the computer. However, such complete integration is not yet well established; there is some delay between the completion of the NC program and the execution of machining. *Rapid prototyping* technology helps accelerate this design process and speed up tooling and machining developments.

Simultaneous rather than consecutive activities for product development, product design, process planning, and production implementation is called *concurrent* (or *simultaneous*) *engineering*. With this technique production lead time is reduced significantly.

24.4 Automatic Operation Planning—Autoprogramming System

NC for Operation Planning

After determining a process route (the sequence of operations), a detailed plan for each operation must be established, such as:

- deciding the order of elemental operations like loading and unloading of workpieces, approaching of cutting tools, a sequence of actual machining operations, dimensions of a part to be machined, etc.;
- selecting proper cutting tools;
- determining machining conditions; and so forth.

These procedures are issued in numerically controlled (NC) machine tools, in most cases, by reading commands from NC digital information via the NC control unit/stored-program computer linked via a machine control unit directly to the NC machine tool.

Autoprogramming

The NC instruction can be generated manually (manual programming) or by an *autoprogramming system*, which is a sort of CAD, and includes the following three principal functions (refer to Figure 24.12).

(1) *Part programming.* This describes the configuration and dimensions of a part to manufacture, motion of the tools, machining conditions, and auxiliary functions of the machine tool to be utilised on a coding sheet using a particular NC auto-programming language. This part program includes:
   (a) statements describing the part,
   (b) statements instructing the path of the cutting tools and a sequence of operations, etc.
(2) *Main processor program.* This processes the input of the part program and calculates a series of coordinates for the tool path, resulting in CL (cutter location) data with supplementary instructions.

COMPUTER-AIDED DESIGN (CAD)

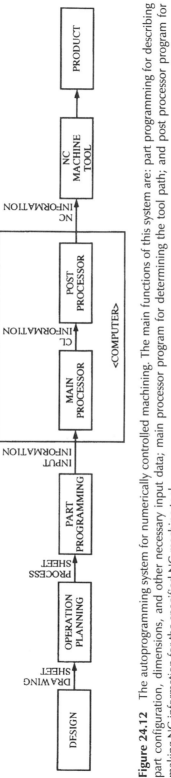

Figure 24.12 The autoprogramming system for numerically controlled machining. The main functions of this system are: part programming for describing part configuration, dimensions, and other necessary input data; main processor program for determining the tool path; and post processor program for making NC information for the specified NC machine tool.

(3) *Post processor program.* This is used to produce NC information from the CL data, for controlling the specific NC machine tool and control unit to be used.

APT

Of the many general- and special-purpose autoprogramming languages which have been developed to date, the most basic and best known is Automatically Programmed Tools (APT). The construction of this system first started with the Servomechanisms Laboratory of the Massachusetts Institute of Technology in 1952, now managed by Computer-Aided Manufacturing-International.

APT is highly complex with a capability of multi-axis (up to five) control, and performs mathematical calculations required to determine the geometrical contours of the tool path and the sequence of operations from the part geometry described (*geometrical processing*). APT has been the basis for other NC autoprogramming systems.

EXAPT

In practical machining, it is required to specify cutting tools and machining conditions. In APT these instructions are made by a part programmer who has thorough knowledge of manufacturing technology. *EXAPT* (extended APT) selects proper cutting tools automatically and determines the optimum machining conditions (*technological processing*) in addition to geometrical processing. This system started in Germany in the mid-1960s, and is now managed by EXAPT-Verein.

The basic EXAPT programming system consists of:

- EXAPT 1 used for programming of point-to-point and single straight line path control, mainly for drilling and simple milling;
- EXAPT 2 concerned with programming of axial turning for both straight-path and contour control;
- EXAPT 3 used for $2\frac{1}{2}$-dimensional contour milling.

Figure 24.13 shows the general flow chart of the EXAPT programs which consist of a geometrical processor and a technological processor. Geometrical statements of a part program are interpreted by the geometrical processor, and the result is produced as CLDATA 1. Based upon this intermediate output and by the use of data in the files for tools, materials, and machine tools, the technological processor interprets the technological statements, producing CLDATA 2, which includes information on the tool path, the sequence of operations, machining conditions, and tools selected. This is then processed by a post processor program to obtain NC information for operating the specific NC machine tool and NC control unit.

NC Languages

In addition to APT and EXAPT, general-purpose NC programming languages have been developed. Some use acoustic or voice programming through an electronic speech recognition system.

Several special-purpose NC programming languages also exist, for instance, for two-axis positioning with some contouring capability, etc.

COMPUTER-AIDED DESIGN (CAD)

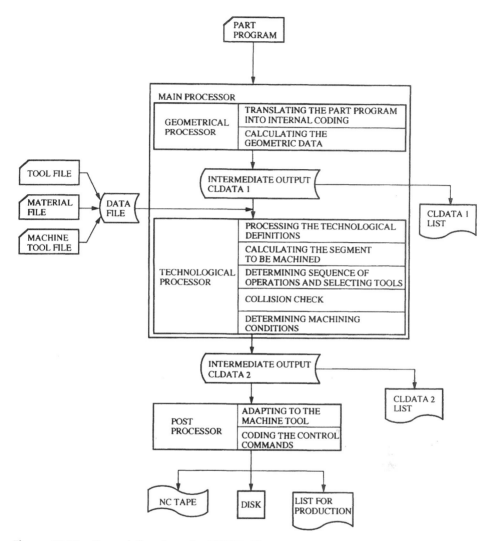

Figure 24.13 General flowchart for EXAPT. The part program is interpreted by, first, a geometrical processor, and then a technological processor with data files, producing intermediate output concerning the tool path, the sequence of operations, machining conditions, and tools selected. This output is converted to the NC control tape for the specific NC machine tools and NC control unit by the post processor.

24.5 Computerised Layout Planning

24.5.1 Scope of Computerised Layout Planning

Computer Use in Plant Layout

A facility layout of a plant can be accomplished with the aid of analytical methods, as explained in Chapter 9. Analytical approaches to layout planning and design have been developed to assist the layout planner in generating alternative layout plans by means of computer programs, which, in most cases, generate block layouts rather than detailed layouts.

Computerised Layout Algorithms

Computer algorithms for plant layout are classified into two (Francis and White, 1974):

- *Construction type.* This constructs the layout by building up a solution from the beginning by successively selecting and placing activities (departments/ workstations) until a satisfactory layout plan is achieved.
- *Improvement type.* This requires a complete initial (often existing) layout, and exchanges locations of activities so as to improve the layout performance.

Typical algorithms in use include Computerised Relationship Layout Planning (CORELAP), Automated Layout Design Program (ALDEP), Richard Muther and Associates Computer I (RMA Comp I), etc. for the construction type, and Computerised Relative Allocation of Facilities Technique (CRAFT) and others for the improvement type. In the following, CORELAP and CRAFT are described briefly.

24.5.2 CORELAP

Outline of CORELAP

CORELAP (Lee and Moor, 1967) is representative of the construction-type layout algorithm. It is concerned with generating a layout based upon closeness ratings determined from an activity relationship (REL) diagram, as described in Section 9.2.

The input data needed by CORELAP include the number of departments N, the area of each department, a REL diagram for all the departments, and weights for activity relationships (which are decided by the layout planner; e.g. $A = 6$, $E = 5$, $I = 4$, $O = 3$, $U = 2$, $X = 1$).

The closeness rating between departments i and j—$V(r_{ij})$ ($i, j = 1, 2, ..., N$)—is based upon the weight assigned to the activity relationship between the two departments. The *total closeness rating* (*TCR*) for department i is defined as:

$$\text{TCR}_i = \sum_{j=1}^{N} V(r_{ij}) \tag{24.1}$$

where $V(r_{ii}) = 0$.

CORELAP Procedure

CORELAP follows the procedure below.

(1) Choose the department having the largest TCR (say department 1–D1) and place it in the middle of the layout drawing.
(2) Scan the REL diagram to find any department having an A rating with D1, and place that department near D1; if no more exist, a check is successively made to find any department having an E rating, followed by an I rating, then

COMPUTER-AIDED DESIGN (CAD) 375

followed by an O rating. In case of ties, the department having the largest TCR is chosen.

(3) If no department has been found so far, choose an unassigned department with the largest TCR. If this happens to be department 2(D2), place it near D1 in the layout drawing.
(4) Scan the REL diagram again to find an unassigned department having an A rating with D1; if not, try to find one with D2. If none exists, search for any unassigned department having an E rating with D1, and then D2, followed by a rating of I, O, and U. For the case that no department is found, choose an unassigned department with the largest TCR. Then place the department next to D1 or D2 (now three departments have been placed in the layout drawing).
(5) Repeat the above search process until all departments under consideration are placed.

---- EXAMPLE 24.5 ----

(CORELAP procedure) Consider the REL diagram given in Figure 9.4. TCR values and sizes of areas of 10 departments to be located are calculated in Table 24.1. A layout scale of 1 m² (10 ft²) is selected as the smallest area element and a single digit on the layout drawing is used to represent this area.

Following the CORELAP algorithm, the order that the departments enter the layout is: 19, 13, 17, 18, 16, 15, 14, 12, 11 and 20. If the plant area is restricted such that the length is no greater than twice the width, we would obtain the CORELAP final solution given in Figure 24.14. □

24.5.3 CRAFT

Outline of CRAFT

CRAFT (Buffa *et al.*, 1964) is representative of the improvement-type layout algorithm. This algorithm is characterised by efficiently allocating the departments in a heuristic way, under the criterion of minimising the materials-handling costs for

16	16	13	13	17	17	12	12	12	0	0
16	16	13	13	19	19	19	19	20	20	20
16	16	18	18	19	19	19	19	20	20	20
14	15	18	18	19	19	19	19	20	20	20
11	0	0	18	19	19	19	19	20	20	20
0	0	0	0	0	0	0	0	20	0	0

Figure 24.14 A sample of the final layout produced by the CORELAP algorithm for the REL diagram given as in Figure 9.4, under the constraint that the length of the plant is no greater than twice the width (also see data calculated in Table 24.1) (after Francis and White, 1974, p. 114, courtesy of Prentice-Hall).

Table 24.1 Sample calculations for the total closeness rating (TCR) used in the CORELAP program, based upon the REL diagram given as in Figure 9.4, together with the sizes of department areas (after Francis and White, 1974, courtesy of Prentice-Hall).

Department	Number of units	Width	Length	TCR
19	14	4	4	33
14	1	1	1	33
18	5	2	3	28
16	6	2	3	26
15	1	1	1	26
13	4	2	2	24
20	13	4	4	23
12	3	1	3	23
17	2	1	2	23
11	1	1	1	23

multiple-product items, where this cost is expressed as a linear function of the transportation distance.

Basically, CRAFT seeks an optimum design by sequentially improving the layout, based upon material-flow analysis, as given in an initial layout specified as either an existing layout or a predetermined spatial array. Input data required include:

- the initial layout;
- material-flow data;
- material-handling cost data;
- the number and location of all fixed departments.

Material-flow and material-handling cost data, in the form of from–to charts, are conveniently used to express the number of items transported per time period and the material-handling cost required to move one unit of distance between any combination of two departments.

CRAFT Procedure

CRAFT calculates the total cost of materials-handling for the initial layout by utilising the from–to charts, regardless of whether any departments have been preassigned as the restricted areas.

Next, CRAFT considers exchanges of locations for those departments which either require the same area or have a common border, by employing pairwise or three-way interchanges. The exchange producing the greatest anticipated improvement, i.e. reduction in total cost, is made. This procedure is repeated until no further improvement can be made.

The transportation distance between pairs of departments in CRAFT is calculated as the rectilinear distance between centroids of departments; that is, $|x_i - x_j| + |y_i - y_j|$ for departments i and j, where (x_i, y_i) and (x_j, y_j) are the coordinates of the two departments.

The CRAFT program does not guarantee to find the least-cost layout, since not all

Table 24.2 Basic data for a CRAFT example problem (after Francis and White, 1974, courtesy of Prentice-Hall). Input data for CRAFT are: material-flow data and materials-handling cost data between pairs of departments, given as from–to charts, together with an initial layout represented in Figure 24.15(a).

To From	A	B	C	D	E	F	G	H	I	J	K	L	Department	Area (ft²)	Number of units[1]
A		1000 3.00	300 1.50										Receiving	600	38
B			1000 1.23	200 0.98									Machining	425	27
C		200 0.98		400 1.23		600 1.10							Grinding	200	13
D					600 2.10								Welding	250	16
E						200 1.80	400 1.50						X-ray	210	13
F							600 1.00	200 1.00		400 1.20			Inspection	175	11
G						400 1.00				200 1.00		600 2.40	Standing	125	8
H							200 1.00		600 1.70				Facing	275	17
I											200 1.30	400 2.10	Painting	285	18
J													Cleaning	150	9
K												200 2.00	Labelling	75	5
L													Storage and shipping	715	45
M													Restricted area	576	36

Note: Figures in upper part of entries: material flow data (in trips per month).
Figures in lower part of entries: material-handling cost data (in units of $1 per 4 ft of travel).
[1] 1 unit = 16 ft².

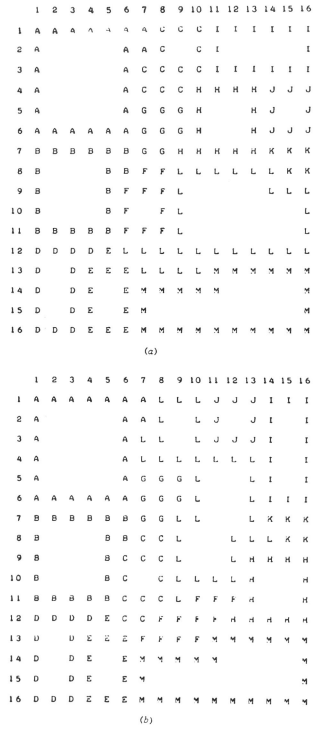

Figure 24.15 An initial layout (a) is improved by the CRAFT algorithm exchanging any pair of departments so as to reduce the total cost, resulting in the final least-cost layout (b) on the fourth iteration (after Francis and White, 1974, pp. 134–138, courtesy of Prentice-Hall).

possible exchanges (four or more at a time) are considered. It does, however, produce a suboptimal solution that cannot be easily improved upon. In CRAFT the final solution is dependent upon the initial layout given. Thus it is advised to obtain alternative solutions from several different starting layouts and then choose the best one.

EXAMPLE 24.6

(CRAFT procedure) Consider basic data for floor space requirements for the departments in terms of a scale of 1.5 m^2 (16 ft^2) per unit and from–to charts for material-flow data and transportation-cost data as given in Table 24.2.

Suppose that an initial layout is given in Figure 24.15(a), with the restricted area designated as department M. Using a pairwise exchange, the CRAFT program exchanged, first, departments H and L, next, C and L, then, C and F, and finally, I and J. Thus, on the fourth iteration the final least-cost layout was obtained as represented in (b) of the figure. The evaluated total-cost value was reduced from $110 696.88 for the starting layout to $87 963.81 for the final layout. □

The final layout plan obtained through the CRAFT program is recommended to be modified into a realistic layout plan via some manual adjustments taking qualitative factors and practical limitations into account.

References

Automated Manufacturing Planning (White Plains, NY: IBM).
BUFFA, E.S., ARMOUR, G.C. and VOLLMAN, T.E. (1964) Allocating facilities with CRAFT, *Harvard Business Review*, **42** (2).
FRANCIS, R.L. and WHITE, J.A. (1974) *Facility Layout and Location* (Englewood Cliffs, NJ: Prentice-Hall), Chap. 3.
HITOMI, K. (ed.) (1993) *Principles of Computer-Integrated Manufacturing* (in Japanese) (Tokyo: Kyoritsu Publishing), Sec. 2.1.
LEE, R.C. and MOOR, J.M. (1967) CORELAP— COmputerized RElationship LAyout Planning, *Journal of Industrial Engineering*, **18** (3).
SUH, N.P. (1990) *Principles of Design* (Oxford: Oxford University Press).

Supplementary reading

BEDWORTH, D.D., HENDERSON, M.R. and WOLFE, P.M. (1991) *Computer-integrated Design and Manufacturing* (New York: McGraw-Hill).
CHOROFAS, D.N. (1992) *Expert Systems in Manufacturing* (New York: Van Nostrand Reinhold).
ROGERS, D.F. and ADAMS, J.A. (1989) *Mathematical Elements for Computer Graphics* (2nd edn) (Singapore: McGraw-Hill).
ZHANG, H.-C. and ALTING, L. (1994) *Computerized Manufacturing Process Planning Systems* (London: Chapman & Hall).

CHAPTER TWENTY-FIVE

Factory Automation (FA), Computer-aided Manufacturing (CAM) and Computer-integrated Manufacturing (CIM) Systems

25.1 Automatic Machine Tools for Mass Production

Progress of Automation for Mass Production

Automation for manufacturing—*factory automation (FA)*—has been developed basically for continuous mass production. As the quantity of parts to be produced increases, the associated machine tools to be employed are more and more automated to increase the productive efficiency and decrease unit production cost—*economies of scale*.

The progression of automation extends from general-purpose machine tools to automatic machine tools, and further to special-purpose machine tools, as demonstrated in Figure 25.1. On the other hand, high-level automation requires a large amount of invested capital. The use of special-purpose machine tools reduces the ability to adapt to changes of product requirements caused by market needs.

General-purpose machine tools are capable of manufacturing a large range of shapes and dimensions of parts and products upon manipulation by the operator. In automatic machines and special-purpose machine tools machining instructions, such as shapes and dimensions for parts to be machined, tool selection and its path, machine motion, etc., are provided for in the hardware and/or software of the equipment. Typical automatic machine tools for mass production are explained below.

Automatic Machines

'Automatic machines' are mechanised to produce components continually without needing any manual operation, except for initial setup, replacement of worn tools and, possibly, loading of bar stocks or workpieces. Shapes and dimensions for the product are adjusted at the initial setup.

The most common of this type is the automatic lathe, sometimes referred to simply as an *automatic*. There are single-spindle and multi-spindle automatics, turret

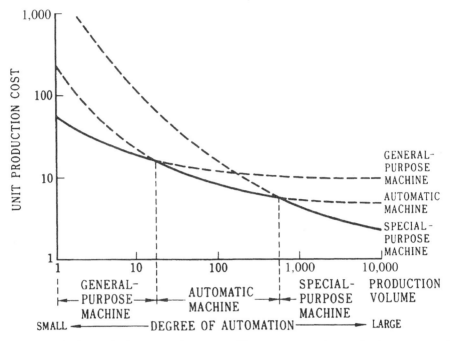

Figure 25.1 Unit production cost decreases with an increase in production volume. An appropriate machine tool is selected, depending upon the production volume. As it increases, the associated machine tool should be more and more automated to increase the productive efficiency.

and multi-tool (mounted radially around the machining region) types, with a variety of work-holding methods.

The multi-spindle automatic is basically a flexible, rotary-transfer type. Each spindle with its work-holding collet or chuck is indexed around the various tooling positions, and a component is finished during each indexing cycle.

If loading and unloading of workpieces are done manually, this machine is 'semi-automatic'.

High-level automatics are equipped with preset program control for each operation sequence by means of a pin or plug board, switch board, cam stepping, etc., and an automatic loading and unloading device. This type of *program control* is limited to straight-line and positioning controls with a limited number of control steps. Program-controlled machine tools are highly accurate and lower in cost for medium-sized and medium-volume production when compared to special-purpose machine tools, transfer machines, and numerically controlled machine tools.

Copying Machine Tools

Each point on a template which replicates the shape of the product is followed by a tracer. A *copying machine tool* (lathe, milling machine) produces the product by mechanical position control with the use of a servo-mechanism, such that the cutting tool follows the same path as the tracer does on the template.

Special-purpose Machine Tools

These are mechanised to produce large quantities of specific identical components with high productivity through fixed instructions. Two types of special-purpose machine tools are:

- *stationary machine*, which processes a workpiece which is set in a fixed position, with tools approaching from multiple directions;
- *indexing machine*, which holds workpieces on an indexing transfer table and machines them automatically as they are transferred from one position to the next.

These are a type of transfer machine which is explained below.

Transfer Machines

In transfer machines a series of simple machining heads, such as milling, boring, drilling, reaming, etc., are allocated as workstations according to the process route for the component, and stations for inspection and assembly operations may also be incorporated. Workpieces are automatically transferred from one machining operation to the next.

Two types of transfer machines are (Boothroyd, 1975):

- *rotary transfer system* via a circular indexing table—Figure 25.2(a);
- *in-line transfer system* along a conveyor—Figure 25.2(b).

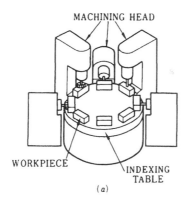

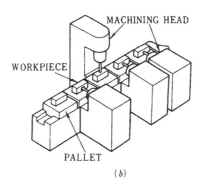

Figure 25.2 Transfer machines: (a) rotary and (b) in-line systems (after Boothroyd, 1975), in which a series of simple machining heads are allocated, and workpieces are automatically transferred and continually machined, resulting in the completion of finished parts at the end.

In the in-line transfer system, workpieces are in many cases held on pallets. Once manually fed into the transfer line, the workpieces receive the required series of operations continually and automatically from the machining heads as they progress along the line. In this way a complete component, such as an engine cylinder block, is produced. After it has been manually removed from the pallet at the end of the line, the empty pallet is returned to the beginning of the line. These transfer machines are basically special-purpose systems for highly productive mass production.

In contrast to the above type where stations on the transfer line are closely connected and dependent upon each other, there is another type called the *independent-transfer system*, in which each station works independently and between stations work-in-process inventories may be held. This is more flexible, but less efficient, than the rotary or in-line transfer system.

Manufacturing Efficiency

The efficiency of operating transfer machines is evaluated by the following two time elements:

- *Cycle time*—the time interval between successive completions of machined component: the maximum actual machining time plus transfer and positioning times. The cycle time is often decided from consideration of the monthly anticipated production volume and operating time, together with the machining processes, machining conditions, accuracy, etc.
- *Manufacturing lead time* for a product—total unit production time to complete a component through all the stations of the transfer line: sum of actual machining times on all the machining heads, transfer and positioning times.

Modular Method for Transfer Machines

Transfer machines are not usually capable of producing a variety of components. Since all the stations are closely related to each other, breakdown of any portion of a transfer line causes the breakdown of the total line, resulting in reduction of productivity.

To overcome this disadvantage, they are frequently constructed in a modular form, where the machining heads are standard items, and the system is equipped with those necessary heads at the appropriate stations. This construction is called the *building-block* (or *modular*) *system*, which was described in Section 2.4. This design raises the interchangeability between stations, thereby simplifying system maintenance, and can cope with versatile production requiring some model changes.

25.2 Numerically Controlled (NC) Machine Tools

25.2.1 Fundamentals of Numerical Control (NC)

Need for NC

NC is useful for low- and medium-volume varied-item production, in which shapes, dimensions, process routes, and machining methods are varied. The use of the conventional general-purpose machine tools is labour-intensive and low in productive efficiency, requiring a large amount of manual time for setup and preparation for machining, setting and approach of the cutting tool, etc.

The 'learning effect' (see Sections 4.3 and 8.3) is not expected where production volume is extremely small. Numerically controlled (NC) machine tools are machines developed to overcome this difficulty.

Development of NC

As the name implies, *numerical control* (*NC*) is a control over machines on the basis of digital information. This was introduced in the area of manufacturing in 1952 when, as previously mentioned in Chapter 22, a three-axis NC milling machine was first demonstrated at the Massachusetts Institute of Technology. Industry in the USA and Europe had developed various prototypes of NC machines before that; it is said that Kearney & Trecker Corporation had an NC milling machine in 1945. In NC machines, numerical information on machine motion, machining conditions, tool path, use of coolant, etc. are provided in the form of punched tape or stored program in the computer. The motion of the tool or workpiece depends upon the digital signal issued to servomotors, which drive the work table or tool holder.

Types of NC Systems

There are two basic types of NC systems as follows:

(I) A *point-to-point* (or *positioning*) *control system* controls, e.g. a drilling machine, a jig borer, a punching machine, etc., so as to position the machine table accurately to a specified location where an operation is to be carried out, but without close control of the path taken during the transition from one point to the next.

(II) A *continuous-path* (or *contouring*) *control system* controls the cutting path continually. Simultaneous motion control of two or more axes for machining of a complicated shape, such as a complex profiled shaft or cam, is performed on NC lathes or milling machines.

The position of a tool relative to the workpiece is specified by a series of coordinates, and a path between these points is interpolated, following either a straight-line path (linear interpolation) or a curved path (circular or parabolic interpolation). A contouring control unit and the NC program are much more complex and more expensive than a positioning control unit.

Advantages/Disadvantages of NC Machining

There are several advantages yielded by numerically controlled machining:

- adaptability to production of a variety of products;
- easy machining of parts with complicated forms;
- increase in accuracy;
- high-productive efficiency even by unskilled workers;
- decrease in manufacturing lead times;
- decrease in work-in-process inventories;
- accurate calculation of production costs, etc.

On the other hand, the invested capital increases, and planning and scheduling operations, including preparation of the necessary production information (NC programming, tape makeup, etc.), are required.

25.2.2 Classification of NC Machine Tools

NC Machine Tools Classified

NC machines are classified according to the level of information processing as mentioned in the following.

Conventional NC/Computer Numerical Control (CNC)

These include machines with a single-purpose operation, such as lathe, drilling machine, boring machine, milling machine, etc. Numerical information as to machining operation is provided by a punched tape (*conventional NC*), or in a control computer with a database storing NC instructions which directly controls the NC machine tool (*computer numerical control (CNC)*, also called a *soft-wired NC* or a *stored-program NC*).

Coded data are transformed to a series of pulse commands, through which the servomechanism, such as a stepping motor that rotates a fixed amount with each pulse, drives a work table and a tool to perform a specified motion and machining by either an open- or closed-loop system.

The CNC controllers can do linear, circular, or parabolic interpolation, which is made in software, and any interpolation routine can be selected with ease. Contouring or positioning modes are similarly selectable, as are absolute or incremental programming. Cutter compensation can be calculated as the work progresses. Known errors can be recorded and added to or subtracted from the position commands to give nearly perfect positioning.

Machining/Turning Centres

Conventional NC machine tools usually carry out single-purpose work with a single tool on a tool post or several tools on a turret. By contrast a *machining centre* (MC), which was originated in 1958 by Kearney & Trecker Corporation, automatically performs complicated multiple operations, such as milling, drilling, boring, facing, reaming, tapping, etc. on all faces of a workpiece except the base at one setup, with several axes of control on a milling-type machine and with several dozen cutting tools held in a magazine, selected, mounted on the main spindle, and exchanged automatically by the automatic tool changer (ATC). Figure 25.3 illustrates a computer numerically controlled machining centre.

Similarly multi-functional NC turning for rotational parts is carried out on a turning centre (Figure 25.4), most commonly of the turret-type.

With a machining/turning centre several production processes are centralised. This simplifies process planning to a great extent, and also scheduling is made with greater flexibility, yielding high utilisation of the machine. Thus, flexible, versatile component machining is automatically performed and production times are accurately estimated, resulting in easier process control. Moreover, the number of

Figure 25.3 Machining centre with a pallet for pre-setup of workpieces (courtesy of Yasuda Industries).

Figure 25.4 Turning centre with an NC display unit (courtesy of Okuma).

setups, setup times, transfer times, and work-in-process inventories are reduced, and the utilisation of the shopfloor space is raised.

On the other hand, a machining/turning centre requires a very large amount of invested capital; hence, high utilisation of the machine is a key requirement. Now that microcomputers are not as expensive as the early conventional NC controllers, CNC systems based on their use have been developed, and machining/turning centres are most commonly equipped with the CNC system. Inputs are read into memory, and thereafter re-read from memory as many times as there are parts to be made. If desirable, the program can be edited or corrected before use, so that a customised part can be made, and the system is therefore capable of flexible manufacturing.

25.2.3 Adaptive Control for Machining

Why Adaptive Control?

In ordinary numerically controlled machining, the cutter path, machining conditions and the sequence of operations are under constant control with commands from a punched tape or a computer, which represent a 'fixed' set of instructions, and cannot be easily changed even when the actual work situation has been changed due to the wear of a cutting tool, increase in tool forces and power, and other causes. The programmer must calculate a 'safe' set of operating instructions to allow for 'worst case' conditions, at the time of preparing NC instructions.

It is desirable to adjust and compensate machining control parameters (machining conditions) quickly to changes of actual work situations (particularly, the gradual wear of the tool as machining progresses), such that a suitable measure of performance, e.g. production time, or cost, or profit, is always optimised.

This technology for automatically optimised machining is *adaptive control* (AC). It offers a means of instantly recognising changes in the actual work situation and makes immediate compensation of machining conditions.

Mechanism of Adaptive Control

Figure 25.5 shows the mechanism of the NC system with adaptive control. The performance of adaptive control depends upon sensors mounted on the machine. The sensors measure dynamic work situations, such as tool forces, spindle torque, motor load, tool deflection, machine and tool vibration, accuracy of the machined workpiece, roughness of the machined surface, cutting temperature, heat deflection of the machine, etc. at short intervals during the machine operation.

Tool wear and life are not easy to monitor directly during actual machining; hence, these are estimated indirectly through the above measures.

These inputs to the adaptive control system are processed in a real-time mode to determine the optimum machining conditions for the next instant, and the spindle speed, feed rate, or slide displacement velocity are finely adjusted through the control unit, thereby yielding real-time optimisation of machining.

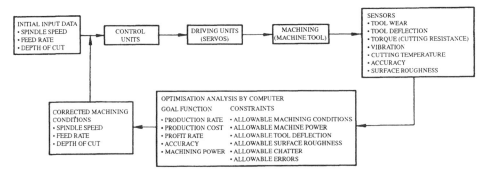

Figure 25.5 The machining system with adaptive control. In place of directly measuring the tool wear and life, they are estimated through an appropriate measure such as tool forces, tool deflection, vibration, accuracy, cutting temperature, etc., and the dynamic optimum machining conditions are determined by the adaptive control system in a real-time mode.

FA, CAM AND CIM SYSTEMS

Advantages/Disadvantages of Adaptive Control

Adaptive control for machining protects the cutting tools by preventing excessive stress, hence, wear; accordingly, tool life is increased. Due to the system's ability to maintain optimum machining conditions, feed and speed programming is simplified for any work material to be machined, and productive efficiency is improved.

However, adaptive control requires an additional fairly large expense for a real-time, highly sensitive sensor system, real-time optimum control for information processing, and a real-time numerical control unit.

Off-line Adaptive Control

Adaptive control for machining in an off-line mode is beneficial to the workshop. As described in Chapter 11, the slope constant n and one-minute tool-life machining speed C in the Taylor tool-life equation play essential roles in optimisation of machining. They are usually determined by machinability tests in the laboratories and provided in the machining database.

Without using such idealised machining data, these parameters must be estimated. Machining conditions will be optimised more accurately and economically by practical data obtained sequentially through real-life production in the workshop, such as the machined length with a cutting tool during its life, the number of machined parts per edge, etc., by the 'optimum-seeking machining (OSM) method', which is discussed in Chapter 29.

25.3 Computer-controlled Manufacturing Systems

DNC

The combination of several single [C]NC machines under the control of a [mini]computer is called *direct* (or *distributed*) *numerical control* (*DNC*), wherein CNC machine tools, industrial robots, transfer equipment, inspection devices, assembly machines, automatic warehousing facilities, etc. are operated through machine control units (MCU), on an on-line, real-time, and time-sharing basis.

'System 24', the 'Omnicontrol System', and the 'Group-control System' mentioned in Chapter 22 are early examples of DNCs.

Advantages/Disadvantages of DNC Systems

The DNC system has benefits as follows:

(1) Expansion of the manufacturing system is possible.
(2) Part programming is made with ease.
(3) All necessary numerical information can be stored in memory, and used at any time if desired.
(4) Production information concerning production volume and progress in the workshop, production time and cost, tool life, etc., are easily collected.
(5) The number of operators is reduced, and even unskilled workers can operate easily.
(6) Practical work based upon the predetermined schedule is possible with an increase in productive efficiency.

On the contrary, the system requires a vast amount of capital investment. Breakdown of the computer causes a great loss to the DNC system; to cope with this a backup for the computer system is required.

25.4 Flexible Manufacturing System (FMS)

25.4.1 Mechanism of Automatic Production

Need for Flexibility

For mass production of a single product, an automatic or a special-purpose machine tool is convenient and in the case of continuous flow production, the transfer machine is very effective, though it lacks in flexibility when the production object is changed.

To cope with the difficulty of multi-product, small-batch production, it is desirable to produce a variety of parts and products automatically.

Four Basic Automatic Activities for High Flexibility

System integration of the following four automatic activities is of vital importance for automation of multi-product, small-batch production, together with allocation of hardware equipment with high flexibility (Hitomi, 1990).

(1) *Fabrication.* In fabricating parts with machine tools, it is necessary to provide many cutting tools, which can be exchanged automatically. Machining centres and turning centres can operate in this way with high flexibility. Industrial robots also have flexibility in product assembly.
(2) *Setup of workparts.* For automatic loading and unloading of parts to and from machine tools or pallets, mechanical hands or robots are available.
(3) *Transfer of workparts.* For short distances mechanical hands or robots may be appropriate. For longer distances conveyors or automated guided vehicles (AGVs) can be used to transfer workparts from the warehouses to the machine tools, between the machine tools, and from the machine tools to the warehouses.
(4) *Storage of workparts.* Automated warehouses or carousels for small-scale works are used to store raw material and final products. Occasionally in-process works are temporarily stocked on the conveyor or pallet.

25.4.2 FMS

Role of FMS

Computer integration of the above-mentioned automated facilities realises a *flexible manufacturing system* (*FMS*) for automated multi-product, small-batch production.

FMSs are the most advanced production equipment in CAM, and play an important role in the construction of a CIM. As demonstrated in Figure 25.6, an FMS is concerned with the operation and working level (conversion of raw materials into the finished products). It works more effectively when linked to a hierarchical large

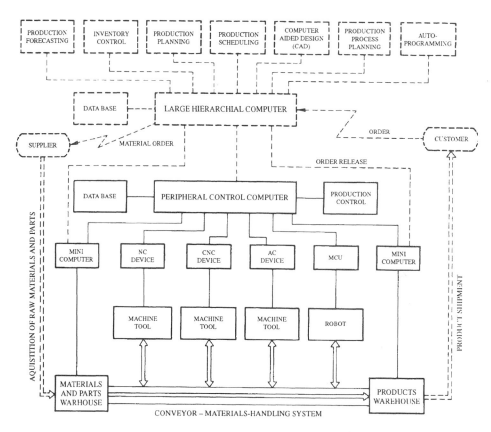

Figure 25.6 Structure of the FMS as well as the computer-integrated manufacturing (CIM) system. In the DNC system a few or several dozen numerically controlled machines are operated by a computer (or computer system) to perform the working level (conversion of raw materials into the finished products), as indicated by solid lines. The computer-integrated manufacturing system links this DNC system to a hierarchical large computer which deals with the management-level works, as demonstrated by the broken lines; thus, the total production information processing as to planning, implementation, and control is made. (In this figure, single lines exhibit information flow and double lines exhibit material flow.)

computer, which interacts with the management level (planning and control), and performs, in addition to auto-design (CAD), process planning (CAPP), autoprogramming, etc., production forecasting, inventory control, production planning and scheduling, etc. Production information thus obtained is directly communicated with low-level control computers for the selection of manufacturing instructions.

CIM and Manufacturing Information Systems

The hierarchical computer systems can then conduct total information processing as to planning, implementation, and control of a manufacturing system. This is a most advanced phase of development of automated manufacturing/production, aiming at the unmanned factory. It is *computer-integrated manufacturing* (CIM), sometimes referred to as *integrated* (or *intelligent*) *manufacturing system* (IMS).

This system is supported by a *manufacturing information system* which deals with information processing and allocation to various functions with the use of computer networks and a common data/knowledge-base.

Kinds of FMSs

There are three kinds of FMSs.

(1) *FMC* (*flexible machining cell*). This is a basic configuration consisting of a machining/turning centre and a robot or a pallet pool.

— EXAMPLE 25.1 —

A separate-robot-type FMC is illustrated in Figure 25.7. This FMC produces a variety of products. Materials handling is carried out by robots and conveyors. □

A larger-scale system is constructed by combining several basic (autonomously distributed) FMC. In general, to install a regular FMS is expensive. Small- or medium-sized firms can better afford FMCs, which generally require less investment.

(2) *Flow* (or *tandem*)-*type FMS*. Several machining/turning centres are laid out according to the work flow in a linear or loop arrangement. This is in some cases called a *flexible transfer line* (*FTL*).

— EXAMPLE 25.2 —

A flow-type FMS is illustrated in Figure 25.8. Bar-type workpieces are fed into the system, which is made up of three stages: turning operation, milling operation (including high-frequency hardening), and finish grinding and inspection, all stages being connected with transfer equipment having automatic loading and unloading devices; various products are completed automatically. □

(3) *Random-access-type FMS*. This type of FMS contains several machining/turning centres, conveyors or AGVs, and automated warehouses. Complicated workpieces with various shapes and operations may be loaded to the system at random. Each workpiece is transferred to the appropriate machine tools, fabricated according

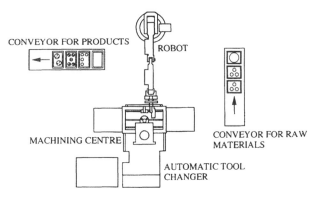

Figure 25.7 Flexible machining cell with a robot for material handling (after Holt, 1981).

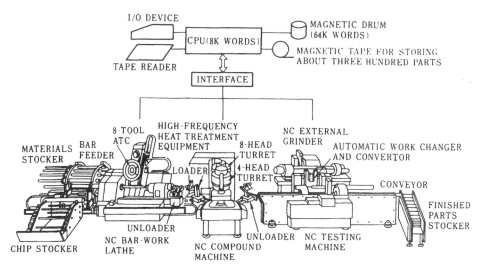

Figure 25.8 An illustration of the tandem-type FMS—'Parts Centre' by Okuma. Bar-type workpieces are fed into the system, which is made up of three stages: turning operation, milling operation (including high-frequency hardening), and finish grinding and inspection, all stages being connected with transfer equipment having automatic loading and unloading devices; thus, various products are completed automatically.

to machining instructions, and unloaded from the machines to be transferred to the exit. Sometimes the system can store in-process workpieces and exchange jigs and fixtures automatically. The system has information-processing functions to generate, store, transmit, and control a variety of information for multi-product, small-batch production. In addition it has managerial functions to generate machining schedules and perform stock and process controls.

—— EXAMPLE 25.3 ——

A random-access-type FMS is shown in Figure 25.9. This system automatically produces various components by transferring raw materials from the automated warehouse to an appropriate machine tool with an AGV. □

Thus this random-access FMS is concerned with consistent control of material flow and information flow in on-line and real-time bases. It is the most flexible automation system for multi-product, small-batch production.

Merits and Problems of FMS

The FMS has several advantages, automating multi-product, small-batch production, and reducing direct labour, in-process works, and lead times.

Although FMSs have very high flexibility to produce a variety of parts, they can handle only those products similar in shape and size. In this sense, they are merely hardware to realise group technology (GT) which aims to produce a variety of parts with similar shapes and dimensions.

The present FMSs are mostly used for parts fabrication; FMSs for sheet metal processing with presses and for forging are also realised.

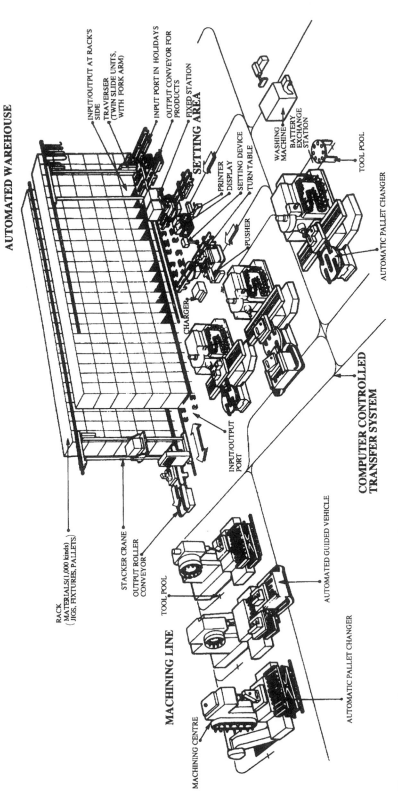

Figure 25.9 Random-access-type FMS (courtesy of Murata Machinery).

Since many machine tools and robots made by many vendors are used in an FMS and operated in a network such as a local area network (LAN), difficulties lie in the lack of standardisation of equipment control. Manufacturing Automation Protocol (MAP) is a useful solution for this, together with Technical and Office Protocol (TOP) for office automation, both being based upon Open Systems Interconnection (OSI), which will be explained in the next chapter.

25.5 Automated Assembly

Assembly Automation

The ultimate goal for assembly operations is automated assembly, in which the employment of mechanised means eliminates manual operations. Achievement of this through the use of the automatic devices to aid partially in the operators' work is *semi-automated assembly*, and when the whole operation is undertaken exclusively with machines, it is *fully automated assembly*.

Advantages of Automated Assembly

Automated assembly brings about advantages such as:

- reduction of assembly time and cost,
- uniformity of assembled products,
- increase in productive efficiency,
- release of operative workers from dangerous work, etc.

Automated assembly also helps to eliminate the use of people for mundane and simple repetitive work on the flow-type conveyor line.

Automated vs. Manual Assembly

However, perfect assembly by machine alone is not easy and super-precision assembly is not performed by machines. Moreover, it may often be economical to assemble details by using human labour, especially in developing countries where labour costs are low. *Cellular assembly systems* by a single worker or by a group of workers can economically and efficiently make versatile products.

Assembling

Assembling is an operation in which two or more parts are brought together and fixed in their specified relationship with either permanent or temporary fastenings. The basic functions of assembling are:

- feeding,
- transferring, and
- fastening,

as indicated in Figure 25.10.

Whilst the assembly function is of a combinative pattern (refer to Figure 8.2(c)), one part, such as the chassis for the assembly of a television set, becomes the

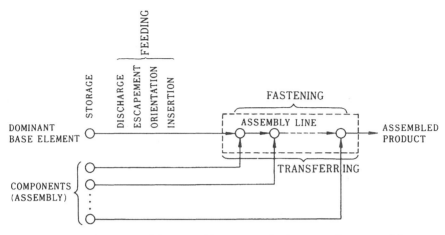

Figure 25.10 Basic functions of the assembly system—feeding, transferring, and fastening.

dominant base element, which is usually set up on a pallet and transferred from station to station and other parts are serially added in the correct order.

Defective assemblies, discovered at inspection points, are withdrawn from the main stream and reworked or rejected.

Parts Supply in Automated Assembly

In automated assembly parts are supplied from feeders:

- vibrating-bowl or rotating-disc parts feeders for small parts, or
- elevating-belt feeders for large parts.

Parts are oriented and fed to and loaded on the specified workstations.

Assembly Machines Classified

Assembly machines are classified as (Boothroyd and Redford, 1968):

- rotary- and in-line transfer systems (as in the transfer machines);
- continuous and intermittent movements;
- fixed-cycle (or synchronous) and free-cycle (or nonsynchronous) transfer systems.

Continuous vs. Intermittent Movement

In 'continuous movement', the assembly work head must move with the same speed as the transfer equipment, but indexing time is eliminated and production rate is high.

In most automated assembly the 'intermittent movement' type is advantageous. There are two kinds:

- The *machine paced fixed-cycle system* repeats operation with a constant speed; however, the breakdown of even one portion stops the entire assembly system.
- In the *free-cycle system*, buffer storage is installed between workstations, and each workstation is run independently to some degree. Hence, this is paced by

the individual operator. This type is considered flexible and suitable for the case of semi-automated manual assembly.

Use of Mechanical Hands/Industrial Robots

A recent trend in automated assembly is the use of mechanical hands or industrial robots, either individually programmed or controlled by computer.

Such robots select a part from some deposit point such as the chute end, conveyor, pick-up stand, etc., and transfer it into a new position, possibly to place it in contact with another part. Also the robots can insert pins, rods, shafts, etc., and fit parts together by threading, sliding, nesting, interlocking, etc.

By employing mechanical hands with multiple degrees of freedom and appropriate jigs and fixtures as modules according to the robot's usage, flexibility of automated assembly by robots will be increased, such as for assembling various products.

There are still many problems to be investigated, such as the hardware and software of robot technology, speed and accuracy of assembly, reliability, etc.

Flexible Assembly System (FAS)

Flexible manufacturing is also done for assembling mixed product types, based upon the group technology (GT) concept. This is the flexible manufacturing system for assembly, which is called the *flexible assembly system* (*FAS*). At present, however, the number of product types that can be assembled is limited and the assembly is usually done in lots.

—— **EXAMPLE 25.4** ——————————————————————————————

A general-purpose FAS for making watches is illustrated in Figure 25.11. □

———

Design for Automated Assembly

It is important that not only assembling operations, but also parts or components and products are suitable for automated assembly. Design of both components and products may not be the same as for manual assembly, but may be made in such a way that the basic operations of feeding, transferring, and fastening are easily and accurately performed on the machine. Noteworthy facts are as follows (Tipping, 1969):

(1) The number of parts should be minimised.
(2) Parts should be standardised.
(3) Parts should be symmetrical or, if not, of a special geometrical feature, such as an additional flat, lug, notch, groove, and so on, for easy orientation and positioning.
(4) Separate fasteners such as screws, bolts, and rivets should be avoided.
(5) Straight plunging motion is preferable to rotational.
(6) Products should be designed so as to be assembled step-by-step, loading parts from the top onto the dominant base component.

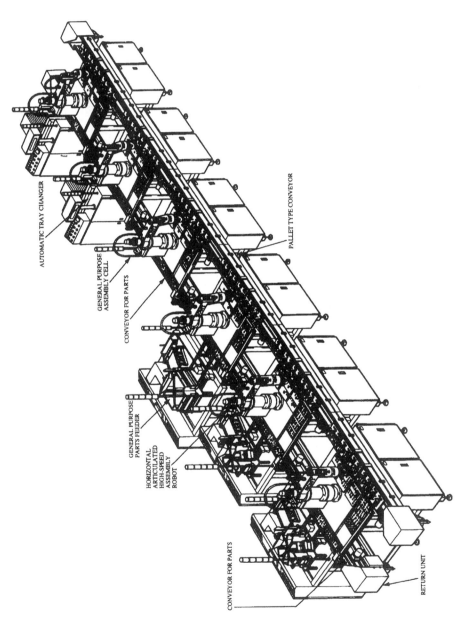

Figure 25.11 General-purpose automatic assembly system (FAS) (courtesy of Sony).

25.6 Automatic Materials Handling

25.6.1 Automatic Handling of Materials

Role of Materials Handling

Whilst the machining operation achieves change in the form of the workpiece, *material[s] handling* is concerned with the change of place and time—movement or transfer and storage—as was discussed in Chapter 6. It may be considered nonproductive, but it is a necessary aspect in manufacturing.

The *movement* is divided into:

- setup or loading the workpiece to the machine tool and subsequent unloading on completion of the operations;
- transferring between machines or stages.

Storage may be necessary:

- temporarily or long term;
- planned or unplanned (delay).

Materials handling of workpieces can involve various operations such as:

- transferring,
- storage,
- orientation,
- dividing,
- escapement, etc.

as demonstrated in Figure 25.12.

Automatic Handling of Materials

Automatic handling of materials is done mostly for mass production; it requires flexibility for varied production.

Automatic loading and unloading of workpieces is performed by the use of gravity, pushing, and grasping. The latter is the most flexible and commonly used. Recently high-level, flexible grasping, including picking up a part, holding it, moving it in any desired direction, and depositing it at the required location, has been developed with manipulators, mechanical arms and hands, or industrial robots.

Automatic Transfer Equipment

'Automatic transfer equipment' is required to provide rapid feeding between workstations and accurate orientation and positioning, together with large flexibility and simple maintenance. It is classified into two, as in automated assembly:

- *Fixed-cycle* (or *synchronous*) *type* feeds in intermittently and simultaneously all the components concerned with a constant cycle time on either a rotary indexing table or an in-line transfer machine.
- *Free-cycle* (or *nonsynchronous*) *type* transfers the workpieces independently of cycle time on the belt, roller, chain, or overhead conveyors, and in some cases, accompanies buffer stock areas between workstations, on the conveyor, in a

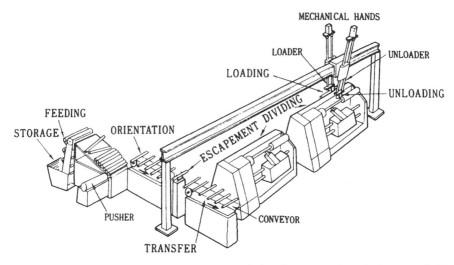

Figure 25.12 An illustration of automatic materials handling. Stored workpieces are fed into the conveyor, oriented, [taken], divided and loaded for machining and unloaded, and transferred.

branched-path, random-pattern conveyor, or in a temporary automatic warehousing facility.

In either type workpieces are fed either directly on the transfer equipment or placed on pallets. The flexible manufacturing system is typical of the free-cycle type, controlled by computer.

AGV

Conveyors occupy space whereas *automated guided vehicles* (*AGVs*) move freely along the wires embedded in the shop floor. AGVs now play major roles in running a flexible manufacturing system.

25.6.2 Industrial Robots

Birth of Robots

The term 'robot' has a wide variety of definitions including its science-fiction connotations originated by K. Čapek in 1920 from the ancient Slavic '*robota*', meaning 'slave labour'.

Industrial robots are easily reprogrammable, automatic, mechanical manipulators operating in several degrees of freedom and conducting a wide range of sequential works corresponding to those of the human arm, hands, and fingers. Industrial robots have developed greatly since the patent on a repetitive playback manipulator was presented by G.C. Devol in 1954. Robots are now equipped with sensory devices with such capabilities as vision (pattern recognition), artificial intelligence, voice response, tactile sense, etc.

Types of Robots

There are four basic types of industrial robot, as shown in Figure 25.13.

(1) *Polar configuration.* The body with vertical motion pivots either horizontally (rotary) or vertically, or a combination of the two.
(2) *Cylindrical configuration.* The body is a column that rotates on a vertical axis and moves the arm up and down.
(3) *Cartesian configuration.* The body moves vertically and horizontally and the arm horizontally.
(4) *Multiple-joint configuration.* The body, the column and the arm flexibly rotate about their respective joints.

In any configuration, the arm attached to the robot's body traverses radially and provides basic positioning while the hand may have a variety of grippers (e.g. jaws,

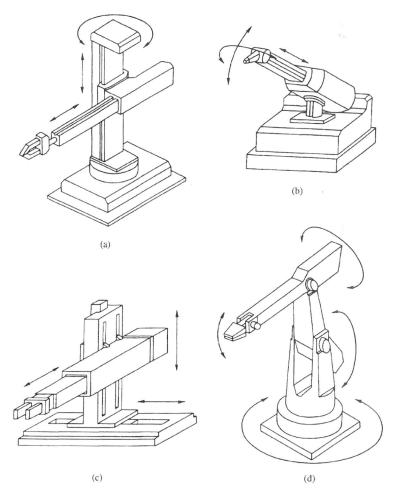

Figure 25.13 Robot configuration: (a) cylindrical, (b) polar, (c) cartesian, and (d) multiple-joint. A robot performs six basic motions (degrees of freedom): vertical, radial (or horizontal), and rotary motions, and wrist swivel, bend, and yaw.

suction pads) or other devices/attachments for carrying out operations (e.g. welding torches or guns). Grippers provide grasping, holding and wrist motions; swivel, bend, or yaw, provide orientation. Thus, a robot has the six basic motions or degrees of freedom.

Typical examples of industrial robots are (in use since 1964):

- Unimate by Unimation Inc.;
- Versatran by AMF.

Components of Industrial Robots

Industrial robots are usually equipped with:

- *manipulator*—performing operations with the use of tools and grippers;
- *driving force*—using power supply for action and movement (hydraulic, pneumatic, electromechanical);
- *control*—regulating the total motion; and in some cases,
- *teaching facility*—for creating the robot path and operations program.

Robot Motion

The motion patterns of robots are:

- *pick-and-place* (or *point-to-point*) *type*—simple loading, unloading, and transfer operations from one distinct point in space to another, with a large number of stopping points available for each motion;
- *continuous-path control type*—providing contour motions in two or three dimensions, such as required for welding or spray-painting operations.

Robots Classified

Robots are classified into:

- *fixed* or *variable sequence robot*—runs according to the paths, sequences, and conditions preset *ex ante*;
- *playback robot*—regenerates the paths, sequences, and conditions which are taught beforehand;
- *NC robot*—runs according to digital instruction;
- *robot system*—a group of robots performs efficient unmanned production.

Industrial Robots in Manufacturing

Industrial robots have been utilised to work in place of human operators in many areas of manufacturing systems, for severe, dangerous, dirty, or simple work, such as welding, die-casting, injection moulding, forge loading, press feeding, paint spraying, transferring, loading and unloading, assembling, etc.

Industrial robots are used in variations—suspended from overhead gantries, running on rails, etc.

The installation of industrial robots [systems] in manufacturing systems can require substantial invested capital; accordingly the cost-effectiveness should be carefully checked.

25.6.3 Automated Warehouse

Role of Warehouse

The modern warehouse plays the role not only of storage for raw materials, parts, end products, tools, jigs and fixtures, and gauges, but also of a dynamic inventory control for a smooth logistic system, such as procurement, production, inventory, distribution and sales, by establishing the information system to update kinds and quantities of stored items.

Role of Automated Warehouse

A warehouse which automatically controls both material and information flows, by using automatic transfer equipment with stacker cranes and conveyors as well as the automatic storage facilities within a high-rise stack building, which is wholly controlled by computer, is called an *automated warehouse*. In this system inventory information is continuously updated, and entering and exiting items and the request for inventory for particular items are easily responded to on a real-time basis.

The function of storage has been modernised by palletised handling of fixed amounts of like items, the unit load or pallet storage and retrieval system—as mentioned in Chapter 6. The sequence of operations is normally undertaken by an operator by means of an appropriate set of push-button controls. By keying the item and its quantity on the computer control console, the load on a standard pallet is moved automatically by a stacker crane. The crane usually consists of a vertical mast structure with platform, arm, and fingers rolling on a rail embedded in the floor or hung from overhead rails. The load is stored in random-access or assigned storage rack or bin, identified by a storage location number, of the storage or warehousing facility.

According to the order picking schedule put into the computer, the crane can collect a pallet from the specified storage rack and transport it to the picking location where the specified number of items are selected and transported to the shipping area. Empty pallets are then moved back to the pallet pool for reuse. If the pallet is not empty it is returned to the storage cell, or remains in the picking area in the event that another pick-up of the same item is already scheduled. Inventory files are updated as changes occur on a real-time basis.

The structure of an automated warehouse is indicated in Figure 25.14.

Control of Automated Warehouse

Automatic control is provided for the automated warehouse in two phases:

- *Control of material flow* includes:
  — operation of stacker cranes,
  — operation of conveyors for entry and exit of the items concerned,
  — search for the location number to pick up the specified item, and
  — instruction for the operations.
- *Control of information flow* includes:
  — scheduling operations of entry and exit of the items,
  — determining the location number,
  — reporting the current inventory status, and
  — processing the request for the particular item.

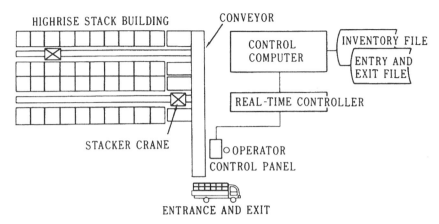

Figure 25.14 Structure of an automated warehouse. Stacker cranes move the loaded pallets automatically to storage, and also pick up the specified pallet and transfer it to the conveyor for picking the required kinds and number of items.

This computer-based system enables elimination of manual errors, improved item location, and increased storage facility utilisation.

Automatic Carousel

Small-scale automatic storage is done by 'automatic carousels', shelves on which the materials are stored rotate and the necessary items are stored or picked up at the entry/exit.

Automatic assortment

Automatic assortment with automatic packing has been developed. After this automatic assortment, trucks transport the packages for shipment.

25.7 Automatic Inspection and Testing

Need of Inspection

Almost all machined or formed parts are inspected for specified accuracy and quality. Appropriate action must be taken in case that quality of the parts is below the specifications. Because of recent trends in high production speed, quick detection of defective products, high labour cost, on-line process control, and so on, automatic inspection is increasingly important.

Inspection in Manufacturing

Major characteristics to be inspected in the manufacturing processes are numbers, dimensions, shape, weight, and defects, and other miscellaneous inspections for colour, chemical composition, etc.

Checking of numbers (counting) is one of the major non-productive operations in many industries, and many types of counting instruments are available, such as the simple mechanical, photoelectric, ultrasonic, and capacitive counters.

Automatic inspection for defects (cracks, flaws, and presence of foreign bodies) is done by inductive, ultrasonic, X-ray devices.

Weight of finished products is checked by conventional beam or spring-type weighing equipment with auxiliary electronic devices providing an indication in digital form.

Dimensional inspection is made automatically by examining parts for very small variations from target figures using gauging devices with pneumatic, inductive, capacitive, nucleonic, or optical sensors. Some of these devices can also be utilised to inspect the geometrical shape or surface features as well.

Automated Measuring Machine

Dimensional inspection is also performed rapidly and precisely with the *automated coordinate measuring machine (CMM)* that has opened the door to computer-aided inspection. The numerically controlled machine head with a movable probe or sensing point measures the surface of the object set up on the table in two- or three-coordinate dimensions. Measurements of the absolute dimensions or any discrepancies between the actual and the design dimensions are displayed by a digital readout device or output to a printer. Alternatively, the output of inspection can be fed to and recorded by a computer for further analysis of the data.

In-process Measurement

In automatic inspection by in-process measurements, objects to be measured are automatically checked at the necessary stages between workstations, or even during machining operations, and checked to ensure that the operation has been correctly performed.

The sensing and measuring device of an automatic inspection machine separates the measured objects into 'accepted' or 'rejected'. The rejected piece is retrieved and, where possible, corrected on a second pass through the machine.

In some cases, by the use of feedback systems, the machine feed is controlled, based upon the measured data, until the measured feature of the component reaches the specified size. In manufacturing sheets of metal, plastic, or paper, it is often required to inspect the thickness of a continuously moving sheet. This in-process measurement is done by radiation, nucleonic, or ultrasonic gauges, with reasonable accuracy and without directly contacting the material.

Transmission of the measured data and their recording and processing for statistical analysis are made quickly and accurately through a computer in an on-line, real-time mode. In some applications of computer-monitored in-process inspection such as in cylindrical grinding, the quality of a finished product is predicted, and based upon the result dynamic control is made by compensating the position of the wheel as it wears during metal removal.

CAT

When the functional capability of an assembly or a final product is evaluated under actual or simulated operating conditions, the process is 'testing' as distinct from 'inspection' (Harrington, 1973).

Typical of *automatic* or *computer-aided testing* (*CAT*) is the testing of aircraft jet engines, which is run by simulating a wide variety of pressure, temperature, load, fuel, etc. Temperatures are measured at critical points in the structure and strain gauges measure stress in components and weigh the thrust generated by the engine. A computer printout records the results as either acceptable or unacceptable.

25.8 Computer-integrated Automation System—Unmanned Factory

Unmanned Factory

The advent of the unmanned factory, which was suggested as the 'factory of the future' in 1946, is no longer a dream from a technological standpoint.

The economic disadvantage of huge capital investment aside, a fully automated manufacturing system could be constructed by integrating various functions and technical capabilities of automation for manufacturing (hard technologies), together with a hierarchical computer system (soft technologies).

--- **EXAMPLE 25.5** ---

(Unmanned factory) Figure 25.15 demonstrates an early version of the computer-integrated manufacturing system for low-volume variety production of small electric motors (Mifune, 1975). Product specifications from customers are read into the hierarchical host computer. Parts explosion is done for each of the products to be made. Stored parts, if any, are assigned, and new parts are ordered to be manufactured. Schedules concerning withdrawal of assigned parts from the warehouse and manufacture of new parts are made up to meet the assembly schedule determined by the due dates set for the ordered products, and these schedules are transmitted to the control computer.

Stored workpieces and parts are automatically drawn by a stacker crane and conveyor according to the withdrawal schedule from the automated warehouse, and delivered automatically to the machining line or the assembly line by the transfer monorail crane. A variety of motor shafts, which are difficult to standardise and depend upon the customers' needs, are machined through the DNC system. Workpieces are transferred to the conveyor, and automatically loaded and unloaded on the NC machine tools by the manipulators (mechanical hands) to receive necessary operations.

The finished parts are directly transferred to the assembly line, where assembling a variety of small motors is performed on the conveyor according to the assembly schedule, by using those finished parts and the parts delivered directly from the automated warehouse. The various completed products are automatically inspected, painted, and packed, and then shipped to the customers. □

Configuration of the Factory of the Future

The above-mentioned is an example of a full automation system for manufacturing. Here it is important to unify both the material flow and information flow completely, which is of major concern in this book.

In the material flow a series of machining, assembly, inspection, transfer, and inventory must be automated, as also must a series of demand forecasting, product

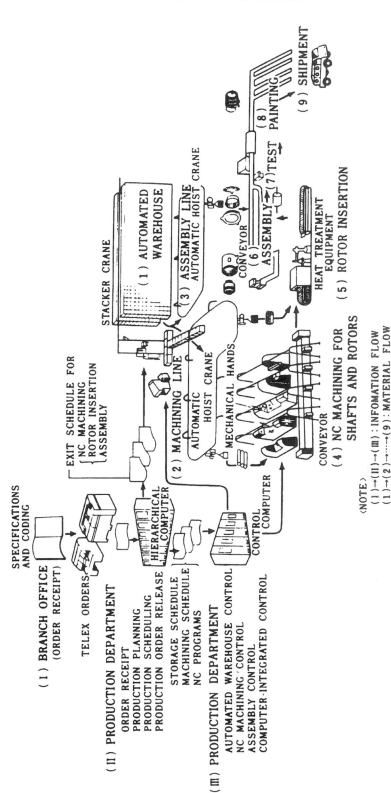

Figure 25.15 A computer-integrated manufacturing system for low-volume variety production of small electric motors at the Narashino plant of Hitachi, Ltd. (after Mifune, 1975). Orders from customers are analysed by the large hierarchical computer, which releases manufacturing schedules. Materials and parts are automatically picked up from the automated warehouse controlled by the control computer, and delivered to the DNC machining line, where materials are automatically machined, and also to the assembly line, where a variety of motors are assembled automatically and in part manually.

design, production planning, process planning, scheduling, operation planning, production control, and sales control in the information flow.

Once automated systems for both flow of materials and information are well integrated, manufacturing processes and production management—planning, implementation, and control—are done automatically. This is the configuration of the unmanned factory or the factory of the future.

References

BOOTHROYD, G. (1975) *Fundamentals of Metal Machining and Machine Tools* (New York: McGraw-Hill), pp. 246–247.

BOOTHROYD, G. and REDFORD, A.H. (1968) *Mechanized Assembly* (London: McGraw-Hill), Chap. 2.

HARRINGTON, J., Jr. (1973) *Computer Integrated Manufacturing* (New York: Industrial Press), pp. 206–208.

HITOMI, K. (1990) Steps toward CIM: System integration of computer-aided design, computer-aided manufacturing and computer-aided management, *Japanese Journal of Advanced Automation Technology*, **2** (2), 7–11.

HOLT, H.R. (1981) *Industrial Robot* (2nd edn) (Dearborn, MICH: Society of Manufacturing Engineers), p. 92.

MIFUNE, T. (1975) Order entry systems for motor manufacturing (in Japanese), *Journal of Japanese Society of Mechanical Engineers*, **78** (685), 49–57.

TIPPING, W.V. (1969) *An Introduction to Mechanical Assembly* (London: Business Books), Chaps. 4 and 5.

Supplementary reading

BOOTHROYD, G. (1992) *Assembly Automation and Product Design* (New York: Marcel Dekker).

CHANG, T.-C., WYSK, R.A. and WANG, H.-P. (1991) *Computer-Aided Manufacturing* (Englewood Cliffs, NJ: Prentice-Hall).

CHOROFAS, D.N. (1992) *Expert Systems in Manufacturing* (New York: Van Nostrand Reinhold).

JHA, N.K. (ed.) (1991) *Handbook of Flexible Manufacturing Systems* (San Diego: Academic Press).

OWEN, A.E. (1984) *Flexible Assembly Systems* (New York: Plenum Press).

REINTJES, J.F. (1991) *Numerical Control* (Oxford: Oxford University Press).

TEMPELMEIER, H. and KUHN, H. (1993) *Flexible Manufacturing Systems* (New York: John Wiley).

PART SIX

Information Systems for Manufacturing

This part describes basic principles and software of manufacturing information processing for automated 'flow of managerial information', in connection with Part III. To begin with, fundamentals of information technology—[management/strategic] information systems (MIS/SIS) and information network techniques are introduced (Chapter 26).

Two effective methods for multi-product, small-batch production with computers are introduced—parts-oriented production information systems for order entry by customers (Chapter 27) and on-line production control (or management) and information systems for efficiently running shop floors under dynamic conditions (Chapter 29).

Computer aids to production scheduling are discussed by introducing interactive group scheduling techniques and computer-aided line balancing (Chapter 28).

Computer-based production management and manufacturing information systems are recognised as a total system with several modules having distinct functions required for planning, implementation and control. The structure and procedure of COPICS (communications-oriented production information and control system) and BAMCS (business and manufacturing control system) are explained (Chapter 30).

CHAPTER TWENTY-SIX

Fundamentals of Information Technology

26.1 Concept of Information

Role of Information in Manufacturing

There is no need to point out the importance of the role of computers and information systems in the field of manufacturing technology and production management. In an attempt to realise this computer control of manufacturing [information] systems, seeking the so-called unmanned factory, it is necessary to manage both the material flow (manufacturing processes) and the information flow (production information processing) by computer.

What is Information?

Information is a system of knowledge that has been transformed from raw 'data', which are the raw materials of information and merely an expression of 'events', into some meaningful form to the recipient, and is of value in current or prospective decision-making at the specified time and place for taking appropriate 'action', thus resulting in evaluation of that 'performance', as exhibited in Figure 26.1. The terms data and information are often used interchangeably, but there is a distinction in that data are the raw material that is processed to provide information, and information is related to decision-making (Davis, 1974). If there is no need for making decisions, information would be unnecessary.

File and Database

A collection of related records in which data/information are assembled, arranged, classified, and coded in a computer is called a *file* (or data set).

Logically related files that fully cover up-to-date data and information are maintained in physical storage media such as disk packs, tapes, etc., and those data/information are timely and quickly updated and easily retrieved. Such an integrated system is called a *database*. The database is generally constructed to cover all the

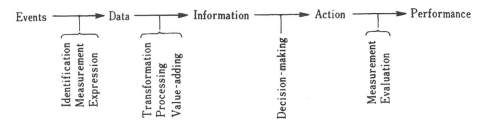

Figure 26.1 Generation and utilisation of information. Information is a system of knowledge transformed from data, and utilised for decision-making for future action.

records and information needed to handle a company's business with structures such as:

- hierarchical;
- relational;
- network;
- object-oriented.

The centralised management of data and information avoids the cost of duplication that would be involved in the maintenance of separate files. This common database is especially important for a CIM environment, as pointed out in Chapter 23.

Database System

The software for effective management and use of the database, such as data expression, storage, maintenance, retrieval, protection, etc., is the *database system*. Computer-aided decision-making, in an on-line and real-time mode at multiple remote decision points of the various subsystems (e.g. an executive's office), is made possible by timely access to the database from the data terminals through a communication channel. This is the information technology approach.

Characteristics of Information

There are several points to be considered in the discussion of information, as follows (Ogasawara, 1969):

(1) *Quantity and quality of information.* Since information is used for effective decision-making for various purposes, both quantity and quality of information are vital by expressing and recording contents of an event or a transaction multidimensionally. Of course, the costs associated with collecting, storing, and retrieving larger amounts of information increase rapidly.
(2) *Up-to-dateness of information.* Needless to mention, as the information utilised becomes more correct and more truly represents a current situation, the more accurate is the resulting decision-making.
(3) *Accuracy and reliability of information.* Accurate and reliable data and information are essential for correct decision-making.
(4) *Response and adaptability of information.* Stored data and information should respond quickly to and adapt to the various inquiries and future use for efficient decision-making.

(5) *Value and cost of information.* Raising the quantity and quality (value) of information for high-level decision-making requires large expenditure, e.g. for installing the information system for decision-making. Thus, both the value and the cost of information increase with its quantity and quality. Theoretically the optimal quantity and quality of information is to be decided by the point at which the difference between the value and the cost of information is the greatest. As mentioned in Section 20.2, at that point the added cost required to increase the information one more unit equals exactly the added revenue which resulted.

26.2 Information Systems

What is an Information System?

An *information system* is a network providing an information flow by processing data and information as inputs, and providing timely accurate information as output to decision-makers, as was also depicted in Figure 1.8.

Structure of Information Systems

The information system is composed of:
- equipment (hardware);
- procedures and programs necessary to make the equipment do the jobs (software);
- information;
- operators.

Hardware of Information Systems

The hardware structure of the information system is usually a computer system, which is made up of the following four units, as shown in Figure 26.2:

- *input device*—transmits the data/information from the outside world into the computer;
- *central processing unit (CPU)*—performs calculations and data processing, stores instructions and data, and monitors and controls the retrieval of data from the storage units;
- *storage unit* (or memory device/database)—stores data and programs that are used by the CPU;
- *output device*—delivers and displays the output from the computer in a form understandable by humans.

The use of these digital electronic devices for processing of data is called the *electronic data processing system (EDPS)*.

Functions of Information Systems

Major processing functions performed by software within the information system are explained below (Emery, 1969) (see Figure 26.2).

(1) *Data collection.* This function captures data about events impinging on the manufacturing system and its environment, serving as the sensory organ, and

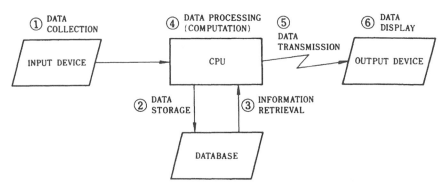

Figure 26.2 Structure and basic functions of the computer system made up of four major units. The input device feeds data collected into the central processing unit (CPU). Data are stored in the database, from which required data and information are retrieved for utilisation in the CPU. The CPU performs data processing or computation, and the result is transmitted to and displayed by the output device in a form understandable by humans.

loads data into the computer by means of keyboard, punched-card or tape reader or from a remote terminal, etc. Collected data are classified and indexed so as to make retrieval of desired information easy.

(2) *Data storage.* This serves the role of storing past data and information for future decision-making. The sum total of stored data and information available is the database as mentioned previously. Access to the necessary data and information contained in the database is provided by the data management function.

(3) *Information retrieval.* This function extracts and utilises the necessary data and information in the database at the right times.

(4) *Data processing*—computation. This consists of all processes within the information system that transform input data into output data.

(5) *Data transmission.* This function involves communication between geographically separated points by transmission of coded information.

(6) *Data display.* This is concerned with the preparation of output information in a form for human perception by means of printed form, temporary display on a CRT display or permanent copy, audio display, and the like.

Types of Data Processing

The following are typical of data processing:

- *Routine processing* performs numeric calculations such as calculation of payrolls and invoices, preparation of accounting data, technical calculation, and the like.
- *Optimising processing* generates optimal plans for well-structured problems by optimising algorithms.
- *Heuristic processing* obtains near-optimal plans for ill-structured problems by heuristic programs.
- *Interactive processing* makes decisions by means of the dialogue between human and computer as follows:
  — a decision-maker makes a plan,
  — based upon the plan the computer calculates and generates a result,

— the result is judged by the human and if it is satisfactory, the plan is adopted for execution; if not, a new plan is generated and the above procedure is repeated until a satisfactory plan is obtained.

Role of Manufacturing Information System

The information system for manufacturing (*manufacturing information system*) is now an absolutely necessary network for the manufacturing system to process accurately a variety of information at the right time, to make proper decisions, and to take appropriate actions, thereby adapting to the dynamically changing external environment. Examples will be described in Chapter 30.

26.3 Management Information System (MIS) and Strategic Information System (SIS)

Material and Information Flows for Effective Manufacturing

For effective and optimum management of a manufacturing system, the appropriately installed information system is vital for optimum decision-making concerning integrated manufacturing problems with regard to the process system, which implements efficient conversion of raw materials into the products (the material flow), and the management system, which deals with planning and control of manufacturing activities (the information flow).

MIS

Based upon the total systems concept mentioned in Sections 2.1 and 2.4, an information[-processing] system for the efficient integrated management of a manufacturing firm is designed so as to provide accurate and timely information for managerial decision within the problem-solving activities of planning, implementation, and control of various divisions and hierarchies in a company's management activities. This is the *management information system* (*MIS*). Managers and decision-makers rely on this system to choose appropriate means and ends for taking future actions.

Whilst simple data processing deals with simple, routine, non-managerial computations of past data mainly from an accounting-oriented standpoint, MIS is concerned with the provision of managerial output resulting from nonroutine, complex computations and futuristic considerations with respect to external as well as internal conditions of the company's management. It will need continuous efforts for the design, development, and maintenance of an effective MIS from the viewpoint of cost-effectiveness.

SIS

The highest level of management is strategic planning, as mentioned in Section 5.1. An information system which deals with this function is called the *strategic information system* (*SIS*). This term is relatively new, coined by C. Wiseman in 1985 (Wiseman, 1988). SIS is mainly concerned with the strategy of the tertiary (service) industries rather than manufacturing industries.

EXAMPLE 26.1

(SIS through POS) Sales data at each store—kinds and quantities of products—are collected through point-of-sales (POS) systems. These data from all the stores are gathered to a centralised computer facility. Based upon the data future actions are analysed and strategically issued (the information flow) in order to transport the correct products and necessary amounts strategically to the appropriate stores/markets (the material flow) (see Figure 26.3).
□

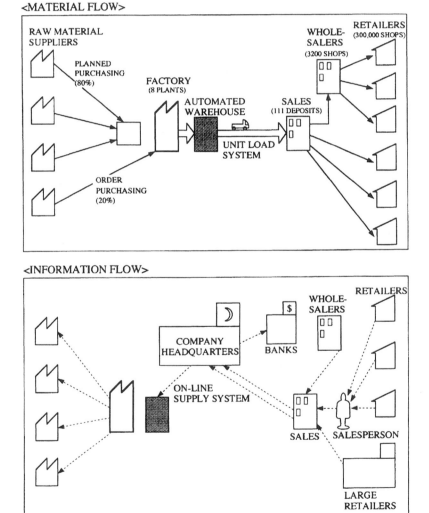

Figure 26.3 The strategic information system (SIS) installed in Kawo, in which sales data are gathered through point-of-sales (POS) systems, and the strategic information flow controls the flow of goods from raw-material suppliers through the plants to retailers (after Ohtsuki).

26.4 Information Networking

Ways of MIS

Currently, input/output and data processing are autonomously done with an engineering workstation (EWS) and geographically separated EWSs are directly connected with communication lines in the form of star, ring, and bus networks.

On-line/Real-time System

Receipt and transmission of the data and information are done in direct mode (*on-line*), and the response of throughput time (the delay between inputting data at the terminal and receiving output messages) is small enough to control the current work by the output (*real time*). This is the *on-line, real-time system*.

This computer system is utilised not only in the field of business management, such as the company's management information system, the order-entry system, the accounting system in the bank, etc., but this mode is also used in technological areas, such as the DNC system in the workshop, CAD in a communication mode, etc.

Ways of Data Network

A data network for direct communications among engineering data workstations, such as CNC machines and computers in a factory, is called a *local area network* (*LAN*).[1] If this configuration is extended to a wider area, it is a *wide area network* (*WAN*).[2]

A network with services to add extra-values, such as line conversion, code conversion, etc. is the *value added network* (*VAN*).

At present communicating among dissimilar systems provided by various vendors is not simple. In order to address the complexity of specifications and accommodate changes, the Open Systems Interconnection (OSI) reference model contains the architecture to segment the specifications into layers, as indicated in Table 26.1 (Bedworth *et al.*, 1991, p. 620).

MAP/TOP

In the area of manufacturing any computerised device installed in a factory can communicate via the Manufacturing Automation Protocol (MAP) which was first developed by General Motors Corporation in 1982. This protocol is a broadband, token-bus network protocol based upon the OSI reference model.

Table 26.1 Seven layers in OSI reference model.

Layer	Name	Function
7	Application	Selects appropriate service for applications
6	Presentation	Provides code conversion and data reformatting
5	Session	Coordinated interaction between end-application processes
4	Transport	Provides end-to-end data integrity and quality of service
3	Network	Switches and routes data
2	Data link	Transfers units of data to other end of physical link
1	Physical	Transmits on to network

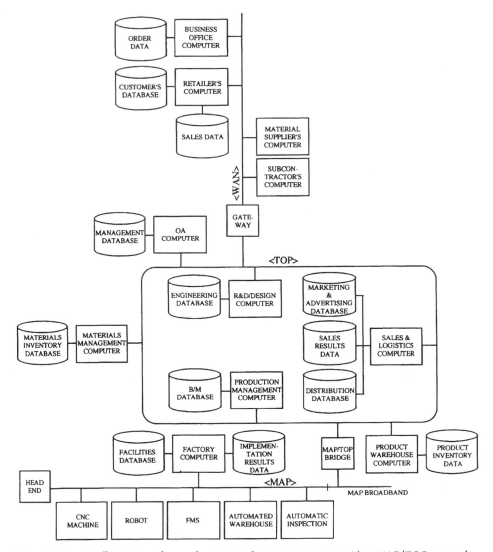

Figure 26.4 An illustration of manufacturing information systems with a MAP/TOP network.

A standard protocol for office LANs has different performance requirements from factory LANs (MAP). A base-band system successfully proposed by Boeing in 1985 is the Technical and Office Protocol (TOP) as a supplement to MAP.

These two networks are compatible. Their success, in that different vendors' protocol executions can correctly and reliably communicate with each other, is made by three types of testing (Bedworth et al., 1991, p. 624):

- conformance—confirms an execution to the specifications;
- interoperability—ensures interoperation among two or more conformant executions.
- functionality—ensures useful work by interoperable executions.

A manufacturing information system with a MAP/TOP network[3] is illustrated in Figure 26.4.

Notes

1. This can at present be performed by 'Intranet'.
2. This is at present supported by 'Internet'. Cooperative business activities through this computer network are called 'CALS' as explained in Section 3.5.
3. The strict configuration of MAP has not been successfully applied.

References

BEDWORTH, D.D., HENDERSON, M.R. and WOLFE, P.M. (1991) *Computer-integrated Design and Manufacturing* (New York: McGraw-Hill).

DAVIS, G.B. (1974) *Management Information Systems* (Tokyo: McGraw-Hill, Kogakusha), p. 33.

EMERY, J.C. (1969) *Organizational Planning and Control Systems* (New York: Macmillan), Chap. 3.

OGASAWARA, S. (1969) Information economics, in *MIS Handbook* (in Japanese) (Tokyo: Japan Management Association), Part 2, Chap. 6.

OHTSUKI, N. A Great Strategy by Kawou—VAN (in Japanese) (Chukei Publishing).

WISEMAN, C. (1988) *Strategic Information Systems* (Homewood, IL: Irwin).

Supplementary reading

BURCH, J.G. Jr., STRATER, F.R. and GRUDBITSKI, G. (1979) *Information Systems* (New York: Wiley).

SCOTT, G.M. (1985) *Principles of Management Information Systems* (New York: McGraw-Hill).

CHAPTER TWENTY-SEVEN

Parts-oriented Production Information Systems

27.1 Concept of Parts-oriented Production Information Systems

Parts-oriented Production System

As mentioned in Section 4.3, an approach to variety production is the *parts-oriented production system*, in which parts are produced according to demand or sales forecast in advance of receiving orders. After receipt of actual orders various products are assembled by suitably combining the on-hand parts. The plant layout for this is shown in Figure 3.1.

Information Systems for Parts-oriented Production

The parts-oriented production is to be supported by a proper information system in order to manufacture standard parts in appropriate amounts, to manufacture special parts rapidly at the right time, and to assemble ordered products within their due dates. The *parts-oriented production information system (POPIS)* is designed and developed for this purpose—producing diversified products, decreasing the manufacturing lead time, reducing the product inventory, and increasing the service level (Hitomi *et al.*, 1972, 1975).

Preconditions for POPIS

The following preconditions are taken into account in the design of POPIS.

(1) Manufacturing parts are made in appropriately large lots for stock replenishment, so that they are supplied on time to the assembly lines.
(2) Parts inventory is effectively utilised to cope with variations in demand for the variety of products.
(3) Products are assembled based upon actual orders.

27.2 Structure of Parts-oriented Production Information Systems

Basic Construction of Parts-oriented Production Systems

The parts-oriented production information system (POPIS) as developed is made up of three basic modules:

(I) the demand forecasting and production planning subsystem;
(II) the parts production subsystem;
(III) the products assembly subsystem.

The flow chart for this information system is represented in Figure 27.1.

Procedures of Parts-oriented Production Information Systems

The outline of functions in the above three modules is described in the following.

Module I *Demand forecasting and production planning subsystem.* This is concerned with forecasting demands and sales, and planning parts production and products assembly. It is the basis of the other two modules.

(1) *Demand forecasting.* Demand forecasting is done for parts, rather than products. Forecast amounts of products resulting from this analysis are referred to as the 'primary information'.
(2) *Receipt of orders.* Actual orders with due dates specified from customers are referred to as the 'secondary information'.
(3) *Determining safety stock.* Appropriate safety stock is established, mainly based upon the potential forecasting error to compensate for discrepancy between primary and secondary information.
(4) *Parts production planning.* This is performed periodically for several consecutive periods, mainly based upon primary information. That is, the parts explosion is made on primary information. Then based upon the calculated required amounts for the parts, the actual amounts to be released for parts production are determined by taking into account adjusted information including released orders on hand.
(5) *Products assembly planning.* This is made periodically for the next time period, based upon secondary information. That is, exact amounts for parts required are determined by parts explosion, based upon secondary information. Then amounts for assembly of required products are decided by taking parts inventory available at the assembly periods into consideration. In case of shortages of required parts, the 'crash' parts production program is initiated for fabricating such parts, and an urgent order release is sent to the workshop.

Module II *Parts production subsystem.* Based upon outputs from the parts production planning of Module I, this second module establishes a production schedule for required parts, which are produced for stock.

(6) *Parts production scheduling.* Based upon the results of parts production planning, machine loading is made for production of the required parts, which are allocated and sequenced according to an appropriate priority, thereby establishing the parts production schedule. In case of overload, proper production smoothing methods, such as overtime work, subcontracting, etc., are used. In case of underload, a schedule is established to produce the most

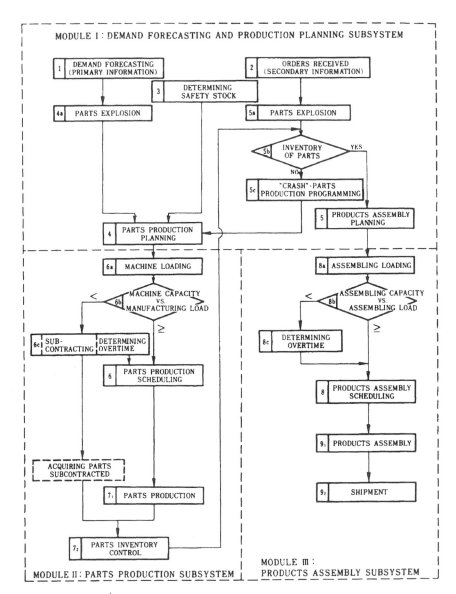

Figure 27.1 General flow chart for the parts-oriented production information system (POPIS). It consists of three modules—demand forecasting and production planning, parts production, and products assembly subsystems.

standardised parts for stock in advance prior to the receipt of definite orders for products, thereby maintaining a stable utilisation of production facilities.

(7) *Parts production and inventory control.* Specified parts are manufactured from supplied raw materials according to the parts production schedule, and the completed parts, together with purchased parts, are stored in the parts storage area or warehouse for control.

Module III *Products assembly subsystem.* Based upon outputs from the products assembly planning of Module I, this third module establishes the assembly schedule for required products, and products assembly is performed by the use of parts supplied according to Module II.

(8) *Products assembly scheduling.* Based upon the results of products assembly planning, loading is carried out for the assembly of the required products, which are allocated and sequenced according to the due-date priority rule, thereby establishing a products assembly schedule. This schedule involves overtime work in case of overload, whilst assembly of the most standardised products is made in case of underload, utilising idle times.

(9) *Products assembly and shipment.* Specified products are assembled according to the products assembly schedule, and then delivered to customers or markets, or stored temporarily in the warehouse as a special rare case.

27.3 Advantages of Parts-oriented Production Information Systems

General Advantages of POPIS

The parts-oriented production information system can provide the following general advantages under a dynamically changing manufacturing environment:

(1) Variation in the amounts ordered for various products and deviation between the primary (uncertain) and the secondary (certain) information do not greatly affect parts production.
(2) Production to order for products can be converted into production for stock for parts constituting the products, thereby resulting in economies of scale for parts production. The effectiveness of parts-oriented production increases as the level of standardisation increases.

--- **EXAMPLE 27.1** ---

The service level increases with the level of standardisation, and it is much higher in parts-oriented production than in conventional products-oriented production, as demonstrated in Figure 27.2. □

---

(3) Manufacturing lead times are reduced or eliminated; hence, a variety of products can be delivered to customers without storing a large amount of products inside the plant and in much less time than in products-oriented production.

The above results emphasise the superiority of parts-oriented production over conventional products-oriented production.

Practical Application of Parts-oriented Production

Parts-oriented production and its information system were employed by a certain manufacturer of domestic electrical appliances which produces 150 different product items giving an overall total of 1 200 000 pieces per annum and involving the combination of about 6000 parts. The plant consists of the parts machine shop, the assembly shop, and the automated parts warehouse.

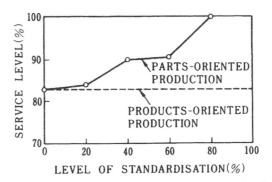

Figure 27.2 The services level increases with the level of standardisation for parts-oriented production, while it remains constant for products-oriented production, much less than that for parts-oriented production.

Parts-oriented production management on an on-line basis by the use of a computer was applied to this plant and executed, to replace the products-oriented production management that had been run so far.

The results obtained showed the following advantages and effectiveness of the application of parts-oriented production to the practical manufacturing situation.

(1) Product inventory was reduced by an amount equivalent to 30 days' shipment, and parts inventory was reduced by the amount required for 6 days' production.
(2) The management cycle of order receipt—fabrication—shipment was decreased by 2 months.
(3) The shortage of product required was reduced and the service level increased by 6%.
(4) Even when trouble occurred in the assembly shop, it was absorbed in a buffer stock that was established in the parts warehouse. Hence, the parts machine shop raised its utilisation by 6% without being affected by that trouble.

Thus, the feasibility and effectiveness of the parts-oriented production and information system and its adaptability to the ever-changing dynamic environmental situation of the manufacturing system have been both theoretically and practically demonstrated.

References

HITOMI, K. *et al.* (1972) Designing parts-oriented production information systems (in Japanese), *IE Review*, **13**(2), 87–96.

HITOMI, K. (1975) *Theory of Planning for Production* (in Japanese) (Tokyo: Yuhikaku), Sec. 16.

Supplementary reading

MURAMATSU, R. and DUDLEY, N.A. (1978) (eds) *Production and Industrial Systems—Future Developments of Industrial and Production Engineering* (London: Taylor & Francis), 509–522.

CHAPTER TWENTY-EIGHT

Computerised Production Scheduling

28.1 Interactive Group Scheduling Technique

28.1.1 Group Scheduling Technique (GST)

Group Scheduling

An important contribution to manufacturing is the concept of *group technology (GT)* which has now found wide acceptance in real-life workshops, as mentioned in Section 7.2. The following advantages are obtained by applying GT to production scheduling.

(1) The total production time is reduced by setting a group setup for a parts family included in a group.
(2) The material-flow pattern can be of flow type by establishing GT layout. Hence, with GT operations scheduling can be simplified as flow-shop scheduling; it is called *group scheduling* (Ham et al., 1985).

Solving Group Scheduling

The optimal solutions for group scheduling—the optimal sequences for groups and for jobs included in each group under the minimum-time criterion—are determined through the branch-and-bound technique, as explained in Chapter 14.

However, this procedure is not practical because of a huge consumption of computer time, especially as the number of jobs and the number of stages increase. This is recognised as an NP-hard problem. Hence a heuristic approach—the Petrov method—is applied to this problem, as explained below.

Modelling for Group Scheduling

In GT jobs (or parts) to be processed in N stages in a flow-shop pattern are classified into K groups. The group index is denoted by i ($=1, 2, \ldots, K$). In group $i(G_i)$, M_i

jobs are included. The job index is denoted by j ($=1, 2, \ldots, M_i$); then $\sum_{i=1}^{K} M_i = M$; that is, the number of jobs is M. Job j in G_i is expressed as J_{ij}.

Suppose that the setup time for G_i on M_k—stage k ($=1, 2, \ldots, N$)—is S_{ik} and the processing time (including a small amount of preparation time for loading and unloading) for J_{ij} on M_k is p_{ijk}. Let the total processing time of G_i on M_k be denoted by P_{ik}; then $P_{ik} = \sum_{j=1}^{M} p_{ijk}$.

Then $S_{ik} + P_{ik} = Q_{ik}$ is the group setup time + job processing times of G_i on M_k.

Petrov's Method Extended to Group Scheduling

In order to obtain fairly good group and job sequences for a multistage flow-shop scheduling, the Petrov method, which was mentioned in Section 14.2, is applied.

(I) *Determining job sequence in the group.* The following values are calculated for J_{ij} ($j = 1, 2, \ldots, M_i$) included in G_i ($i = 1, 2, \ldots, K$):

$$\left.\begin{aligned} A_{ij} &= \sum_{k=1}^{h} p_{ijk} \\ B_{ij} &= \sum_{k=h'}^{N} p_{ijk} \end{aligned}\right\} \quad (28.1)$$

where $h = N/2$ and $h' = h + 1$ if N is even; otherwise, $h = h' = (N+1)/2$.

Then Johnson's algorithm is applied for the A_{ij}s and B_{ij}s to determine the job sequence in the group.

(II) *Determining group sequence.* The following values are calculated for G_i ($i = 1, 2, \ldots, M$).

$$\left.\begin{aligned} A_i &= \sum_{k=1}^{h} Q_{ik} \\ B_i &= \sum_{k=h'}^{N} Q_{ik} \end{aligned}\right\} \quad (28.2)$$

Group sequence can be determined by application of Johnson's algorithm.

(III) *Calculating the makespan.* After both group and job sequences have been determined by the above procedures, the makespan—$F_{\max}$—is calculated as follows (Hitomi, 1978):

$$F_{\max} = \sum_{i=1}^{K} \left(Q_{\langle i \rangle N} + \sum_{j=1}^{M_i} D_{\langle i \rangle \langle j \rangle N} \right) \quad (28.3)$$

where

$$D_{\langle i \rangle \langle j \rangle k} = \begin{cases} F_{\langle i \rangle \langle 1 \rangle k-1} - F_{\langle i-1 \rangle \langle M_{i-1} \rangle k} - S_{\langle i \rangle k} > 0 \text{ for } j = 1 \\ F_{\langle i \rangle \langle j \rangle k-1} - F_{\langle i \rangle \langle j-1 \rangle k} > 0 \text{ for } j = 2, 3, \ldots, M_i \\ 0, \text{ otherwise} \end{cases} \quad (28.4)$$

$$F_{\langle i \rangle \langle j \rangle k} = \sum_{h=1}^{i-1} \left(Q_{\langle h \rangle k} + \sum_{j=1}^{M_h} D_{\langle h \rangle \langle j \rangle k} \right) + S_{\langle i \rangle k} + \sum_{h=1}^{j} (D_{\langle i \rangle \langle h \rangle k} + P_{\langle i \rangle \langle h \rangle k}). \quad (28.5)$$

In the above equations the symbol $\langle \rangle$ in the subscript is used to signify the order in the sequence; that is, $G_{\langle i \rangle} = G_\xi$ means that G_ξ (group ξ) is processed in the ith position in the group sequence, and $J_{\langle i \rangle \langle j \rangle} = J_{\xi \eta}$ (job η in G_ξ) means that J_η is processed in the jth position in G_ξ.

28.1.2 Interactive Group Scheduling Technique

Inputs/Outputs for Group Scheduling

Interactive human–computer systems are applied to group scheduling based upon the solution procedures mentioned above (Hitomi, 1988).

Inputs for interactive group scheduling (IGS) are:

- group setup times: S_{ik};
- job processing times: p_{ijk}

for $i = 1, 2, \ldots, K$, $j = 1, 2, \ldots, M_i$ and $k = 1, 2, \ldots, N$.

Then, the following graphical outputs are provided through IGS:

- Gantt chart drawn according to the group and job sequences obtained;
- makespan (total production time): $F_{\max}$;
- critical path, on which there is no time slack;
- operation sheet representing start and finish times of group setups and job processings.

Interactive Mode of Group Scheduling

To begin with, the number of stages (N), the number of groups (K), the number of jobs in each group (M_i), group setup times (S_{ik}), and job processing times (p_{ijk}) ($i = 1, 2, \ldots, K$; $j = 1, 2, \ldots, M_i$; $k = 1, 2, \ldots, N$) are put into the computer as basic data on an interactive mode.

--- **EXAMPLE 28.1** ---

This interactive process of group scheduling is demonstrated in Figure 28.1. ☐

After all the basic data have been given to the computer, they are expressed on the display as a table.

--- **EXAMPLE 28.2** ---

An illustration of the input table—group setup and job processing times—is shown in Table 28.1. ☐

Then the computer asks whether or not these basic data are correct. If there are any corrections, the table is interactively modified. If the human answers that the table is correct, then the computer determines group and job sequences, and displays the Gantt chart based upon the specified group and job sequences.

Table 28.1 Basic time data for group scheduling.

Group	Job	Stage 1	Stage 2	Stage 3	Stage 4	Stage 5
G_1	S_1	30	15	25	30	10
	J_{11}	41	65	39	79	52
	J_{12}	75	75	68	71	61
	J_{13}	32	25	62	73	54
G_2	S_2	10	20	15	30	25
	J_{21}	50	41	22	41	55
	J_{22}	30	28	41	48	64
	J_{23}	70	20	56	54	62
	J_{24}	48	34	48	29	52
G_3	S_3	15	25	30	20	10
	J_{31}	29	55	46	37	31
	J_{32}	26	20	37	51	28
	J_{33}	72	66	40	47	62
G_4	S_4	25	30	10	25	35
	J_{41}	47	71	29	38	24
	J_{42}	27	69	42	75	57
	J_{43}	78	45	73	74	29
	J_{44}	22	42	35	68	17

```
INPUT THE NUMBER OF THE GROUPS AND STAGES ?
4, 5
INPUT THE NUMBER OF JOBS IN GROUP 1 ?
3
INPUT SETUP TIMES AT THE STAGES FOR GROUP 1 ?
30, 15, 25, 30, 10
INPUT PROCESSING TIMES AT THE STAGES FOR JOB 1 ?
41, 65, 39, 79, 52
INPUT PROCESSING TIMES AT THE STAGES FOR JOB 2 ?
75, 75, 68, 71, 61
INPUT PROCESSING TIMES AT THE STAGES FOR JOB 3 ?
32, 25, 62, 73, 54
INPUT THE NUMBER OF JOBS IN GROUP 2 ?
4
INPUT SETUP TIMES AT THE STAGES FOR GROUP 2 ?
10, 20, 15, 30, 25
INPUT PROCESSING TIMES AT THE STAGES FOR JOB 1 ?

    ... Ready ...
```

Figure 28.1 The group scheduling technique (GST) inputs basic data for operations scheduling interactively.

EXAMPLE 28.3

The Gantt chart which resulted from GST for the basic data indicated in Table 28.1 is demonstrated in Figure 28.2. In this figure, a series of setups and jobs marked with an asterisk are on a critical path. No delay in completing setups or jobs is allowed on this path. □

The operation sheet representing the starting and finishing times of group setups and job processing can also be displayed. This sheet is convenient for the operators to employ in the workshop.

EXAMPLE 28.4

An operation sheet by group scheduling is demonstrated in Table 28.2; setups and jobs marked with an asterisk are on the critical path. The makespan for the example is 1091 hours, as indicated in this table and Figure 28.2. Setups and jobs marked with an asterisk must not be delayed. □

Relation Between Group Scheduling and Flow-shop Scheduling

The conventional flow-shop scheduling is a special case of group scheduling, in that only one job is contained in each group. Hence, by setting $M_i = 1$ ($i = 1, 2, \ldots, K$) and by replacing $S_{ik} + p_{ilk}$ with processing time, p_{ik}, for the ith job on stage k (=1, 2,

Table 28.2 The operation sheet obtained through group scheduling for the basic data of Table 28.1.

Group	Job	Stage 1 Start	Stage 1 Finish	Stage 2 Start	Stage 2 Finish	Stage 3 Start	Stage 3 Finish	Stage 4 Start	Stage 4 Finish	Stage 5 Start	Stage 5 Finish
G_2	S_2	*0	10	20	40	53	68	79	109	132	157
	J_{22}	*10	40	40	68	68	109	109	157	157	221
	J_{21}	*40	90	90	131	131	153	157	198	221	276
	J_{23}	*90	160	160	180	180	236	236	290	290	352
	J_{24}	*160	208	*208	242	*242	290	290	319	352	404
G_1	S_1	208	238	255	270	*290	315	347	377	440	450
	J_{13}	238	270	270	295	*315	377	*377	450	450	504
	J_{11}	270	311	311	376	377	416	*450	529	529	581
	J_{12}	311	386	386	461	461	529	*529	600	600	661
G_4	S_4	386	411	461	491	529	539	*600	625	661	696
	J_{44}	411	433	491	533	539	574	*625	693	696	713
	J_{42}	433	460	533	602	602	644	*693	768	768	825
	J_{43}	460	538	602	647	647	720	*768	842	842	871
	J_{41}	538	585	647	718	720	749	*842	880	880	904
G_3	S_3	585	600	718	743	749	779	*880	900	941	951
	J_{32}	600	626	743	763	779	816	*900	951	951	979
	J_{33}	626	698	763	829	829	869	*951	998	*998	1060
	J_{31}	698	727	829	884	884	930	998	1035	*1060	1091

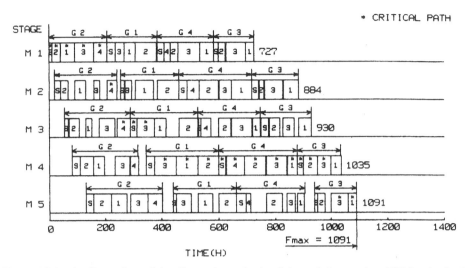

Figure 28.2 An illustration of the Gantt chart obtained through interactive GST for the basic data of Table 28.1. A chain of jobs marked with an asterisk indicates a critical path.

..., N), the above procedure can be interactively applied to the conventional operations scheduling.

28.2 Computer-aided Line Balancing (CALB)

Computer-aided line balancing developed by the Illinois Institute of Technology Research Institute includes both single-item and multiple-item line balancing (IITRI, 1969).

Inputs for CALB

Inputs for the CALB program are as follows:

(1) *Element definition deck.* Work element, its content, preceding work elements, required restriction, are described.
(2) *Restriction definition deck.* Name and kind of restriction, priority, are described.
(3) *Line definition deck.* Number of workstations, work position, number of workers, allowance time as station specification; the upper and lower limits of operation time as cycle specification; and change of operation time as unit position change allowance, are specified.
(4) *Model mix specification.* Name of item and production quantity are described.
(5) *Element/model catalogue.* Name of work element and name of item (or model) in which the work element is required, are described.

For mixed-model line balancing the following input decks are additionally employed.

(6) *Model operation.* This deck is produced automatically, indicating item name and production quantity as the first card; operation time for each item for each workstation.
(7) *Station definition.* Length of workstation, limits of the upstream/downstream working distance for workers, losses incurred in each workstation, are described.
(8) *Conveyor speed card.* Movement of conveyor per unit time is specified.

Table 28.3 A summary of the output through computer-aided line balancing (CALB) programs.

Number of stations	2
Number of operators	2
Number of elements	25
Total time assigned	2509
Total idle time	5
Total available time	2514
Average time/operator	1254.50
Cycle time limits	1250–1260
Cycle time	1257
PCT balance delay	.20

(9) *Sequence definition*. This consists of model-mix specification deck, cycle time card, and partition specification deck (item name and percentage production quantity incurred by partition).

Outputs of CALB

The outputs of CALB are:

- list of input data
- result of assigning work elements on each workstation
- summary of the assignment result.

— **EXAMPLE 28.5** —

Table 28.3 shows an example of the summary of the assignment result (Kuroda, 1984), in which

$$\text{percentage (PTC) balance delay} = \frac{\text{total idle time}}{\text{total available time}} \times 100\%. \tag{28.6}$$

□

References

HAM, I., HITOMI, K. and YOSHIDA, T. (1985) *Group Technology—Applications to Production Management* (Boston: Kluwer-Nijhoff), Chaps. 7 and 8.
HITOMI, K. (1978) *Production Management Engineering* (in Japanese) (Tokyo: Corona Publishing), p. 121.
HITOMI, K. (1988) Group scheduling with interactive computer graphics, *Manufacturing Systems*, **17** (4).
IITRI (1969) *Advanced Manufacturing Methods—Computer Aided Line Balancing*, Vol. I: User's manual.
KURODA, M. (1984) *Line Balancing and Its Application* (Tokyo: Business and Technology Newspaper Company).

Supplementary reading

CHOW, W.-M. (1990) *Assembly Line Design—Methodology and Applications* (New York: Marcel Dekker).
HITOMI, K. and HAM, I. (1977) Group scheduling technique for multi-product, multi-stage manufacturing systems, *Journal of Engineering for Industry, Trans. ASME*, Series B, **99** (3).

CHAPTER TWENTY-NINE

On-line Production Control Systems

29.1 Concept of On-line Production Control

On-line Production Control Systems

It is not easy, because of complicated situations of production activities, to control the manufacturing systems, especially in jobbing and variety production modes. Installation of an automated process control system is thus a necessity for the production management system, if the plant requires the intense control called for in a modern manufacturing environment by the construction of practical scheduling and control systems.

This system is quite effective, if the computer system is constructed in an on-line, real-time mode by connecting the control computer and the terminals (or factory computers) located in workstations. Production data generated at each of the workstations are collected on an on-line basis from the terminal to the control computer. Then the computer processes the input data to create new production instructions. This type of planning and control system for manufacturing is called *on-line production* (or *processs*) *control* (or *management*) *system,* as described in Section 4.3. It is an effective methodology to cope with jobbing—multi-product, small-batch—production.

Computerised Production Information Systems

Several types of on-line production management systems have been presented; the *computer-aided production information system (CAPIS)* (Akiba *et al.*, 1972; Hitomi *et al.*, 1974), which is discussed in this chapter, is a computerised production information system for controlling the multi-product, multi-stage manufacturing system by the timely establishment of optimal working conditions and operation schedules which adapt to dynamic changes in manufacturing situations and environment.

CAPIS provides methods of collecting information relating to practical operations in an on-line, real-time computer system, and of determining the optimal working conditions and schedules for all jobs to be processed which are then dispatched to appropriate workstations for further processing.

29.2 Structure of On-line Production Control

Functions Involved in CAPIS

The general flowchart of CAPIS is represented in Figure 29.1. CAPIS processes the following functions.

(1) *Collecting production data.* By means of terminals (or factory computers) located in workstations various data of actual production, such as the job–machine–operator combination, the production quantity, etc., are provided via keyboard and dials, card and badge reader, or bar codes, and transmitted to the computer on an on-line, real-time basis.

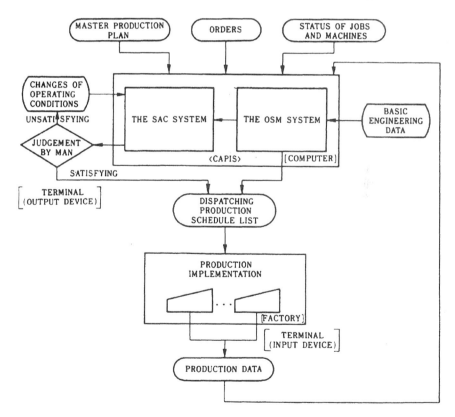

Figure 29.1 General flow chart of the computer-aided production information system (CAPIS). Practical data are collected through terminals located in work centres with on-line communication and processed through the scheduling and control (SAC) system and the optimum-seeking machining (OSM) system to obtain optimal operation schedules and the optimal machining conditions, which are dispatched to the workshop as subsequent instructions. Optimality of the operation schedules is evaluated by human operators.

(2) *Processing production information.* Based upon input data collected via terminals and with basic information (order information, master production plan, and engineering data such as status of jobs and machines and working conditions, processing times and costs for processing jobs), the computer carries out the following information processing as two core functions of CAPIS.
  (a) *Planning and controlling operation schedules.* Simulation is performed by the scheduling and control (SAC) system to obtain optimal operation schedules, which are judged by a production planner.
  (b) *Determining optimal working conditions.* The optimal working (or machining) conditions, especially optimal machining speeds, which are suitable to current manufacturing practice, are determined by the optimum-seeking machining (OSM) system, based upon practical machining data collected successively from production. Furthermore, optimal production time and cost data, as well as the optimal working conditions, are recorded in a file as basic engineering data.
(3) *Evaluation.* Optimality or feasibility of operation schedules and working conditions is judged by a decision-maker. If such results are not satisfactory, a new trial is made by setting another appropriate priority rule and operating conditions. This is done effectively by the use of a display device in human–computer interactive mode.
(4) *Dispatching the production instruction.* Optimal schedules and working conditions thus determined are displayed at each workstation by means of a visual display or printed instruction sheet. The next production will be executed in the workshop according to the instructions provided.

29.3 Scheduling and Control of On-line Production

Main Routine of Scheduling and Control of On-line Production

The detailed operation plan (process route and time schedule) is most conveniently made up by means of simulation based upon the human–computer communication involved in the scheduling and control (SAC) system of CAPIS. As indicated in the main routine of this simulation program shown in Figure 29.2, basic events treated are:

- *date change*—change of date after the completion of a day;
- *job-move completion*—completion of transfer of a job to the subsequent stage;
- *machine release*—occurrence of readiness for the next processing on a machine.

Other events such as start and completion of the use of jigs and tools, composition and decomposition of each individual job, etc., are treated as dependent events of the above basic events.

Outline of Scheduling and Control of On-line Production

The outline of the simulation process is as follows:

- *Initialisation.* Generate and set the start status of the simulation from the practical information.

ON-LINE PRODUCTION CONTROL SYSTEMS

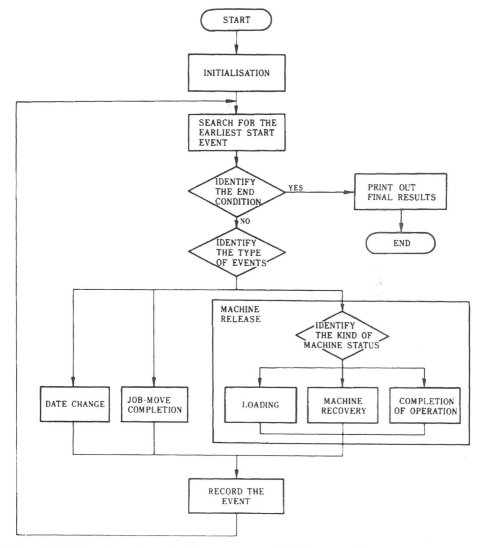

Figure 29.2 Flowchart of the scheduling and control (SAC) system of CAPIS. It provides the detailed operation plan (process route and time schedule) by means of simulation in human–computer communication mode.

- *Searching for the earliest start event.* Search for the earliest start event, and check completion of that event.
- *Output of the final results.* Generate the instruction, i.e. schedules for each combined job, machine and operator for every workstation.
- *Identifying the type of events.* Check the type of events, perform the event, as follows, and return the program to the next event processing:
  — *Date change module.* List possible jobs to be processed on the current day, and set the date as the next day.
  — *Job-move completion module.* After a job is transferred to the subsequent machine and its move is completed, add that job to the queue prior to that machine.

- *Loading module.* By loading a job with the highest priority on each individual workstation or machine, the job completion time is calculated.
- *Machine-recovery module.* Any machine is made available for use after its preventive maintenance or repair.
- *Operation completion module.* The move-completion time of a finished job is calculated.

29.4 Optimum-seeking Method for On-line Production Control

Optimum-seeking Machining

The optimal working conditions, particularly optimal machining speeds to be issued to the workshops, are theoretically determined by equations (11.23), (11.25), and (11.27). However, any reliable tool-life equation usable with those equations is usually unknown at the start of manufacturing.

To cope with this situation, an attempt is made to improve the current tool-life equation by the use of machining data obtained in practical work, so that it becomes increasingly accurate and reliable. It has been proved experimentally that in several iterative procedures close optimal machining speeds are achieved. This is the concept called the *optimum-seeking machining* (or *method*) (*OSM*) (Hitomi, 1970).

OSM Procedure

The OSM program depicted in Figure 29.3 is operated as follows: every time new machining data (in particular, tool life or number of workparts produced in the life time of a cutting edge) are obtained via the terminals or the factory automation computers located in the workstations. This program can provide an optimal machining condition—machining speed or feed rate for turning, milling, and drilling operations.

OSM Algorithm

The optimum-seeking machining proceeds as follows:

- *Initialisation.* An appropriate machining condition is set as an initial value, based upon past experience and conveyed to the workshop.
- *Measuring the tool-life value.* A tool-life value is obtained by conducting the machining operation at the initial machining speed.
- *Estimating the tool-life equation.* The two parameters, n and C, of the tool-life equation are estimated with the data obtained up to the present, using the method of least squares, by the following procedure.

    (1) First stage: Let the tool life be T_1 when conducting actual machining at the initial machining speed v_1 (e.g. $= 250$ m/min). Supposing that $n = n_1 = 0.25$,

    $$C = C_1 = v_1 T_1^{0.25}.$$

    Calculate an optimal machining speed v_2^o for n_1 and C_1 by using equation (11.23), or (11.25), or (11.27) according to the evaluation criterion adopted.

ON-LINE PRODUCTION CONTROL SYSTEMS   437

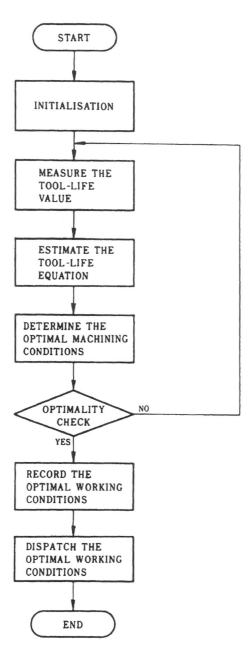

Figure 29.3 Flow chart of the optimum-seeking machining (OSM) system of CAPIS. This determines the optimal working conditions, particularly optimal machining speed by the step-by-step procedure of revising the tool-life parameters, based upon practical data obtained sequentially in the workshop.

(2) Second stage: Conduct actual machining at machining speed $v_2 = v_2^o$, and measure the new tool life T_2. Then with two pairs of actual data: (v_1, T_1) and (v_2, T_2), we obtain:

$$n_2 = \log(v_1/v_2)/\log(T_2/T_1), \qquad C_2 = v_2 T_2^{n_2}.$$

Based upon these n_2 and C_2, a new optimal machining speed v_3^o can be calculated.

⋮

(k) kth stage ($k = 3, 4, \ldots$): Conduct actual machining at machining speed $v_k = v_k^o$ (repeated as $k = 3, 4, \ldots$), and obtain a new tool life T_k. Then with k pairs of actual data: $(v_1, T_1), (v_2, T_2), \ldots, (v_k, T_k)$, we calculate new values for n and C by the method of least squares from the following formulae:

$$\left.\begin{array}{l} \log C_k \sum_{i=1}^{k} \log T_i - n_k \sum_{i=1}^{k} (\log T_i)^2 = \sum_{i=1}^{k} (\log v_i)(\log T_i) \\ k \log C_k - n_k \sum_{i=1}^{k} \log T_i = \sum_{i=1}^{k} \log v_i \end{array}\right\}. \quad (29.1)$$

Table 29.1 Outputs of CAPIS: Time schedules for (a) each job and for (b) each machine, produced by the scheduling and control (SAC) system, and (c) optimal machining speeds based upon criteria of the maximum production rate, the minimum cost, and the maximum profit rate, provided from the optimum-seeking machining (OSM) system.

(a) Job schedule

JOB NO	PART NO	GT CODE	NO OF LOT	NO OF OP	RELEASE	DUE DATE
10	10	9021	75	5	74-1-9	74-1-14

OP SEQ NO	MCN NO	UC NO	HOT CODE	JIG CODE	PCS	START DAY	TIME	COMPLETION DAY	TIME	LOT SET UP TIME	MCN TIME	QUEUE STAY TIME	LATENESS
10	9	21	0	1	1	74-1-9	8.0	74-1-10	9.2	42	450	0.	
20	15	23	1	2	1	74-1-10	9.6	74-1-11	9.1	30	360	0.	
30	10	21	2	2	7	74-1-11	9.6	74-1-11	15.7	48	270	0.	
40	14	23	3	1	1	74-1-11	15.9	74-1-16	10.6	30	240	0.	
50	16	24	0	4	1	74-1-16	13.4	74-1-17	10.6	42	270	50.6	• 3

(b) Machine schedule

MCN NO	UC NO	PRTY	SHIFT
3	12	1	2

JOB NO	PART NO	GT CODE	NO OF LOT	OP SEQ NO	HOT CODE	JIG CODE	PCS	START DAY	TIME	COMPLETION DAY	TIME	LOT SET UP TIME	MCN TIME	QUEUE STAY TIME
7	8	9025	10	11	0	3	1	74-1-8	8.0	74-1-8	18.4	84	540	0.
4	5	9025	25	11	1	0	4	74-1-8	18.8	74-1-10	10.8	30	1800	0.
2	3	9024	40	21	1	4	0	74-1-10	18.0	74-1-14	15.4	84	360	0.
19	9	8134	15	11	1	0	2	74-1-14	15.8	74-1-19	15.5	84	600	0.2
15	14	8126	50	30	0	9	1	74-1-19	15.6	74-1-23	19.2	66	450	0.
20	12	9028	30	31	1	0	0	74-1-23	19.8	74-1-25	15.4	30	360	0.
34	3	8322	50	30	1	2	1	74-1-25	15.8	74-1-29	15.8	84	2700	49.6
41	13	7210	90	21	0	8	0	74-1-29	10.8	74-2-4	10.6	66	150	53.1
13	15	9612	10	30	0	0	2	74-2-4	19.4	74-2-14	14.6	30	360	0.
36	15	7452	15	40	2	1	1	74-2-10	13.3	74-2-18	8.3	60	600	0.
		7241	25	60										

(c) Optimal machining speeds

JOB NO	PART NO	GT CODE	NO OF LOT	NO OF OP	OP SEQ NO	MCN NO	UC NO
10	10	9021	75	5	60	14	23

MIN-TIME CRITERION

ITERATION NO	CUT SPEED	CUT DEPTH	FEED RATE	TOOL LIFE	SLOPE CONST	1-MIN TOOL LIFE SPEED	OPT CUT SPEED	UNIT PROD TIME	UNIT PROD COST	UNIT PROFIT RATE
1	120	2.00	0.30	57.3	0.10	180	129	7.11	217	581
2	129	2.00	0.30	39.7	0.19	260	160	7.10	217	582
3	160	2.00	0.30	14.8	0.22	286	171	7.09	220	582
4	171	2.00	0.30	11.2	0.22	260	172	7.09	220	582
5	172	2.00	0.30	10.5	0.22	258	172	7.09	220	582

MIN-COST CRITERION

ITERATION NO	CUT SPEED	CUT DEPTH	FEED RATE	TOOL LIFE	SLOPE CONST	1-MIN TOOL LIFE SPEED	OPT CUT SPEED	UNIT PROD TIME	UNIT PROD COST	UNIT PROFIT RATE
1	129	3.00	0.30	58.7	0.10	180	107	7.09	217	581
2	107	2.00	0.30	99.4	0.21	283	114	7.11	217	580
3	114	2.00	0.30	73.8	0.21	283	114	7.11	216	581

MAX-PROFIT-RATE CRITERION

ITERATION NO	CUT SPEED	CUT DEPTH	FEED RATE	TOOL LIFE	SLOPE CONST	1-MIN TOOL LIFE SPEED	OPT CUT SPEED	UNIT PROD TIME	UNIT PROD COST	UNIT PROFIT RATE
1	130	2.00	0.30	56.9	0.10	179	124	7.11	217	581
2	165	2.00	0.30	11.2	0.20	305	153	7.10	220	582
3	153	2.00	0.30	16.7	0.20	269	152	7.10	212	582

Based upon these n_k and C_k, a new optimal machining speed v_k^o is calculated. Repeat this stage until the difference between v_k^o and v_{k+1}^o is within a small range (e.g. 5 m/min), then stop. The current values for n and C are reliable enough from a practical standpoint.

- *Determining the optimal machining conditions.* The optimal machining conditions are calculated, based upon the evaluation criterion of either the minimum production time, or the minimum production cost, or the maximum profit rate, using the tool-life parameters determined above. Then additional data such as unit production time and cost are also calculated.
- *Optimality check.* Whenever the difference between two successively obtained machining conditions comes within a small range, by repeating the above steps, then the last values obtained are the optimal machining conditions.
- *Recording the optimal working conditions.* The reliable tool-life equations, the optimal machining conditions, and other basic data are recorded in the basic engineering data file.
- *Outputting the optimal working conditions.* Optimal working conditions are output to workstations for practical use, or they are used as basic engineering data in the SAC program.

── **EXAMPLE 29.1** ──────────────────────────

(CAPIS procedure) Several results obtained through execution of CAPIS for planning and controlling a job shop on an on-line, real-time basis are illustrated in Table 29.1. The effectiveness of this on-line production control system is obvious; thus, CAPIS is capable of adapting to dynamically changing environmental situations of manufacturing.   □

References

AKIBA, H., HITOMI, K. and NISHIKAWA, N. (eds.) (1972) *On-line Production Control for Job Shops* (in Japanese) (Tokyo: Japan Management Association).

HITOMI, K. (1970) Economical and optimum-seeking machining, in *Metal Processing and Machine Tools* (Vico Canavese: Instituto per le Ricerche Tecnologia Meccanica), pp. 137–153.

HITOMI, K. *et al.*, (1974) Design of the Computer-aided Production Information System (CAPIS), *Proceedings of the International Conference on Production Engineering*, Tokyo, Part I.

Supplementary reading

HITOMI, K. (1987) *Multi-product, Small-batch Production—A Text* (in Japanese) (Tokyo: Business & Technology Newspaper Company), Chap. 11.

CHAPTER THIRTY

Computer-based Production Management Systems

30.1 Computerised Production Management

Use of Computers for Production Management

Production management by computers requires optimum control of manufacturing information for integration of planning, implementation, and control of manufacturing. The information needed for this management decision at the right time at the lowest possible cost is adequately provided by means of the management information system (MIS) supported by a total systems approach.

System Integration and Total Systems for Computerised Production Management

In constructing a large-scale system, such as computer-based production management systems, several modules with distinct functions required for planning, implementation, and control of manufacturing are integrated and connected to the computer based upon 'system integration' mentioned in Section 2.1.

This is the concept of *total systems*, which has been defined in Section 2.4.6 as 'an advanced approach in the design of management information systems for timely, optimum integration of administrative information, and this concept is closely associated with the use of electronic computers and data communications devices as "systems tools" to process large quantities of data, converting them into useful, timely information for managerial decision-making in complex modern business organisations. These are conceived of as an integrated entity composed of inter-related systems and subsystems' (Dickey and Senensieb, 1973).

MIS/DSS

Based upon the total systems and systems integration concepts, the information processing system for efficient production management is designed so as to provide accurate and timely information for managerial decisions within the problem-solving activities of planning, implementation, and control of various divisions and hierarchies in a company's management.

This is the *management information system (MIS)*, discussed in Section 26.3. Since this helps top management's strategic decisions, this decision support is called the *decision support system (DSS)*.

30.2 Computerised Manufacturing Information Systems

COPICS

A most widely used management information system for manufacturing or manufacturing information system for computers is the Communications Oriented Production Information and Control System (COPICS) developed by IBM (*Communications Oriented Production Information and Control System*, 1972). It is concerned with an approach to computer-integrated manufacturing (CIM) seeking a management (planning–implementation–control) oriented system by data communication through the use of computers, display terminals, shop-floor terminals, etc., in an on-line, real-time mode.

COPICS deals with establishing a production plan based upon a demand forecast, implementing productive activities from the purchase of raw materials, through production, to the shipment of the finished products, and allocating and managing production resources such as production facilities, labour force, and materials.

Modules/Functions Involved in COPICS

COPICS is composed of the following 12 interrelated modules via a system database, as shown in Figure 30.1.

(1) *Engineering and production data control*. This module creates, organises, and maintains basic engineering data, which are used by other modules of the system, in one central location accessible to all departments of the business as a database. These data consist of bill of materials, process routes for making products, production facilities, and additional data such as engineering drawing, product history, etc.

The basic functions in 'engineering and production data control' include:

(a) coordinating the acquisition of all data needed to describe a new item and creating a new bill of material;
(b) structuring the bill of materials by the bill-of-material processor;
(c) controlling engineering change;
(d) tracing the location and status of engineering drawings;
(e) creating and maintaining technical documents;
(f) applying CAD to product design;
(g) engineering information retrieval.

(2) *Customer order servicing*. This module links the sales information (order entry) to production. It covers rapid and accurate servicing and control of the customer's order from receipt of the quotation request, entry, and inquiry of the order to shipment of the finished products. This is achieved by the use of telecommunication systems.

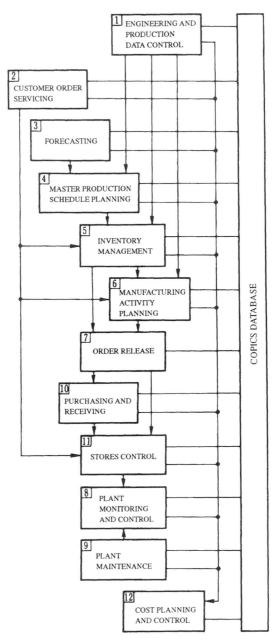

Figure 30.1 Structure and applications covered by the communications-oriented production information and control system (COPICS) composed of 12 interrelated modules via the system database. This is an approach to a computer-integrated management information system by data communication in an on-line, real-time mode.

Functions of 'customer order servicing' are:
(a) identifying the customer and the product ordered and pricing of the products;
(b) checking whether the requested due date is met, and if not, quoting the attainable delivery date;

(c) controlling unshipped orders from the stage of design through manufacturing to shipping;
(d) processing order information for answering customer inquiries and for evaluating the order backlog;
(e) monitoring the status of open quotations.

(3) *Forecasting.* This provides mathematical techniques to project finished product demands in the coming production periods and establish management standards to control manufacturing activity. The objective of this module is to find a best forecast model to represent consistent patterns of demand, thereby improving forecast accuracy. Thus obtained long-range future demand quantity is modified by human judgement to reduce the forecast error, thereby allowing a reduction in buffer stock. The result of this module is used in the development of the succeeding module, especially 'master production schedule planning'.

The basic functions of 'forecasting' include:

(a) conditioning historical data;
(b) selecting the forecast model to fit consistent patterns of demand, modifying it based upon economic and other external factors, and monitoring its applicability;
(c) projecting the future demands for time periods required by using the forecast model;
(d) applying life-cycle curves according to the history of similar items to modify the long-range projections and improve the forecast accuracy;
(e) correction of the forecast by human judgement, e.g. for the effect of one-time occurrence known in advance.

(4) *Master production schedule planning.* This is a statement of future product requirements specified by date and quantity which is determined, based on plant capacity, by reflecting management inventory policy as well as customer demand. This is a result of aggregate production planning. Short-term requirements for all subassemblies, components, and raw materials are determined by parts explosion so as to provide input to 'inventory management', and production resources such as production capacity are estimated for long-range periods. This function is *manufacturing resource planning* (MRPII), incorporating material requirements planning (MRP) and capacity requirements planning (CRP). In the case of varying seasonal demands, the system can be directed to stabilise the utilisation of production facilities by applying production smoothing. A realistic master production schedule obtained as a result of this module, is used for further detailed planning.

Basic steps of 'master production schedule planning' are:

(a) creating a net change caused by a revised forecast or customer order to the current production schedule;
(b) calculating the net change to the machine load, based on the above changes and a product load file expressing the load imposed on production resources by each of the types of end products;
(c) modifying the plan by examining overload and underload that would result from the above schedule change.

(5) *Inventory management.* This determines the quantities and timing of each item to be ordered at the level of raw materials (purchased), work-in-process goods, and

finished products, in order to meet the requirements of the master production schedule. This module plans and controls those inventories which enhance customer delivery service as well as reduce investment.

'Inventory management' provides two major functions. One is *inventory accounting* concerning inventory status records and transaction processing for all types of inventory, and the other is *inventory planning and control*, which is made up of:

(a) entry of end item gross requirements;
(b) determining net requirements as planned orders by subtracting on-hand inventory plus released orders from end item gross requirements;
(c) calculating safety buffer stock for absorbing demand fluctuations and safety lead time for forward adjustment in delivery date against purchased items;
(d) determining the economic order size to balance the costs of acquisition and carrying inventory;
(e) exploding planned orders for assemblies to develop requirements for the associated components and raw materials by the bill of materials;
(f) scheduling release of planned orders to cover remaining net requirements by determining the release date caused by the normal manufacturing or purchasing lead time. Requirements for subassemblies, components, and raw materials for the 'parent' assembly item are scheduled to be available on the release date of that item;
(g) inventory classification for determining items to be controlled stringently and accurately;
(h) pegging component requirements by allocating available and planned inventory of that component to requirements generated from various sources;
(i) predetermining the effect of a proposed schedule change, such as indicating materials to be expedited and the probable effect on other orders;
(j) projecting inventory investment by calculating the on-hand inventory in every time period.
(k) material requirements planning (MRP) for interrelated multi-plant situations by rapid replanning based upon net changes to master production schedules, together with high-speed data transmission between interrelated plants, resulting in reduction of total manufacturing lead time;
(l) inventory control for branch warehouses.

(6) *Manufacturing activity planning*. This module implements MRPII. It plans the details of capacity requirements indicated by 'master production schedule planning', and adjusts the release date and schedule of planned order established by 'inventory management' to be consistent with plant capacity, so as to meet due dates. Its objective is to achieve a reasonable level load (machines and manpower) as well as to minimise work-in-process inventory, idle machine time, and manufacturing lead time.

'Manufacturing activity planning' deals with the following three major functions:

(a) capacity requirements planning (CRP) in detail for covering production requirements in the coming time periods by analysing workloads and establishing planned capacity with consideration of overtime work, extra shifts, subcontracting, etc.;
(b) order release planning to the shop floor by adjusting the release dates and levelling the planned load on each workstation, together with the subsequent

operations sequencing. The sequence of release of orders is also determined by priority rules, and the start and completion times for each order are estimated;
(c) operations sequencing to produce a job-sequence list by considering the latest conditions of the shop so as to minimise machine idle times.

(7) *Order release.* This is concerned with the connection between the manufacturing planning and execution phases; that is, 'planned' orders, which are the primary output of the planning phase and indicate a release date and quantity, changed to 'released' ones for either purchase or production on the appropriate dates. It may be difficult and expensive to alter an order after release.

The basic functions of this module, 'order release', are:

(a) periodical review of the planned order release date;
(b) check of availability of materials and machines: the result of this check might prevent the release of orders. If they are available at order release time, their requisitions are generated and orders are allocated. Raw materials and workpieces in short supply are issued to high-priority orders;
(c) preparing the shop order documentation for production of released jobs;
(d) generating requisitions for purchase orders.

(8) *Plant monitoring and control.* This implements the production plan established by 'inventory management' and 'manufacturing activity planning', and serves to reduce the delay in the progress of each order by effective monitoring based upon feedback of manufacturing and shop-floor status data to provide better coordination between production and supporting activities such as maintenance, material handling, and inspection, and to improve the efficiency of production and machine utilisation.

Basic functions of 'plant monitoring and control' are:

(a) Attendance of actual manpower is quickly reported at start of shift.
(b) Delivery of materials required at other than the first operation is made to reduce inventory on the shop floor.
(c) Shop documentation is controlled, and required changes are incorporated quickly, if possible, through the use of terminals.
(d) Jobs are assigned based upon the work-sequence list prepared in 'manufacturing activity planning' and by the foreman's knowledge of individual workers' skills, machine tolerances, etc.
(e) A new job is dispatched with all required materials and tools immediately after the previous job has been reported complete from a machine.
(f) Manufacturing activity reporting of the completed job or any interruptions to normal production is made by the worker on an on-line, real-time basis through the shop terminal located at or near his/her work area.
(g) Inspection and testing on the shop floor. Earlier correction of off-quality production problems is made by analysis of data fed directly into the computer via terminals.
(h) Material handling of raw materials, work-in-process workpieces, and jigs and tools is performed to reduce idle times caused by their non-delivery.
(i) Jig and tool control by receiving advance notice of their requirements.
(j) Direct machine monitoring and control by continuous feedback of data of machine status to the computer to solve critical problems. Direct numerical control (DNC) of machines is effective in this action.

(9) *Plant maintenance*. This module addresses maintenance manpower planning, work order dispatching and costing, as well as preventive maintenance scheduling, so as to minimise the manufacturing costs by reducing machine breakdown and maintenance costs, whilst improving machine efficiency.

Some functions included in this module, 'plant maintenance', are:

(a) establishing labour standards for repetitive maintenance;
(b) automatic scheduling of preventive maintenance;
(c) determining the scheduled maintenance intervals;
(d) dispatching maintenance jobs on a priority basis;
(e) costing maintenance jobs;
(f) establishing the maintenance labour force.

(10) *Purchasing and receiving*. The objective of this module is to obtain the required quantity of raw materials by the date specified on the requisition, at the lowest possible price consistent with the required quality level. This assures purchasing concerned with relevant vendor selection, order release, and follow-up activities, receives ordered materials through counting and inspection by neglecting off-quality materials to be rejected, and routes the accepted ones to the warehouse or directly to the manufacturing area to be used.

The major functions of 'purchasing and receiving' are:

(a) preorder activities such as creating material supplier and quotation data, and evaluation and rating of individual supplier on the basis of price, on-time delivery, and quality level based upon purchased item history;
(b) order placement activities such as automatic creation of requisitions generated by the 'order release' module and their control, proper selection of supplier with a good rating at the time of order placement, and automatic generation of purchase orders and approval;
(c) post-order activities such as follow-up of purchase orders including changes to orders after their placement;
(d) receipt of orders by identifying incoming materials, validating their quantities against delivery documents, and expediting critical items and shortages;
(e) determination of routing of the goods—i.e. to the warehouse, inspection, the area to be used, reclamation or rework, or back to the supplier;
(f) inspecting and reporting on the quality of items received, and transmitting the results to 'purchasing', 'stores control', and 'inventory management' modules.

(11) *Stores control*. This regards inventory (measured in terms of quantity in 'inventory management') in physical terms with characteristics such as weight, volume, and physical location. The module maintains close and accurate physical control and record-keeping in respect of order receiving, storing, and issuing stocked items, and reduces the storage cost. Automatic warehousing techniques are used to reduce both picking time and picking errors, to eliminate manual errors, to improve item location, and to respond quickly to inquiry from remote places.

'Stores control' involves the following functions:

(a) Basic disciplines necessary for effective control over physical inventories are established.
(b) Location of stocked items is controlled with the aim of decreasing material handling within the warehouse and increasing space utilisation.

(c) Needs for order and requisition are filled so as to improve the efficiency of picking activities.

(12) *Cost planning and control.* This is addressed particularly to the financial executive and provides techniques whereby the information created and maintained for manufacturing purposes can be used for managing cost and accounting data and budgeting. All the modules presented in the above are connected to 'cost planning and control', as seen in Figure 29.1, and provide information for its specific functions.

Topics that this module provides are:

(a) direct labour and material costs and other direct costs;
(b) overhead costs, their apportioning to departments and products, and planning and control of divisional expense budgets;
(c) long-range planning of capital expenditure and investment.

In summary COPICS aims at computer-integrated production management on an on-line and real-time basis, through the total systems approach to planning, implementation, and control of manufacturing with rapid processing of production

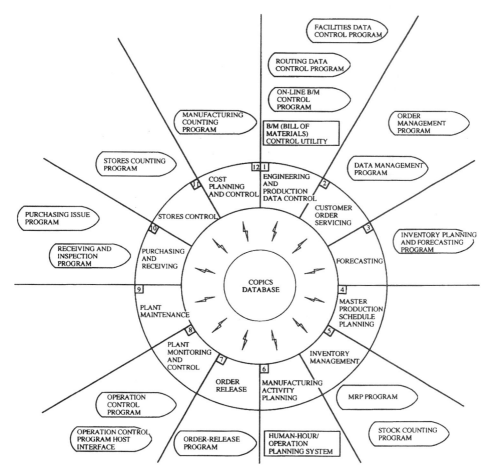

Figure 30.2 Programs supporting COPICS (International Business Machines Corporation).

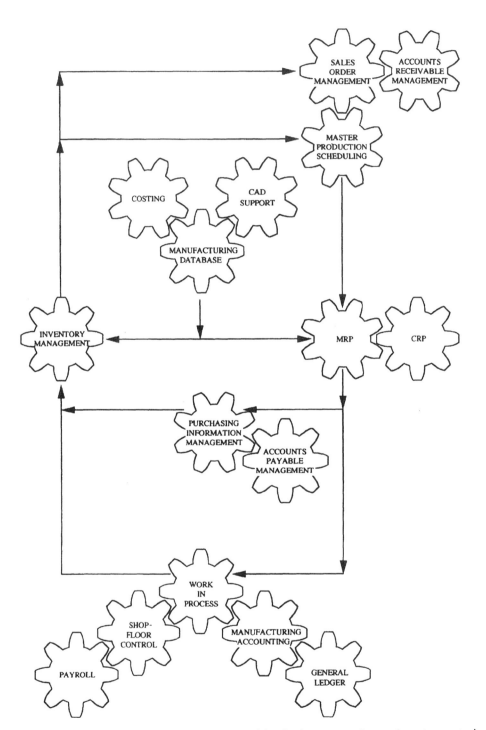

Figure 30.3 Configuration and functions covered by the business and manufacturing control system (BAMCS) composed of 16 interrelated modules (UNYSIS).

information by computer. It handles the major problems from development of a forecast, through creation of the production schedule, to the actual shipment of the finished products. Major benefits expected from this are:

- high productivity through effective utilisation of production facilities;
- reduction in inventories;
- improvement of service level;
- realistic planning and allocation of production resources including machines, manpower, and capital;
- reduction of production times and costs;
- timely and appropriate decisions to various changes of manufacturing environments, etc.

COPICS programs have been supplemented since their first presentation in 1972; Figure 30.2 shows the programs in comparison with the 12 modules explained above.

BAMCS

UNISYS developed the Business and Manufacturing Control System (BAMCS) (*Business and Manufacturing Control Systems*, 1987) in which 17 modules cover the widest functions of computerised production management systems. Based upon MRP, these modules are as follows (see Figure 30.3):

- *business foundation.* System support facilities, manufacturing database, inventory management, CAD support.
- *customer order management.* Sales order management, accounts receivable management.
- *production planning and operations management.* Master production scheduling, MRP, CRP, work in process, shop-floor control.
- *purchasing management.* Purchasing information management, accounts payable management.
- *financial management.* Costing, manufacturing accounting, general ledger, payroll.

References

Business and Manufacturing Control Systems (1987) User's Manual (UNISYS).
Communications Oriented Production Information and Control System (1972) Vols. I–VIII (White Plains, NY: International Business Machines Corporation).
DICKEY, E.R. and SENENSIEB, N.L. (1973) Total systems concept, in Heyet, C. (ed.), *The Encyclopedia of Management* (2nd edn) (New York: Van Nostrand Reinhold), p. 1050.

Supplementary reading

SKIVINGTON, J.J. (1990) *Computerizing Production Management Systems* (London: Chapman & Hall).

PART SEVEN

Social Systems for Manufacturing

This part describes basic principles of the social aspect of production and manufacturing systems. Production is viewed in connection with consumption and inventory to start with. The historical development of production modes is mentioned (Chapter 31).

The strategic aspect of manufacturing is vital in an age of worldwide competition. Essentials of manufacturing strategy, especially the strategic side of computer-integrated manufacturing (CIM) is discussed (Chapter 32). A further strategic point is international production. Basic theories of globalisation and the present situation of global manufacturing, including export and import of industrial products are mentioned (Chapter 33).

The industrial pattern shifts from primary industry (agriculture) to secondary (manufacturing), and further to tertiary (service), as the economies of a nation grow. Historical trends of industrial development in Japan are presented, and the industrial efficiencies—particularly the manufacturing efficiency—are discussed for advanced and developing countries (Chapter 34).

The influences among industrial sectors in a nation and among industrial countries are obtained by input–output analysis. The basics of this method are explained with real-life examples (Chapter 35).

In the last chapter of this book (Chapter 36), a concept of 'manufacturing excellence' is proposed for future production perspectives to cope with the great dilemma of today's manufacturing, and to enhance the elegance of goods production. Several methods for this purpose are discussed; green (environment-preserving) production is especially important to construct the recycle-oriented manufacturing system and to evaluate the product's life-cycle assessment. As an ultimate form of manufacturing excellence 'socially appropriate production/manufacturing' is introduced to save the earth from the destruction caused by excess production and consumption accompanied by the huge disposal of useful resources.

CHAPTER THIRTY-ONE

Social Production Structure

31.1 Social Manufacturing Systems

Production and Consumption

In economics 'production' is discussed in connection with 'consumption'. Further inclusion of 'sales' (distribution and exchange) builds a chain of production–distribution–exchange–consumption. This series of activities was recognised as early as 1825 by J. McVickar.

Production/Manufacturing in Social Systems

From the social systems viewpoint, as discussed in Section 3.1, manufacturing (production) systems are recognised beyond each firm:

- The *production system* forms the socially spatial interaction structure together with the settlement system and the public system (Nijkamp, 1977).
- The *production structure* plays a part in constructing the international structural power in the world system, interacting with the security structure, the finance structure, and the knowledge structure (Strange, 1988).

Relationship Between Production and Inventory

Usually 'inventory' should be avoided or minimised from the viewpoint of the minimum-cost principle. It reflects the economic and business trends of a state/country/society in connection with 'production'.

── EXAMPLE 31.1 ──────────────────────────────

Figure 31.1 shows the relationship between the increase of mining and manufacturing in percentage over the previous year and that of inventory in Japan from 1955 to 1990 (Shinohara, 1994). ☐

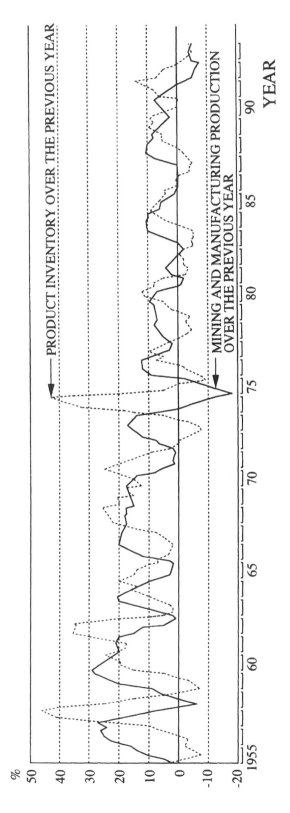

Figure 31.1 Relationship between the increase of mining and manufacturing in percentage over the previous year and that of inventory in Japan (1955–90).

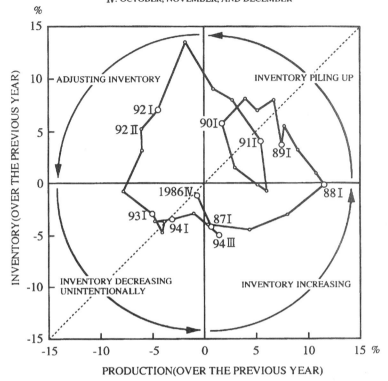

Figure 31.2 Trend of inventory circulation in Japan (1990–94).

--- **EXAMPLE 31.2** ---

Figure 31.2 shows the relationship between production and inventory (percentage)—trend of inventory circulation from 1990 to 1994 in Japan. (*Asahi Newspaper*, 29 October 1994, p. 3). □

31.2 Social Production Modes

General Trends of Production Modes

As mentioned at the start of this book, the mode of production/manufacturing has changed from the stage of production directed towards nature (collection, hunting, fishing, cultivating, breeding) to production for the market (manufacturing, construction), and further to utility production (servicing, information processing). With this production progress, the production means have become highly developed, which raises the ability of material production—'production power' (*Produktionskraft*)—with the enhancement of labour-force (skills, production techniques).

On the other hand, the relation among humans implementing production by establishing a society is said to be the 'production relation' (*Produktionsverhältnis*).

This is the mode of ownership of production means which has been developed, such as primitive communism, slavery, feudalism, capitalism and socialism.

Marx' materialism is the social production process to integrate the above two factors—*Produktionskraft* and *Produktionsverhältnis*—by establishing the social relation dealing with material conversion.

Historical Development of the Mode of Production

The 'mode of production' (*Produktionsweise*) has been historically traced as follows (Shizuta, 1962).

(1) *Primitive communism*—the age of instrument before the 3rd or 4th century BC The production means were owned in common.
(2) *Pre-capitalism*—the age of slavery (the 3rd or 4th century BC to the 5th century AD) and feudalism (6th–17th centuries AD). This mode was developed as:
   (a) *house work* (*Hauswerk*): primitive mode of industrial production which was not separated from cultivation;
   (b) *handicraft*: first mode of industrial production separated from agriculture;
   (c) *putting-out system*: manufacture subjected by the wholesale capital owned by the merchants.
(3) *Capitalism*—appeared through world trade in the 16th century and developed through the Industrial Revolution in England in the latter part of the 18th century. This is a mode of private possession of production facilities and employment of waged labour—separation of the capitalist (bourgeoisie) and the working class (proletariat):
   (a) *co-operation*: mode of a number of labourers working together according to a production plan/schedule;
   (b) *manufacture*: labour-intensive production by the division of labour—transitional stage to the great industry;
   (c) *great industry*: capital-intensive production by machines—the highest stage of the capitalistic mode of production.
(4) *Socialism*—appeared through the Russian Revolution in 1917, but almost gone in 1991; this production mode is based upon public or state-owned production facilities, and the earnings are divided among the workers according to their labour contribution. Two types of this production mode are:
   (a) *central planning mechanism*: production plans are all decided by the central agency of the socialist country;
   (b) *self-managed market mechanism*: individual firms are self-managed and operated by the market principles. Presently this type is common in socialist countries.

31.3 Establishment of Management Systems

Recognition of Management

A *management system* for manufacturing is a procedure consisting of strategic management, aggregate production planning, production process planning, production scheduling, and production control, as described in Chapter 3 (see Figure 3.5). This is a systematic production management procedure which concerns the 'flow of managerial/strategic management information'.

The production management [system] is a high-level concept in Russia in that it plays a central role in the aggregate social structure which includes the economic structure, the political structure, the law structure, the ideological structure, etc. (Lavikov and Koritsky, 1982).

Historical Development of Management

Management or control is said to have originated in the age of Babylonia in the 2nd century BC (Gilbreth and Jaffe, 1960). The modern mode of production management has been developed since the latter part of the 19th century as follows (Mohri, 1966):

(1) *Drifting management.* This is the early stage of production management which accompanied the efficiency-promotion movement in the latter part of the 19th century. It rationalised the wage-payment system.
(2) *Task management.* This is production control based upon industrial engineering disciplines, often called *Scientific Management* or the *Taylor system (Taylorism)*. A 'fair day's work'[1] or a *task* is set for each worker and the wage rate is determined according to the performance of the task. This improved labour efficiency and got rid of organisation soldering (mending lazy work) which had been a big problem in the age of drifting management. The 'management department' was established to enhance the activity of factory management or control.
(3) *Management by synchronisation.* This means productivity increase of a production line (or process) by the use of 'standardisation' and 'division of labour', often called the *Ford system (Fordism)*. This contributed to establishing the mass-production mode, which resulted in reduction of production costs and rise in the wage rate.[2]
(4) *Management by self-control.* This is automatic control of a production line by the use of automatic machines equipped with numerical control (NC), thereby resulting in reduction or elimination of direct labour, often called the *automation system*. This enabled a systematic production management for mass production of a single product by stabilising the product quality, reducing the production cost, manufacturing the required amount of the product by the delivery time, etc.
(5) *Management by systems.* This is integrated management for manufacturing a variety of products in the just-in-time mode to meet the dynamically changing market situation through full automation or computer-integrated manufacturing (CIM) systems, often called the 'manufacturing-management system' (Hitomi, 1978), or 'neo-Fordism'. This functions as manufacturing strategy from the viewpoint of total corporate management, being the most advanced mode of today's production management (Hounshell, 1984).

Notes

1. This was pointed out by C. Babbage in the early 19th century.
2. The wage rate was raised to $5 a day from $2–3; H. Ford instituted the 'five dollar day' on 5 January 1914.

References

GILBRETH, L.M. and JAFFE, W.J. (1960) Management past—a guide to its future, *ASME Paper*, No. 60-W-67.
HITOMI, K. (1978) *Production Management Engineering* (in Japanese) (Tokyo: Corona-sha), pp. 5–6.
HOUNSHELL, D.A. (1984) *From the American System to Mass production, 1800–1932* (Baltimore, ML: Johns Hopkins University Press), p. 11.
LAVIKOV, Yu. and KORITSKY, Z. (1982)/MIYASAKA, J. (1984) *Fundamentals of Russian Management* (translated into Japanese) (Tokyo: Sugiyama-Shoten), p. 158.
MOHRI, S. (1966) *Principles of Management* (2nd edn) (in Japanese) (Tokyo: Chikura-Shobo), Chap. 1.
NIJKAMP, P. (1977) *Theory and Application of Environmental Economics* (Amsterdam: North-Holland), p. 302.
SHINOHARA, M. (1994) *Business Cycles for Fifty Years after the War in Japan* (in Japanese) (Tokyo: Japan Economic Press), p. 146.
SHIZUTA, H. (1962) *Modern Industrial Economics* (in Japanese) (Tokyo: Yuhikaku), Chap. 2.
STRANGE, S. (1988) *States and Markets* (London: Pinter), p. 27.

Supplementary reading

BARRO, R.J. and SALA-I-MARTIN (1995) *Economic Growth* (New York: McGraw-Hill).
HITOMI, K. (1994) *Introduction to Today's Advanced Manufacturing* (in Japanese) (Tokyo: Dobunkan).

CHAPTER THIRTY-TWO

Manufacturing Strategy

32.1 Strategy and Tactics

What are Strategy and Tactics?

Strategy means acts/procedures taken to achieve an aim. *Tactics* denotes actions to execute an individual task based on the strategy.

Features of Strategy/Tactics

Characteristics or features of strategy and tactics are as follows (Hitomi, 1991):

(1) Strategy puts a priority on external issues of the system, and tactics on internal issues. In other words, strategy examines the relation between a system (for example, a manufacturing firm) and an external environment (for example, markets) and makes an appropriate decision. Tactics solve problems inside the system (for example, a manufacturing system).
(2) Strategy deals with issues sustained for a long time period (for example, several years), and tactics treat problems in the short term (several months).
(3) Strategy is, therefore, futuristic rather than oriented for the present time and is involved in decision-making for issues in the future. Tactics give priority to decision-making for present problems.
(4) Strategy treats issues in the wide territory of the system environment, whilst tactics deal with problems of the narrow area inside the system.
(5) Strategy is concerned with the system environment where matters are uncertain, whereas tactics treat only matters of certainty.
(6) From the viewpoint of decision-making, strategy has no routine method/means for obtaining solutions; experience and intuition are helpful in most cases. On the other hand, the decision-making in tactics is rather programmed, and optimum solutions are obtainable.

32.2 Corporate Strategy

Aim of Corporate Strategy

Enterprises determine the optimum use of available resources in order to achieve their targets (profit, market share, turnover, rate of return, growth, etc.) and execute the decisions. *Corporate strategy* deals with the strategic aspect of this corporate activity; extensive arguments have been presented since 1965 (Ansoff, 1965).

Two Aspects of Corporate Strategy

Two aspects exist in corporate strategy as follows:

(1) *Manufacturing strategy* means a strategy for the production function of manufacturers concerning material production.
(2) *Management strategy* means a strategy for the management function of non-manufacturing industries which are not involved in material production. *Strategic information systems* (*SIS*), coined by Wiseman (1985) in 1988, play a role for this purpose. In this system, also, the coincidence or unification of the information flow and the material flow (e.g. goods transportation, as Figure 26.3 represents) is vital.

32.3 Manufacturing Strategy

Meaning of Manufacturing Strategy

Manufacturing strategy is a corporate strategy, dealing with the production function, as mentioned previously.

This terminology emerged in the early 1980s (e.g. Hill, 1985), as the importance of global manufacturing became recognised worldwide and factory automation (FA)/computer-integrated manufacturing (CIM) have become significantly developed.

As Japan has succeeded in establishing highly efficient manufacturing systems such as quality circles, just-in-time (JIT) production and supply, and other systems, to become one of the world-class industrial countries, manufacturing is now considered as an important management strategy for each individual manufacturing firm.

Why Manufacturing Strategy?

Awareness of the environment, manufacturing flexibility and attention to the role of manufacturing managers are regarded as important requirements in manufacturing strategy (Swamidass and Newell, 1987). These are especially important for world-class manufacturers to win against international business competition (Gunn, 1987).

Essentials of Manufacturing Strategy

Manufacturing strategy consists basically of (Buffa, 1984):

- the minimum-cost/high-availability strategy, and
- the highest-quality/flexibility strategy.

In addition the following six basics need to be integrated into the management system:

(1) positioning of the production system;
(2) capacity/location decisions;
(3) product and process technology;
(4) workforce and job design;
(5) strategic implications of operating decisions;
(6) vertical integration of suppliers.

32.4 Computer-integrated Manufacturing as a Corporate Strategy

Essentials of Computer-integrated Manufacturing (CIM)

As mentioned in Chapter 23, *Computer-Integrated Manufacturing* (*CIM*) is a computerised system which integrates the computer aids of the following three different functions with a common database (see Figure 23.1):

(I) computer aid to the production function—automated flow of materials (*CAM*);
(II) computer aid to the design function—automated flow of technological information (*CAD*);
(III) computer aid to the management function—automated flow of managerial information (*CAP*).

Strategic Aspects of CIM

The strategies of CIM covering the three basic regions, CAM, CAD and CAP, together with *office automation* (*OA*) for the automated strategic management function (automated flow of strategic management information: long-term management planning/factory planning) (see Figure 5.4), are explained below (Hitomi, 1989):

(1) *Strategy of CAM*—unmanned small-lot production for a variety of products. An ideal CIM is intended to eliminate direct workers. Since the direct labour cost accounts for more than 10% (in 1993 12% of the total cost/15% of the manufacturing cost in the case of Japanese small- and medium-sized manufacturers: see Figure 19.1), it is expected to reduce production costs by at least 10%, bringing about a remarkable benefit.

CIM is not required for single-item mass production, which can be implemented efficiently by automatic lathes, special-purpose machine tools, transfer machines, and low-cost automation (LCA) systems. However, multi-product, small-batch production reflecting today's market needs for large variety, calls for CIM to automate this production mode. Specifically, flexible manufacturing systems (FMSs) are employed as the most advanced production facility of CAM, as mentioned in Chapter 25.

(2) *Strategy of CAD*—quick design and product development. The most advantageous functions of CAD are quick design and drawing of parts and products. CAD ensures design and drawing of parts and products with appropriate accuracies and qualities for JIT delivery as demanded by the market for large-variety, small-batch production. Benefits from CAD are also found in substantial reduction of design lead time and automatic preparation of process design (computer-aided process planning, CAPP) and numerical control (NC) programs. In other words, the CAD and CAM functions are integrated, as mentioned in Section 24.3.

Innovative products are generated from research and development (R&D), to which quick simulation and virtual factory, technical computation and computer graphics by computer-aided design/engineering (CAD/CAE) contribute greatly.

(3) *Strategy of computer-aided production management*—reduction of lead time and flexible production management. Computer-aided production management is characterised by quick production planning and scheduling for producing a variety of products to meet the market needs. Reduction of production lead time and quick JIT deliveries are also expected, resulting in improvement of service to customers and giving an edge in severe business competition.

(4) *Strategy of OA*—sales promotion, corporate automation, and global production. A practical object of CIM is to integrate production and sales. The strategic management function to make such an integration effective deals with *strategic planning*. OA helps computerisation of this activity. In addition, automation of the marketing and sales functions based upon this OA is expected to result in a 10–30% increase in sales (Moriarty and Swartz, 1989).

The integration of CIM or FA and OA is regarded as a 'corporate (or enterprise) automation'. It automates total systems of strategic planning, management control, operation control, and production implementation (Wakuta and Hitomi, 1983).

Through corporate automation, product orders are transmitted via real-time communication networks between the corporate office and sales branches, quickly analysed by the inter-office information network, and processed for design and production. Then the purchasing requests are immediately given to external suppliers of raw materials and parts. CALS, mentioned in Section 3.5, is expected to raise manufacturing system productivity at light speed in the future, through the Internet.

Additionally, a step towards an effective international production by structuring, implementing and administrating a global CIM will thus be established to meet the requirements for international industrial competition.

References

ANSOFF, H.I. (1965) *Corporate Strategy* (New York: McGraw-Hill).
BUFFA, S. (1984) *Meeting the Competitive Challenge* (Homewood, IL: Irwin).
GUNN, T.G. (1987) *Manufacturing for Competitive Advantage* (Ballinger).
HILL, T. (1985) *Manufacturing Strategy* (London: Macmillan).
HITOMI, K. (1989) *Essentials of CIM* (in Japanese) (Tokyo: Ohm-sha), Chap. 4.
HITOMI, K. (1991) Strategic integrated manufacturing systems—the concept and structures, *International Journal of Production Economics*, **25** (1–3).
MORIARTY, R.T. and SWARTZ, G.S. (1989) Automation to boost sales and marketing, *Harvard Business Review*, **67** (1).
SWAMIDASS, P.M. and NEWELL, W.T. (1987) Manufacturing strategy, environmental uncertainty and performance, *Management Science*, **33** (4).
WAKUTA, H. and HITOMI, K. (1983) *FA and OA—towards corporate automation* (in Japanese) (Tokyo: Business & Technology Newspaper Company).
WISEMAN, C. (1988) *Strategic Information Systems* (Homewood, IL: Irwin).

Supplementary reading

ANSOFF, I. and MCDONNELL, E. (1990) *Implanting Strategic Management* (2nd edn) (Hemel Hempstead: Prentice-Hall).

MOODY, P.M. (ed.) (1990) *Strategic Manufacturing* (Homewood, IL: Dow-Jones Irwin).
RUE, L.W. and HOLLAND, P.G. (1989) *Strategic Management* (Singapore: McGraw-Hill).
SAMSON, D. (1991) *Manufacturing and Operations Strategy* (Englewood Cliffs, NJ: Prentice-Hall).
SKINNER, W. (1978) *Manufacturing in the Corporate Strategy* (New York: Wiley).
VOSS, C.R. (ed.) (1992) *Manufacturing Strategy—Process and Content* (London: Chapman & Hall).

CHAPTER THIRTY-THREE

Global Manufacturing

33.1 Movement Towards Globalisation

Globalisation

Today's manufacturing activity is not confined to domestic production by a factory of a manufacturing firm in the native country; rather manufacturing has been conducted offshore, i.e. globally and internationally, by realising the optimum combination of resources of production available throughout the world. This international business activity, unhampered by the boundaries of a country, is the movement towards *globalisation*.

Multinational Enterprise

Firms which conduct business and/or manufacturing activities by establishing business and/or manufacturing sites in various countries across the world are called *multinational enterprises* or *international corporations*.

The movement towards international corporation started in the 1950s, seeking to capitalise on the low wage rate of developing countries. As a result, developed countries have been faced with *de-industrialisation* or *industrial hollowness*, which incurs a decline in domestic production as well as an increase in unemployment rate.

Theories Concerning Globalisation

The phenomenon of globalisation or international corporation/manufacturing has been explained by several theories, as follows.

- *Product-cycle model* (Vernon, 1966). A cycle is repeated: new product development in a developed country–domestic production–export of the product–foreign investment toward developing countries–production in the developing country–counter-import from the developing countries.

- *Wild-geese flying model* (Akamatsu, 1965). A cycle is repeated: import of products—domestic production—export of the product—foreign investment and production.
- *Eclectic paradigm* (Dunning, 1988). This selects from among possible alternatives of product export, licensed production, and foreign production by direct investment.

33.2 International Production

International and World-class Manufacturing

Manufacturing activities which maximise total profits by installing production sites (factories) in various countries are called *international production/manufacturing* or *foreign production*. This is world-class manufacturing called *global manufacturing*. It is now recognised as manufacturing strategy as discussed in the previous chapter.

—— EXAMPLE 33.1 ——

The beginning of foreign production:

- German chemical industry—Bayer acquired American aniline factory in 1865;
- A.B. Nobel of Sweden installed an explosives factory in Germany in 1867.   □

—— EXAMPLE 33.2 ——

In 1988 a Japanese automobile company made engines in Japan, chassis in Thailand, doors in Malaysia, transmissions in the Phillipines, wheels in Australia, and car radios in Singapore, assembled them into cars in Thailand, and exported the products to Canada.   □

Why Global Manufacturing?

An essential characteristic of global manufacturing is the transfer of internal management resources to foreign countries with low labour cost or those exporting a great amount of products in order to win internationally competitive advantages through 'economies of network'.

Japanese manufacturers have recently installed foreign production sites to reduce trade surplus, resulting in creation of employment in the foreign countries, contribution to technology transfer, internationally vertical/horizontal division of labour, etc. This phenomenon incurs the risk of de-industrialisation or industrial hollowness in the native country.

Foreign Production Rate

The ratio of foreign production over total production (=domestic production + foreign production) is called the *foreign* (or *offshore*) *production rate*.

—— EXAMPLE 33.3 ——

The foreign production rate is: Japan 8.9% in 1995, USA 25.2% in 1993, Germany 18.2% in 1992.   □

Table 33.1 Comparing Japanese and American global manufacturing performances (1990).

Item	Japan	Japan/USA(%)	USA
Domestic production			
• Labour population	15.05 million	71.0	21.19 million
• Added value created	93 118 billion yen		$846.8 billion
(exchange rate*)	$643.1 billion	75.9	
(PPP†)	$485.0 billion	57.3	
Global production			
• Foreign production rate	6%		25%
• Added value created	$676.9 billion	60.0	$1129.1 billion
(PPP†)	$510.5 billion	45.2	

*$1 = 144.79 yen.
†Purchasing power parity: $1 = 192.00 yen.

Global Manufacturing Performance

The performance of global manufacturing is important in the age of globalisation. Table 33.1 shows the comparison between the United States and Japan for 1990 in this respect (Hitomi, 1994).

The Japanese labour force is approximately 71% of that of the USA. Domestic production in Japan created added value equal to 76% of that of the USA in terms of the exchange rate, but only 57% in terms of the purchasing-power parity.

From the global-manufacturing standpoint, Japan created added value equivalent to only 60% of that of the USA in terms of the exchange rate and merely 45% in terms of the purchasing power parity. This demonstrates the USA's strong power in global manufacturing.

33.3 Export and Import of Industrial Products

Export-driven Economy

Japan, Germany, and most developing countries have export-driven economies, and aspire to trade surplus. Japan's surplus (135 billion US dollars in 1995) raised the value of Japan's money (the yen) against the US dollar.

— EXAMPLE 33.4 —

$1 = 360 yen in 1949, 308 in 1971, 227 in 1980, 145 in 1990, below 100 (even 79) in 1995. ☐

Comparison of Japanese, US, and German Trades

The United States is top for both export and import, Germany is second, and Japan is third. International comparison of export and import together with trade balance in 1991 is shown in Table 33.2 for those three countries.

Table 33.2 Comparing Japanese, American and German export/import and trade balance (1991).

Country	Population GDP per capita	Export per capita over GDP	Import per capita over GDP	Import — Export	Trade balance per capita over GDP
Japan	124 million $3386 billion $27 323 (19 211*)	$315 billion $2541 9.3%	$213 billion $1720 6.3%	67.7%	$102 billion $821 3.0%
USA	253 million $5678 billion $22 468	$422 billion $1669 7.4%	$491 billion $1942 8.6%	116.4%	−$69 billion −$273 −1.2%
Germany	64 million $1566 billion $24 427	$403 billion $6288 25.7%	$359 billion $5593 22.9%	88.9%	$45 billion $695 2.8%

* In terms of purchasing power parity.

The ratios of export and import over GDP are extremely large in Germany, i.e. over 20%, whilst these are less than 10% in the US and Japan. Both export and import per capita are much higher in Germany than in other countries. Japan's trade surplus is only 3% of GDP, almost the same as Germany. The US's trade deficit is a little over 1% of GDP.

Japan's special feature is that the export of industrial products contributes 97%, whilst the import contributes only 53%. These figures are 76% and 79% in the United States, and 90% and 82% in Germany.

Export/Import over GDP

The ratio of export over GDP is usually large in developing countries.

--- **EXAMPLE 33.5** ---

The ratio of export over GDP is: Korea and China 35%, Hong Kong 120%, Singapore 150%. □

On the other hand, the ratio of import over GDP is usually larger for countries with less population. Those countries are import-driven.

--- **EXAMPLE 33.6** ---

The ratio of import over GDP is: Belgium (population: 10 million) 70%, Holland (11) 50%, Germany, England, France (50–60) 20–25%, Japan (120) 6%, USA (240) 9% (*Japan Economic Newspaper*, 10 February 1987, p. 22). □

References

AKAMATSU, K. (1965) *Theory of World Economy* (in Japanese) (Tokyo: Kunimoto-Shobo), Chap. 10.

DUNNING, J.H. (1988) *Explaining International Production* (London: Unwin Hyman), Chap. 2.

HITOMI, K. (1994) *Introduction to Today's Advanced Manufacturing* (in Japanese) (Tokyo: Dobunkan).

VERNON, R. (1966) International investment and international trade in the product cycle, *Quarterly Journal of Economics*, **80** (2).

Supplementary reading

HODGETTS, R.M. and LUTHANS, F. (1994) *International Management* (2nd edn) (New York: McGraw-Hill).

HOWELLS, J. and WOOD, M. (1993) *The Globalisation of Production and Technology* (London: Belhaven Press).

IETTO-GILLIES, G. (1992) *International Production* (Cambridge: Polity Press).

RUGMAN, A.M., LECRAW, D.J. and BOOTH, L.D. (1985) *International Business* (New York: McGraw-Hill).

CHAPTER THIRTY-FOUR

Industrial Structure and Manufacturing Efficiency

34.1 Industrial Structure

Industrial Classification

As mentioned in Section 1.1, industries are usually classified into three categories (A.G.B. Fisher):

(I) *primary industry*—production relying on nature—agriculture; forestry and fishing; and mining (which is often included in the next category);
(II) *secondary industry*—production of tangible goods (products) through conversion of raw materials—manufacturing; construction; and electricity, gas and water (which are often included in the next category);
(III) *tertiary industry*—intangible service production—transportation and communication, wholesale and retail trade, finance, insurance and real estate, services, and government (and nonclassified establishments).

Changes in Industrial Structure

As economies of a nation have grown, the industrial pattern shifts from I to II, and further from II to III (*Petty and Clark's law*). A country in which the percentage of labour population in the tertiary industry has exceeded 50 is recognised as an economically advanced country that has entered into the post-industrial or information society.

—— EXAMPLE 34.1 ——————————————————————————

In Japan the share of labour population in the secondary (manufacturing) industry peaked at 36.8 (27.1)% in 1973 (1969) (34.3 (23.2)% in 1994); since then it has been decreasing. The share in the tertiary industry exceeded 50% in 1977, and has been increasing ever since (58.2% in 1994). □

—— EXAMPLE 34.2 ——————————————————————————

The labour-population shares (%) in the primary, secondary (manufacturing), and tertiary industries in 1990 were: USA (labour population: 118 million) 3.5, 25.8 (18.0), 70.7; Britain

(27 million) 3.0, 28.1 (20.3), 68.9; Germany (29 million) 4.5, 39.1 (31.6), 56.4; Japan (66 million) 9.4, 33.1 (23.4), 57.6; Korea (18 million) 18.7, 34.3 (26.9), 47.0; China (567 million) 60.4, 21.4 (17.1), 18.2. □

34.2 Japan's Manufacturing and Industry

34.2.1 Preliminaries to Japan's Manufacturing and Industry

Review of Japan's Manufacturing and Economy

Since the *Meiji Restoration* in 1868, Japan has made a great effort to achieve rapid economic growth. Although Japan's manufacturing industry was completely destroyed during World War II, it recovered due to special demands generated by the Korean War in 1950/51. Japanese products were considered to be 'cheap, but poor' for a long time; presently, however, their quality became quite superior, probably the best in the world. Today, Japan, a small country with land mass only 0.3% of the world (377 750 km^2) and a population of 124 million (2.3% of the world population), is widely recognised to be one of the great economic nations; in 1990 her gross national product (GNP) was $2963 billion or 13.7% (10% in 1970/15.6% ($3669 billion) in 1992) of the world's GNP—second only after the United States at $6020 billion or 25.6%. Japan's nominal GNP per capita exceeded that of the United States in 1987; in 1993 it was $34 077—third in the world (top, $33 764, from the standpoint of gross domestic product (GDP)) (but it is estimated at around 80% of that of the US: $24 364—fifth in the world—from the viewpoint of the purchasing power parity).

Historical Trend of Japan's Industry

The shares of labour population (LP) and of gross domestic product (GDP) from 1955 to 1994 in Japan's industry are represented in Figures 34.1 and 34.2, respectively. During this period the total labour population increased by 63.7% from 40.7 million in 1955 to 66.6 million in 1994. The percentage of manufacturing, labour population peaked in 1969: 27.1% (14.5 million). In 1994 it was 23.2%; the population was 15.4 million with an increase of 106.6%, compared to that in 1955.

The year when LP in the tertiary industry exceeded 50% was 1977; it was 58.3% in 1994. Among this industry the growth of the service sector was large, as is common in the advanced countries; its population share was 22.7%.

LP in the primary industry has decreased significantly; it is presently 7.5%.

During the period 1955–94 nominal GDP increased by 58 times (9 times in real GDP) from 8665 billion yen ($24 billion) in 1955 to 499 013 billion yen ($4881 billion) in 1994 ($1 = 102.23 yen). During this period the manufacturing sector increased by the same amount (21 times in real value) from 2381 billion yen ($6.6 billion) to 117 151 billion yen ($1146 billion).

The share of GDP by the manufacturing sector (and also by the secondary industry) peaked in 1970, amounting to 34.9%; it is presently 27.5%.

The year when the share of GDP by the tertiary industry exceeded 50% was 1971; the share is presently 61.3%, among which the service sector shares 16.1%.

The sector of agriculture, forestry and fishing contributes only 2.2%.

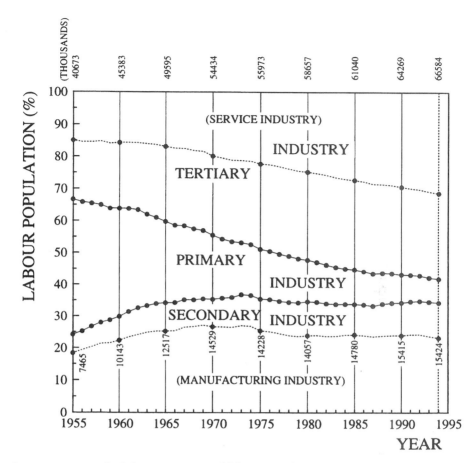

Figure 34.1 Trend of the percentages of labour population in Japan's industries, 1955 to 1994 (data source: *Annual Reports on National Accounts* by Japan's Economic Planning Agency).

Growth of Japan's Industry

Setting as 100 for the year 1955, the growth of the total labour population, population in the manufacturing sector, GDP, added values created by agriculture, manufacturing, and the tertiary industry, as well as the growth of the national (GDP per person employed) and the manufacturing productivities are shown until 1990 in Figure 34.3 (monetary figures based upon the real values).

The growth of the manufacturing industry as well as the growths of several sectors contained in manufacturing industry are represented in Figure 34.4. The increase in the machinery and chemical sectors are tremendous, over one hundred times. The increase in the iron and steel sector shows almost the average growth of the total manufacturing; that is, 21 times. The textile sector did not increase much during this period.

The above results lead to a conclusion that manufacturing industry is the key to creation of a nation's wealth.

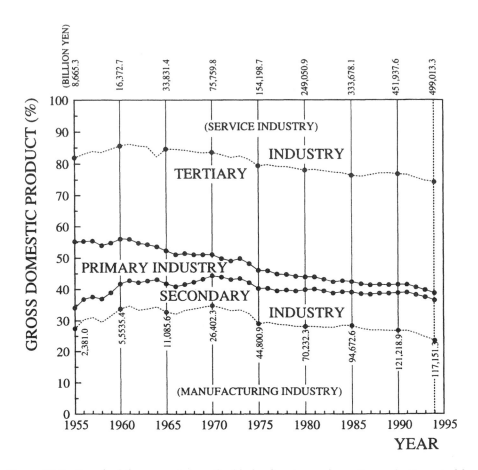

Figure 34.2 Trend of the percentages of added value (gross domestic product) created by Japan's industries, 1955 to 1994 (data source: *Annual Reports on National Accounts* by Japan's Economic Planning Agency).

34.2.2 Japan's Industrial and Manufacturing Efficiencies

Japan's Industrial Power

The labour population, output (production), and added value in the classified industrial sectors for the year 1990 are shown in Table 34.1. In this year (and in 1994) the total labour population in Japan was 64 269 [66 584] thousand, the nominal total output was 865 435.7 [909 565.7] billion yen ($5976 [8897] billion), and the nominal GDP was 451 937.6 [499 013.3] billion yen ($3121 [$4881] billion).

Japan's Manufacturing Performance

Japan's manufacturing industry produced an output of 39.3 [34.5]% with a labour force of 24.0 [23.2]%; hence at a glance Japan's manufacturing industry seems to be very efficient. This aspect is observed in the time series as represented in Figure 34.5.

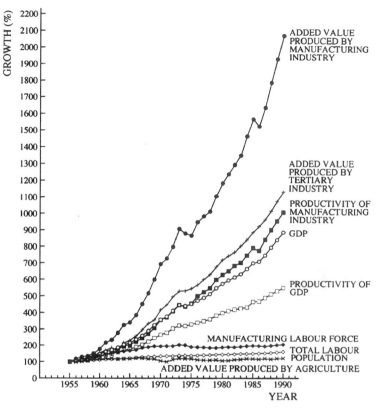

Figure 34.3 Trends of the growths of GDP, agriculture, manufacturing, and tertiary industry in terms of added value in real figures and their productivity trends in Japan, 1955 (as 100) to 1990 (data source: *Annual Reports on National Accounts* by Japan's Economic Planning Agency).

However, from the viewpoint of added value (=output−value paid to the outside) which is a factor more important than the output, manufacturing industry contributes 26.8 [23.5]%; it is hard to say that Japan's manufacturing is very efficient.

Measures of Efficiency

The following two kinds of measures are introduced to evaluate the efficiency of an industrial sector (Hitomi, 1993):

(1) *yield rate*—rate of the added value against the output for an industrial sector, expressed in percentage;
(2) *efficiency index*—rate of the GDP share against the labour-force share for an industrial sector.

Yield rate, G_i, of an industrial sector, i, is expressed:

$$G_i = Y_i/O_i \times 100 \tag{34.1}$$

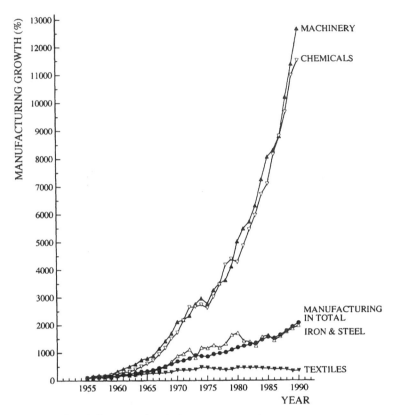

Figure 34.4 Trends of the growth of the sectors made up into Japan's manufacturing industry, 1955 (as 100) to 1990 (data source: *Annual Reports on National Accounts* by Japan's Economic Planning Agency).

where O_i is the output produced by sector i and Y_i is the added value created by sector i. The larger G_i, the more efficient that industrial sector (maximum: 100%).

On the other hand, the efficiency index of sector i is expressed as:

$$E_i = \frac{Y_i/Y}{R_i/R} \tag{34.2}$$

where R is the total resource employed, R_i is the resource employed in sector i, and Y is the GDP in a specified year under consideration. An industrial sector with the efficiency index greater than 1 is efficient; the greater this index, the more efficient that industrial sector. If this figure is less than 1, that industrial sector is not efficient from the viewpoint of creating added value.

Two basic factors taken into consideration as resources are labour and capital; hence we have:

- *labour (human-based) efficiency index*[1]—efficiency figure based upon human resource;
- *capital [-based] efficiency index*—efficiency index based upon capital stock especially tangible fixed assets for direct labour, or simply production facilities.

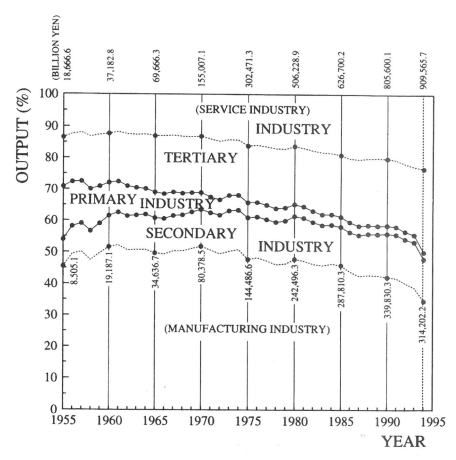

Figure 34.5 Trend of the percentages of output produced by Japan's industries, 1955 to 1994 (data source: *Annual Reports on National Accounts* by Japan's Economic Planning Agency).

Labour Efficiency of Japan's Industry

It is found from Table 34.1 that in 1990 [1994] the yield rate of Japan's manufacturing industry is 35.7 [37.3]%, which is the lowest among all the industries. This figure has recently risen a little, but is much lower than the average for all the industries, even lower than those of agriculture and of commerce, which are usually considered to be low in productivity.

Regarding the efficiency index (man-year base), Japan's manufacturing industry exhibits 1.12 [1.01], which is superior among industrial sectors. Excepting 13.70 for the petroleum refining and coal products, 2.74 for the chemical industry, and 2.17 for the primary metal industries including iron and steel, the efficiency of Japan's manufacturing industry is not extremely superior as is commonly believed. The efficiency index of the textile mill products is very low, 0.32. As shown in Figure 34.6, the trend shows that the efficiency index of the manufacturing industry was high up until the first half of the 1970s, then became almost constant, and recently it has been decreasing; 1.01 in 1994.

Table 34.1 Japan's industrial power and efficiency (1990).

Industry	Labour force		Output			Gross domestic product			Yield rate	Efficiency index
	Thousand[1]	% (a)	Billion yen[†1] (A)	Thousand yen/employee	%	Billion yen[†1] (B)	Thousand yen/employee	% (b)	% (B/A)	(b/a)
Primary	5732	8.9	20 981.3	3660	2.4	12 042.1	2101	2.7	57.4	0.30
Agriculture, forestry and fishing	5631	8.8	18 819.0	3342	2.2	10 920.5	1939	2.4	58.0	0.28
Mining	101	0.2	2162.3	21 409	0.2	1121.6	11 105	0.2	51.9	1.58
Secondary	22 017	34.3	449 888.2	20 434	52.0	175 888.4	7989	38.9	39.1	1.14
Manufacturing	15 415	24.0	339 830.3	22 045	39.3	121 218.9	7864	26.8	35.7	1.12
Foods	1596	2.5	34 403.0	21 556	4.0	12 321.7	7720	2.7	35.8	1.10
Textiles	1106	1.7	7423.8	6712	0.9	2514.0	2273	0.6	33.9	0.32
Pulp and papers	370	0.6	9888.9	26 727	1.1	3366.3	9098	0.7	34.0	1.29
Chemicals	486	0.8	26 512.3	54 552	3.1	9375.2	19 291	2.1	35.4	2.74
Petroleum and coal products	43	0.1	11 073.2	257 516	1.3	4143.0	96 349	0.9	37.4	13.70
Stone, clay and glass	603	0.9	10 140.6	16 817	1.2	4381.7	7267	1.0	43.2	1.03
Primary metals	621	1.0	35 590.5	57 312	4.1	9466.1	15 243	2.1	26.6	2.17
Fabricated metals	1175	1.8	16 223.7	13 807	1.9	7157.8	6092	1.6	44.1	0.87
General machinery	1752	2.7	40 369.5	23 042	4.7	15 901.8	9076	3.5	39.4	1.29
Electric/electronic machinery	2407	3.7	52 163.3	21 671	6.0	19 386.2	8054	4.3	37.2	1.15

Transportation machinery	1456	2.3	42 854.3	29 433	5.0	11 820.0	8118	2.6	27.6	1.15
Precision machinery	332	0.5	4874.9	14 683	0.6	2203.8	6638	0.5	45.2	0.94
Miscellaneous	3468	5.4	48 312.6	13 931	5.6	19 181.3	5531	4.2	39.7	0.79
Construction	6200	9.6	91 157.1	14 703	10.5	43 427.5	7004	9.6	47.6	1.00
Electric, gas and water	402	0.6	18 900.8	47 017	2.2	11 242.0	27 965	2.5	59.5	3.98
Subtotal (goods production)	27 749	43.2	470 869.5	16 969	54.4	187 930.5	6773	41.6	39.9	0.6
Tertiary (service production)	36 519	56.8	394 566.0	10 804	45.6	264 007.1	7229	58.4	66.9	1.03
Wholesale and Retail trade	11 037	17.2	87 538.9	7931	10.1	58 358.0	5287	12.9	66.7	0.75
Finance and insurance	2139	3.3	38 165.3	17 843	4.4	25 545.6	11 943	5.7	66.9	1.70
Real estate	937	1.5	51 475.4	54 936	5.9	46 792.2	49 938	10.4	90.9	7.10
Transportation and communication	3515	5.5	42 792.5	12 174	4.9	28 474.8	8101	6.3	66.5	1.15
Services	13 418	20.9	114 758.3	8553	13.3	63 624.2	4742	14.1	55.4	0.67
Government and non-profit services	5473	8.5	59 835.6	10 933	6.9	41 212.3	7530	9.1	68.9	1.07
Total	64 269	100	865 435.7	13 466	100	451 937.6	7032	100	52.2	1

† 144.81 yen = $1 for 1990.
¹ Source: 1: 1996 *Annual Report on National Accounts* by *Economic Planning Agency*, Government of Japan, pp. 148–165.

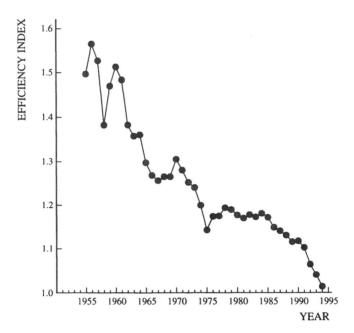

Figure 34.6 Trend of the efficiency index for Japan's manufacturing industry, 1955 to 1990 (data source: *Annual Reports on National Accounts* by Japan's Economic Planning Agency).

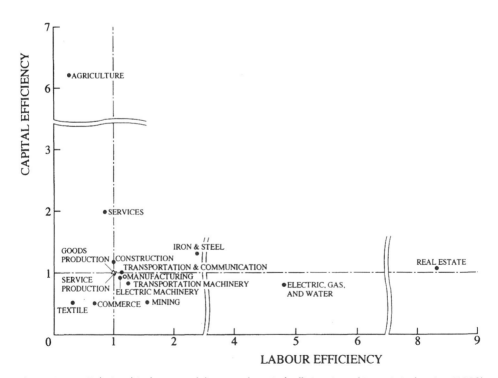

Figure 34.7 Relationship between labour and capital efficiencies of Japan's industries (1989) (data source: *Annual Reports on National Accounts* by Japan's Economic Planning Agency and *Economy*, No. 326 (1991), p. 114).

Labour and Capital Efficiencies

By taking labour and capital efficiency indices as abscissa and ordinate, respectively, Japan's labour and capital efficiencies in 1989 for industrial sectors are plotted as shown in Figure 34.7.

The agriculture, forestry and fishing sector is excellent in capital efficiency, but poor in labour efficiency. The real estate and the electric, gas and water sectors are particularly excellent in labour efficiency. The service and the iron and steel sectors show better efficiency-index values than other sectors.

34.3 Industrial Efficiency in Advanced Countries

The US's Industrial Power and Efficiency

America's industrial power and efficiency in 1990 is depicted in Table 34.2. Her share of world GNP in this year was 24.1% (US$ 5465 billion), top in the world, as mentioned in the previous section.

The manufacturing sector shares 18% of the total labour population, 118 million, producing 22.1% of GDP, $3829 billion. Hence, the labour efficiency index shows 1.23, which is slightly higher than Japan's, 1.18 for the same year.

Table 34.2 The US's industrial power and efficiency (1990).

Industry	Labour force (Thousand)[1]	(%) (a)	Gross domestic product ($ Billion)[1]	($ Thousand per employee)	(%) (b)	Efficiency index b/a
Primary	4085	3.5	135.2	33 097	3.5	1.02
Agriculture, forestry and fishing	3355	2.9	97.1	28 941	2.5	0.89
Mining	730	0.6	38.1	52 191	1.0	1.61
Secondary	30 464	25.8	1174.0	38 548	30.7	1.19
Manufacturing	21 184	18.0	846.8	39 974	22.1	1.23
Construction	7696	6.5	234.4	30 457	6.1	0.94
Electric, gas and water	1584	1.3	92.8	58 586	2.4	1.81
Subtotal (goods production)	34 549	29.3	1309.2	37 894	34.2	1.17
Tertiary (service production)	83 365	70.7	2519.9	30 227	65.8	0.93
Wholesale and retail trade	24 269	20.6	655.6	27 014	17.1	0.83
Transportation and communication	6552	5.6	236.2	36 050	6.2	1.11
Finance, insurance and real estate	13 346	11.3	679.8	50 937	17.8	1.57
Services*	39 198	33.2	948.3	24 193	24.8	0.75
Total	117 914	100	3829.1	32 474	100	1

*Excludes the Government services ($657.9 billion).

Source: 1: Bank of Japan, *Comparative Economic and Financial Statistics*, 1992, pp. 45–46, 107–108.

Table 34.3 British industrial power and efficiency (1990).

Industry	Labour force		Gross domestic product			Efficiency index
	(Thousand)[1]	(%) (a)	(£* Million)[1]	(%) (b)	(£* Thou/employee)	b/a
Primary		3.0		1.5	12.5	0.70
Agriculture, forestry and fishing	792	2.1	7102			
Mining	569	0.8				
Secondary	223	28.2				
	7480					
Manufacturing	5396	20.3	106 995[†]	22.6	19.1[‡]	1.03[‡]/1.12
Construction	1805	6.8	36 085	7.6	20.0	1.12
Electric, gas and water	279	1.1	24 334[†]	5.2	83.9[‡]	4.72[‡]/4.90
Subtotal (goods production)	8272	31.1	174 516	36.9	21.1	1.19
Tertiary (service production)	18 305	68.9	298 287	63.1	16.3	0.92
Wholesale and retail trade	5445	20.5	70 151	14.8	12.9	0.72
Transportation and communication	1546	5.8	34 031	7.2	22.0	1.24
Finance, insurance and real estate	3143	11.8	117 979	25.0	37.5	2.11
Services	8171	30.8	76 126	16.1	9.3	0.52
Total	26 577	100	472 803	100	17.8	1

* £ (pound) 1 = $1.78 for 1990.
[†] Includes a part of added value created in the mining sector.
[‡] Cases where labour population for the mining sector is partitioned among the manufacturing and the electric, gas and water sectors according to the labour population ratios for both sectors.
Source: 1: Bank of Japan, *Comparative Economic and Financial Statistics*, 1992, pp. 45–46, 107–108.

Table 34.4 German industrial power and efficiency (1990).

Industry	Labour force (Thousand)[1]	(%) (a)	Gross domestic product (M* Billion)[1]	(M* Thou/employee)	(%) (b)	Efficiency index b/a
Primary		4.7			2.1	0.55
Agriculture, forestry and fishing	1299	3.8	38.0	36.6		
Mining	1039	0.9				
	260					
Secondary	10 855	39.2				
Manufacturing	8736	31.5	689.6[†]	76.7[‡]	37.2[‡]	1.15/1.18
Construction	1849	6.7	114.3	61.8	6.2	0.92
Electric, gas and water	270	1.0	69.7[†]	250.7[‡]	3.8[‡]	3.75/3.86
Subtotal (goods production)	12 154	43.8	911.6	75.0	49.1	1.12
Tertiary (service production)	15 572	56.2	944.4	60.6	50.9	0.91
Wholesale and retail trade	4093	14.8	187.5	45.8	10.1	0.68
Transportation and communication	1573	5.7	123.6	78.6	6.7	1.17
Finance, insurance and real estate	2181	7.9	274.8	126.0	14.8	1.88
Services	7725	27.9	358.5	46.4	19.3	0.69
Total	27 726	100	1856.0	66.9	100	1

* M (mark) 1 = $0.62 for 1990.
[†] Includes a part of added value created in the mining sector.
[‡] Cases where labour population for the mining sector is partitioned among the manufacturing and the electric, gas and water sectors according to the labour population ratios for both sectors.
Source: 1: Bank of Japan, *Comparative Economic and Financial Statistics*, 1992, pp. 47–48, 107–108.

British and German Industrial Powers and Efficiencies

British and German industrial powers are presented in Tables 34.3 and 34.4, respectively. The manufacturing efficiency indices are 1.03 to 1.12 for Britain and 1.15 to 1.18 for Germany.

As a great industrial country, Germany's manufacturing sector shares 31.5% of the total labour population, and 37.2% of GDP, both top among economically advanced countries. But the manufacturing efficiency index is not very good.

Russian Industrial Power and Efficiency

Russia's (old USSR's) industrial power and efficiency are indicated in Table 34.5, though raw data are not so exact. The manufacturing sector creates over 40% of GDP, larger than that for Germany, and its efficiency index is 1.46, highest among the advanced countries. However, the efficiency index of service production is very low—0.54; underdevelopment of the tertiary industry is seen in Russia.

International Comparison of Labour and Capital Efficiencies

Industrial efficiencies for Japan, the United States, and Germany in terms of labour and capital for primary, secondary and tertiary industries as well as the manufacturing and the commerce sectors are represented in Figure 34.8, taking the labour and the capital efficiencies as abscissa and ordinate.

Table 34.5 Russian industrial power and efficiency (1989).

Industry	Labour force (Million)[1]	(%) (a)	Gross domestic product (Rbl.* Billion)[1]	(%) (b)	Efficiency index b/a
Primary					
Agriculture	23.183	18.2	158.4	23.5	1.29
Secondary	49.598	39.0	368.4	54.7	1.40
Manufacturing	36.414	28.7/29.3[†]	282.0	41.9	1.46/1.43[†]
Construction	13.184	10.4	86.4	12.8	1.24
Subtotal					
(goods production)	72.781	57.3	526.8	78.2	1.37
Tertiary (service production)	51.431	40.5	146.9	21.8	0.54
Commerce	9.877	7.8	109.0	16.2	
Services	31.331	24.7	37.9	5.6	
Transportation	10.223	8.0			
Unidentified	2.845	2.2	0.0	0.0	
Total	127.057	100	673.7	100	1

*Rbl (ruble) 1 = $1.59 for 1989 ($0.5 for 1991).
[†] Figures in cases where labour population for nonclassifiable establishments is excluded.
Source: 1: Bank of Japan, *1991 Foreign Economic and Financial Statistics*, 1992, p. 473.

As capital investment is rather large in Japan,[2] Japan's manufacturing is low (below 1) from the viewpoint of capital efficiency, whilst this efficiency is higher in the US and in Germany (above 1).

The labour efficiency of Japan's manufacturing is above 1, better than the German, but inferior to that of the US.

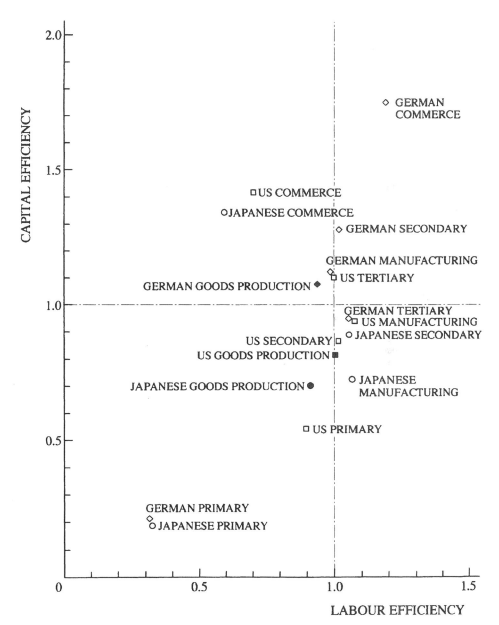

Figure 34.8 Comparison of industrial efficiencies in terms of labour and capital for the United States, Germany, and Japan (1987) (data source: *Comparative Economic and Financial Statistics* (1991) and *Investigation*, No. 140 (1990), Bank of Japan).

It is further found from Figure 34.8 that

- Japan's goods production is inferior to that for the other countries.
- German commerce is particularly high in both capital and labour efficiencies.
- American primary industry is much superior to Japanese and German primary industries.

34.4 Industrial Efficiency in Developing Countries

General Trends of Asian Industries

In recent years economic growth in Asia has been tremendous. Japan's annual growth rate was over 10% in the 1960s, still 5% in the 1980s, and is presently (in the 1990s) low, namely 1–3% (−0.2% in 1993/0.6% in 1994), owing to the economic recession. The newly industrialising economies (NIEs)—Korea, Hong Kong, Taiwan, and Singapore (often called 'four dragons')—showed an economic growth rate of 7–10% in the 1980s, even now it is 6–8%. The Association of Southeast Asian Nations (ASEAN): Indonesia, Malaysia, Singapore, Thailand, the Philippines, Brunei and Vietnam, especially Thailand and Malaysia, have taken off economically with a growth rate of 8–9%. The socialist countries of China and Vietnam are also expanding their economies by introducing market-economic systems as in capitalist countries; their annual growth rate shows nearly 10%. China expanded industrial production by 19.3% during the first half of 1992.

As mentioned in Section 34.2, in 1990 Japan produced an added value of 2963 billion US dollars—13.7% of the world's GNP. Other south-east Asian countries with a population of 438 million produced 335 billion US dollars. China with a population of 1134 million and huge land resources, 9.6 million km^2, produced 416 billion US dollars.

In 1990 Japan's GNP per capita indicated 25 125 US dollars, sixth in the world. GNP per capita is 5 to 15 thousand US dollars in NIEs—5569 in Korea with a population of 43.2 million, 7997 in Taiwan with 20.4 in 1991, 14 186 in Hong Kong with 5.8, and 14 471 in Singapore with 2.8. This figure is even less in other countries; that is, 500–1500 in ASEAN, 370 in China in 1990.

Labour wage is still low in Asia. With Japan's hourly wage in 1990 of US$ 10.74 (cf. the United States, $11.66) as standard of 100, Korea is about 40 Taiwan 30, China, Malaysia and Thailand only 2–6. With Japan's yearly wage in 1993 (4 040 000 yen = $27 900) as standard of 100, Germany was 135, the US 88, Korea 29, Thailand 7.5 and China 1.5 (US Department of Labor, *BLS Report*).

Asian manufacturing industries deal with large amounts of export. In 1991 Japan exported US$315 billion with a surplus of $102 billion ($135 billion in 1995) (see Table 33.2). NIES, ASEAN and China's exports were $220, 63 and 60 billion, respectively (cf. USA 422, Germany 403). The ratio of export against GNP exceeds 100% in Singapore and Hong Kong, namely 148 and 121%; Japan 9, Korea 35, China 36.

Thus the economic growth in Asia is tremendous; however, it is said that Japan and NIEs are now at the turning point, in that the industrial structure is changing for these countries. That is, the shares of the labour force and of the Gross Domestic Product (GDP) in the tertiary industry, which concerns services (non-manufacturing activities) are increasing; in 1990 those figures (%) were 58.2 and 60.8 in Japan, 47.0 and 42.8 in Korea, and 46.3 and 49.4 in Taiwan. This phenomenon might lower the international–competitive power of the nation's manufacturing sector.

Korean Industrial Power and Efficiency

As a representative of Asian NIEs Korean industrial power and efficiency are indicated in Table 34.6. In 1990 Korea employed 26.9% of the labour force in the manufacturing sector. This industry produced an output of 51.7% and created a value added of 32% of GDP. Hence the yield rate is 28.3%, which is the lowest among all the industrial sectors, as in Japan. The efficiency index is 1.19, slightly larger than Japan's—1.18 (but in 1989 this indicator showed 1.13, lower than Japan's 1.24).

Thailand's Industrial Power and Efficiency

As a representative of ASEAN Thailand's industrial power and efficiency in 1990 are indicated in Table 34.7.

ASEAN countries are agricultural excepting Singapore. The labour force in the primary industry is 30–70%; hence the manufacturing labour force shares less than 10%, except Malaysia's 16%. However GDP share created by the manufacturing sector exceeds 25%, except Indonesia's 17%. The manufacturing efficiency index is 3 for Thailand, 2.6 for the Philippines, 1.5 to 2 for Malaysia and Indonesia; manufacturing efficiency is rather good in ASEAN countries, compared to Japan and Korea.

China's Industrial Power and Efficiency

A large socialist country, China (population: 1.2 billion) produced US$ 363.81 billion: 1.6% of the world's GNP in 1990.[3]

Although statistical data are not complete, China's industrial power and efficiency in 1990 are indicated in Table 34.8. The manufacturing sector holds 17.1% of the total labour population, produced 62.9% of the output, and created 38.8% of GDP. The yield rate was 28.7%, lower than Japan's and Korea's; however, the efficiency index was 2.27, much more efficient than those Asian countries.

34.5 International Comparison of Manufacturing Efficiency

Manufacturing Efficiencies

Labour-force ratio, GDP share and efficiency indices for manufacturing are compared internationally in Table 34.9. In general the efficiency in primary industry is low and the manufacturing efficiency is superior to efficiencies for primary and tertiary industries.

The manufacturing efficiency is usually higher in the developing countries and in the socialist countries than in the advanced countries.

Table 34.6 Korean industrial power and efficiency (1990).

Industry	Labour force		Output		Gross domestic product		Yield rate	Efficiency index
	(Thousand)¹	(%) (a)	(W† Billion)¹ (A)	(%)	(W† Billion)¹ (B)	(%) (b)	(%) B/A	b/a
Primary	3373	18.7	22 783.8	6.7	16 371.6	10.5	71.9	0.56
Agriculture forestry and fishing	3292	18.3	21 337.7	6.2	15 583.5	10.0	73.0	0.55
Mining	81	0.4	1446.1	0.4	788.1	0.5	54.5	1.12
Secondary	6186	34.3	216 739.7	63.5	72 778.8	46.7	33.6	1.36
Manufacturing	4847	26.9	176 503.8	51.7	49 894.7	32.0	28.3	1.19
Construction	1339	7.4	40 235.9	11.8	22 884.1	14.7	56.9	1.98
Subtotal (goods production)	9559	53.0	239 523.5	70.1	89 150.4	57.2	37.2	1.08
Tertiary (service production)	8477	47.0	102 039.8	29.9	66 782.7	42.8	65.5	0.91
Others*	8477	47.0	102 039.8	29.9	66 782.7	42.8	65.5	0.91
Total	18 036	100	341 563.3	100	155 933.1	100	45.7	1

† W (won) 1 = $0.0014 for 1990.
* Excludes activities by the Government and non-profit organisations. Includes electricity, gas and water.
Source: 1: Bank of Korea, Economic Statistics Yearbook, 1992, p. 266, 310.

Table 34.7 Thailand's industrial power and efficiency (1990).

Industry	Labour force		Gross domestic product			Efficiency index
	(Thousand)[1]	(%) (a)	(B* Billion)[1]	(%) (b)	(B* Thou/employee)	b/a
Primary	19 620	66.6	298.041	20.5	15	0.31
Agriculture, forestry and fishing	19 577	66.5	250.384	17.3	13	0.26
Mining	43	0.1	47.657	3.3	1108	22.50
Secondary	3283	11.1	492.432	33.9	150	3.05
Manufacturing	2461	8.4	373.326	25.7	152	3.08
Construction	702	2.4	84.791	5.8	121	2.45
Electric, gas and water	120	0.4	34.315	2.4	286	5.81
Subtotal (goods production)	22 903	77.7	790.473	54.5	35	0.70
Tertiary (service production)	6555	22.3	660.108	45.5	101	2.05
Wholesale and retail trade	2897†	9.8	332.977‡	23.0	—	—
Transportation and communication	641	2.2	106.697	7.4	166	3.38
Finance, insurance and real estate	—	—	124.012	8.5	—	—
Services	3017‡	10.2	96.422	6.6	—	—
Total	29 464	100	1 506.977§	100	51	1

* B (baht) 1 = $0.0395 for 1990.
† Includes finance, insurance and real estate.
‡ Includes restaurants and hotels.
§ Includes the Government services (B56.397 billion)/B2 381.0 billion for 1991.

Source: 1: Bank of Japan, *1991 Foreign Economic and Financial Statistics*, 1992, p. 360, 364.

Table 34.8 China's industrial power and efficiency (1990).

Industry	Labour force		Output			Gross domestic product			Yield rate	Efficiency index
	(Million)[1]	(%) (a)	(¥* Billion)[1] (A)	¥/employee	%	(¥ Billion)[1] (B)	(¥/employee)	(%) (b)	(%) B/A	b/a
Primary	342.77	60.4			20.7	501.70	1468	28.4		0.47
Agriculture			766.2							
Secondary	121.58	21.4	2696.7		70.9	771.74	6348	43.6	28.6	2.04
Manufacturing	6.97	17.1	2392.4	24 672	62.9	685.80	7072	38.8	28.7	2.27
Construction	24.61	4.3	304.3	12 365	8.0	85.94	3492	4.9	28.2	1.12
Subtotal (goods production)	464.35	81.8				1273.44	2742	2.0		0.88
Tertiary (service production)	103.05	18.2				494.69	4800	8.0		1.54
Commerce	29.37	5.2	187.1		4.9	83.70		4.7	4.7	0.91
Transportation & Communication	14.69	2.6	153.5		4.0	111.76		6.3	72.8	2.44
Total	567.40	100	3803.5		100	1769.53[†]	3119	100	46.5	1

*¥ (yuan) 1 = $0.21.
[ ]: Data/calculated figures for 1988.
[†] Not coincident with the sum of figures for primary, secondary and tertiary industries.

Table 34.9 Labour-force and GDP shares and efficiency of the manufacturing industry—international comparison.

Country	Labour-force share (%)	GDP share (%)	Manufacturing efficiency
USA	18.0	22.1	1.23
Canada	15.9	21.8	1.37
Britain	20.3	22.6	1.12
France	21.0	25.2	1.20
Germany	31.6	36.5	1.15
Italy	22.2	26.1	1.18
Russia	28.7	41.9	1.46
Japan	23.4	27.5	1.18
Korea	26.9	32.0	1.19
Hong Kong	27.7	18.1	0.65
Taiwan	32.0	37.3	1.17
Singapore	28.9	27.3	0.94
Indonesia	10.2	21.7	2.14
Malaysia	15.5	25.5	1.64
Thailand	8.4	25.7	3.08
Philippines	9.7	25.2	2.60
Pakistan	12.7	15.2	1.20
Australia	15.3	14.9	0.97
New Zealand	17.1	18.3	1.07
China	17.1	38.8	2.27

Notes

1 The 'labour efficiency index' was first introduced by Kuznets (1971). He did not use this terminology, but defined 'the ratio of product per worker in the sector to product per worker for the country' as:

$$Y_i L : Y L_i = (Y_i/L_i) : (Y/L). \qquad (34.3)$$

His definition is different from the definition for the efficiency index stated in this chapter, but the above equation completely coincides with (34.2).

2 For example, in 1988 Japan invested $498 billion, compared to USA $488 and Germany $148.

3 In this year China exceeded Japan in terms of GDP calculated from the viewpoint of the purchasing power parity; it is estimated that China will exceed the US in the same term in 2002 (World Bank).

References

HITOMI, K. (1993) Efficiency analysis of manufacturing industry (in Japanese), *Office Automation*, **14** (1).
KUZNETS, S. (1971) *Economic Growth of Nations* (Belknap Press), p. 208.

Supplementary reading

CAVES, R. and BARTON, D. (1990) *Efficiency in U.S. Manufacturing Industries* (Cambridge, Mass.: MIT Press).

HITOMI, K. (1993) Manufacturing technology in Japan, *Journal of Manufacturing Systems*, **12** (3).

HITOMI, K. (1993) Efficiency of Asian manufacturing, *Proceedings, The International Conference of Office Automation and Information Management*, Beijing.

HITOMI, K. (1994) *Introduction to Today's Manufacturing* (in Japanese) (Tokyo: Dobunkan), Chap. 4.

CHAPTER THIRTY-FIVE

Industrial Input–Output Relations

35.1 Relationships Between Industrial Activities

Influences Among Industrial Sectors

Close relations exist among various industrial sectors, and each industrial sector influences the others nationally and internationally.

─── **EXAMPLE 35.1** ───────────────────────────────

(The ripple effect) In 1980 a demand of 1 trillion yen ($1 = 227 yen in 1980) from the Japanese transportation machinery sector (automobiles, railway cars, airplanes) generated metal production of about 300 billion yen, producing more than 100 billion yen for each of the general machinery sector, industrial goods, and commerce. This further created about 9 billion yen for each of the agriculture and fishery and the mining sector, and more than 0.3 billion yen for pulp, papers and woods. This ripple effect finally washed over the transportation machinery sector again to the tune of about 345 billion yen (Miyazawa, 1987). □

Hence production/manufacturing should be recognised as an integrated 'open system', rather than a 'closed system', covering the relationships among all the industrial sectors and the surrounding overall national and international economies.

Input–Output Analysis

An effective solution procedure—analytically, numerically, and with tables—for nationally/internationally related manufacturing/industrial activities is the *input–output analysis* with the use of input–output tables (a modern version of F. Quesnay's economic table of 1758) which W. Leontief developed in 1936.

35.2 Industrial Input–Output Analysis

35.2.1 Input–Output Tables

Structure of Input–Output Tables

An *input–output table* summarises the economic relationships between 'input' (demands from the industrial sectors) and 'outputs' (production performances) for the trades among industrial sectors in a country or in several related countries.

Basic Formulae for Input–Output Tables

The input–output table (in physical units) is represented in Table 35.1. In this table the following condition of *demand–supply equilibrium* holds for each row:

total output = total demand = intermediate demands + final demand   (35.1)

Hence, the total output for sector i is expressed:

$$X_i = \sum_{j=1}^{n} X_{ij} + F_i \qquad (i = 1, 2, \ldots, n) \qquad (35.2)$$

where X_{ij} is the intermediate demand of sector j from production by sector i, F_i is the final demand toward product[ion] i; n is the number of industrial sectors (usually 100 to 500).

Input–Output Tables in Monetary Units

Input–output tables for a country and for internationally related countries are usually indicated in monetary units.

--- **EXAMPLE 35.2** ---

(Japan's input–output table) The condensed (originally 13 sectors—see Management and Coordination Agency, 1994) input–output table in terms of the primary, secondary, and tertiary industries in Japan for 1990 is represented in Table 35.2.   □

Table 35.1 Input–output table in physical units.

From industrial sector	To industrial sector						Final demand	Output
	1	2	...	j	...	n		
1	X_{11}	X_{12}	...	X_{1j}	...	X_{1n}	F_1	X_1
2	X_{21}	X_{22}	...	X_{2j}	...	X_{2n}	F_2	X_2
⋮	⋮	⋮	...	⋮	...	⋮	⋮	⋮
i	X_{i1}	X_{i2}	...	X_{ij}	...	X_{in}	F_i	X_i
⋮	⋮	⋮	...	⋮	...	⋮	⋮	⋮
n	X_{n1}	X_{n2}	...	X_{nj}	...	X_{nn}	F_n	X_n

Table 35.2 Japan's input–output table in 1990 (condensed into three industries) (unit: trillion yen ($1 = 145 yen)).

Industry	Primary	Secondary	Tertiary	Final demand	Input	Output
Primary	2.3	23.2	1.1	4.8	−11.5	20.0
Secondary	3.2	180.1	50.0	240.3	−25.1	448.6
Tertiary	3.0	75.8	87.3	246.8	−9.3	403.6
Value added	11.4	169.5	265.2	(Total: 446.1)		
Input	20.0	448.6	403.6			Total: 872.2

── **EXAMPLE 35.3** ──

(Global input–output table) Table 35.3 shows the outline of the 1985 Japan–USA–EC–Asia input–output table (Research and Statistics Department, 1993). □

Input Coefficients

The *input coefficient* of sector i for production j is defined as follows:[1]

$$\begin{cases} a_{ij} = X_{ij}/X_j \geq 0 & (i, j = 1, 2, \ldots, n; i \neq j) \\ a_{ii} < 1 \end{cases} \quad (35.3)$$

Then (35.2) is further expressed:

$$X_i = \sum_{j=1}^{n} a_{ij} X_j + F_i \quad (i = 1, 2, \ldots, n). \quad (35.4)$$

This shows the causal ripple effects between the industrial sectors and calculates the production ultimately required for each sector.

35.2.2 Leontief Models

Static Leontief Model

Equation (35.4) can be rewritten with vector expression as follows:

$$(\mathbf{I} - \mathbf{A})\mathbf{X} = \mathbf{F} \quad (35.5)$$

where $\mathbf{X}$ and $\mathbf{F}$ are column vectors for X_is and F_is ($i = 1, 2, \ldots, n$), respectively; $\mathbf{A} = (a_{ij})$ is the input-coefficient matrix; and $\mathbf{I}$ is the $n \times n$ unit matrix.

The above equation is called the 'static' *Leontief model*.[2] $(\mathbf{I} - \mathbf{A})$ is the Leontief matrix.

Calculating Total Production

Total output (or production) is derived from (35.5) as follows:

$$\mathbf{X} = (\mathbf{I} - \mathbf{A})^{-1}\mathbf{F} = \mathbf{BF} \quad (35.6)$$

Table 35.3 Global input–output table among Japan, the USA, the EC, and Asia in 1985 (unit: billions of US dollars).

	Intermediate demand				Domestic final demand				Export to rest of the world	Production value
	Japan	US	EC3	Asia8	Japan	US	EC3	Asia8		
Japanese products	992.0	26.5	5.7	21.2	1260.0	40.9	7.9	15.2	81.7	2451.3
US products	18.3	2206.0	14.6	14.4	5.6	3882.9	12.9	7.7	191.3	6353.8
Products of EC3	3.6	20.2	843.9	8.4	2.2	22.2	1345.3	6.0	296.2	2547.9
Products of Asia8	23.9	20.2	4.4	506.4	3.7	23.5	4.8	595.8	72.1	1254.7
International transportation costs, insurance	6.0	4.2	1.4	5.1	1.0	4.7	1.3	2.2		
Import from rest of world	77.4	122.9	180.9	47.9	10.4	116.0	91.5	16.5		
Customs, etc.	4.3	4.8	5.1	8.2	1.4	9.4	3.1	7.8		
Total for intermediate input	1125.4	2404.9	1056.1	611.6	1284.3	4099.7	1466.9	651.2	Total: 128077	
Total for value added	1325.9	3948.9	1491.8	643.2						
Domestic production value	2451.3	6353.8	2547.9	1254.7						

where

$$\mathbf{B} = (b_{ij}) = \left(\frac{\partial X_i}{\partial F_j}\right). \tag{35.7}$$

Equation (35.7) is called the Leontief inverse matrix; its table is the inverse matrix coefficient table. This helps calculate the incremental output (production) required directly/indirectly in sector i when a unit increase of final output in sector j has emerged. Then the required production in sector i can be calculated:

$$X_i = \sum_{j=1}^{n} b_{ij} F_j \qquad (i = 1, 2, \ldots, n). \tag{35.8}$$

X_i can be obtained under the following conditions:

- *Hawkin–Simon* (necessary and sufficient): all of the principal cofactors of determinant $|\mathbf{I} - \mathbf{A}|$ are positive;
- *Solow* (sufficient):

$$\sum_{i=1}^{n} a_{ij} < 1. \tag{35.9}$$

35.3 Requirements for Input–Output Analysis

Notes for Input–Output Analysis

The following requirements should be noted when employing input–output analysis:

- *Uncombined production.* An industrial sector produces only one kind of product.
- *Fixed input coefficients.* Constant returns for economies of scale are held.
- *Static model.* Time scale is not permitted; long-term, dynamic cases cannot be analysed.

Notes

1. If $a_{ii} \geq 1$, production is meaningless.
2. Based upon L. Walras' general equilibrium theory (1874).

References

Management and Coordination Agency, Government of Japan (ed.) (1994) *1990 Input–Output Tables* (Explanatory Report) (in Japanese) (Tokyo: National Statistics Association), pp. 4–5.

MIYAZAWA, K. (1987) *Economics of Industries* (2nd edn) (in Japanese) (Tokyo: Toyokeizai-shimpo-sha), pp. 89–90.

Research and Statistics Department, Ministry of Industrial Trade and Industry, Japan (ed.) (1993) *The 1985 Japan–U.S.–EC–Asia Input–Output Table* (Tokyo: Tsusan Statistics Association), p. 96.

Supplementary reading

LEONTIEF, W. (1986) *Input–Output Economics* (2nd edn) (Oxford: Oxford University Press).

CHAPTER THIRTY-SIX

Manufacturing Excellence for Future Production Perspectives

36.1 Importance and Dilemma of Today's Manufacturing

Manufacturing Matters

As mentioned in the beginning of this book, manufacturing—the production of tangible goods or products—is a basic activity with a history of several thousand years. Three points of vital importance are:

- It is a basic means of human existence.
- It is a major contributor to the creation of the wealth of a nation.
- It contributes to human happiness and to world peace.

Manufacturing Dilemma Today

Today's manufacturing in many economically developed countries is faced with several issues constituting a considerable dilemma (Hitomi, 1994):

(1) *Maturity and saturation of industrial products.* Many industrial products have been and are being manufactured and supplied to the established markets in excess; hence, in many cases there is no longer a strong demand.

— **EXAMPLE 36.1** ————————————————————————

In Japan most households possess more than two colour televisions, more than one electric cleaner, refrigerator, washing machine, air conditioner, etc.   □

——

At present production capacities are in excess of demands and so it is hard to maintain high factory utilisation rates.

— **EXAMPLE 36.2** ————————————————————————

'Japan's factories are plodding along at only 74% of capacity.' (*Business Week*, 6 February 1995).   □

——

(2) *De-industrialisation* or *industrial hollowness*. As the labour wage becomes increasingly high in the developed (or advanced) countries compared to developing countries, domestically produced non-sophisticated products can no longer compete on price with those produced in the developing countries.

—— EXAMPLE 36.3 ——

Setting Japan's manufacturing labour wage ($30 641 in 1991) at 100, the USA is 92.8, Germany is 113.1, whilst Korea is 36.4, Thailand 7.4 and China 1.5. □

Accordingly, manufacturing industries of the developed countries have established factories outside their countries. This has resulted in the decline of the domestic manufacturing labour force and production activities. This phenomenon is 'de-industrialisation' or 'industrial hollowness', as mentioned in Chapter 33.

—— EXAMPLE 36.4 ——

Japan's overseas production in 1993 was 74% of colour television sets, 64% of microwave ovens, and 50% of videotape recorders. □

---

(3) *Unattractiveness of employment in manufacturing*. The younger generation tends not to be attracted to work in factories. In Japan they are now said to be '3K', that is, 'Kitanai' (dirty), 'Kitsui' (hard work), and 'Kiken' (dangerous). There are fewer skilled workers/craftmen in workshops, and it is increasingly difficult to attract suitable people to be trained in the required skills and techniques.

—— EXAMPLE 36.5 ——

Advanced countries cannot compete with developing countries in production skills; the number of gold medals obtained in the 'Skills Olympics' in 1993 was 18 for Taiwan and 12 for Korea, compared with 3 for Germany, 2 for France and Japan, and 0 for the United States and the United Kingdom. □

---

(4) *Environmental impact*. Excess production threatens to exhaust natural resources, materials and energy, and contaminate the environment. As a result, increasing attention must be given to recycling of manufacturing material, energy efficiency in the manufacturing process and the control of environmental pollution.

—— EXAMPLE 36.6 ——

The world's energy consumption has increased more than sixfold between 1950 and 1990. The world's energy resources will last only about 50 years for petroleum and natural gas, about 75 years for uranium. Coal will last more than 200 years. □

—— EXAMPLE 36.7 ——

Car production consumes a great amount of resources: 17% of the iron and steel produced, 36% of bearings, 27% of machine tools, more than 40% of dies, etc. (*Nikkei Mechanical*, 6

September 1993, p. 21). Around 10% of the cars being used are dumped every year; (more than 5 million in Japan). □

36.2 Concept of Manufacturing Excellence

Proposal for Manufacturing Excellence

Manufacturing excellence (or *elegance*) is now proposed as a keyword expressing the enhancement of the conditions of goods production for pursuing human happiness, whilst eliminating the 'hollowing' of production, '3K', and environmental problems. These words have been employed in literature without any clear definition. Hall (1987) suggests that they consist of 'just-in-time', 'total quality', and 'total people involvement'.

Manufacturing Culture

Cultural orientation of production was pointed out prior to manufacturing excellence (Sahlins, 1976).

A word 'culture' appeared in 1480[1], meaning worship. Its fundamental meaning, 'cultivate', which comes from its origin (Latin: *colere*) appeared in 1620; incidentally *'agriculture'* appeared earlier in 1603. In some cases culture means produces/crops; however, nowadays *culture* is also an abstract noun meaning intellectual/artistic activities such as philosophy, history, literature, art, music, sculpture, drama, etc. (Williams, 1976).

The term *manufacturing culture* ('industrial culture' or 'corporate/enterprise culture') implies a combination of intellectual, conceptual, philosophical activities and ideas with human relationships, motivations and involvements. In addition, another expression, *manufacturing renaissance* means the restructuring of goods production by constructing 'amenity' factories for humane working conditions.

Culture usually holds a positive meaning, although it may be used negatively, e.g. monoculture, mass culture, throw-away culture, etc. In this respect 'manufacturing culture' might be misunderstood as throw-away culture through mass production/mass consumption and even mass disposal. This is a reason why 'manufacturing excellence' is used to describe the enhancement and elegance of goods production, rather than 'manufacturing culture'.

Cultural/Civilised Production

There exists a term *cultural production*. It means immaterial production such as art, literature, etc. (Bourdieu, 1992). It should be noted that Aristotle (384–22 BC) established an academic subject of producing things but production (*poiēsis*) based upon production technique (*poiētikē technē*) was poesy, not tangible goods (Ide, 1972).

The expression *civilised production* exists in China. Often 'culture' is region-restricted, whilst 'civilisation' (which appeared in 1704) means a mode or a state of human activity applicable all over the world (Kimura, 1992). Occasionally civilisation means materialistic civilised production; these two words are often employed synonymously. Civilised production therefore can mean simply a clean operation in a workshop, or it can mean manufacturing based upon modern science and technology.

36.3 Approaches to Manufacturing Excellence

Methods of Manufacturing Excellence

The following methods are effective to moving towards manufacturing excellence.

(1) *Automated production/CIM systems*. A way to get rid of '3K' is to establish 'amenity' factories through factory automation and CIM (computer-integrated manufacturing) systems, as discussed in Chapters 23 and 25.

(2) *Flexible/human-centred production*. Manufacturing a variety of products of super-high quality with the use of high technologies but still drawing heavily on skills of workers is what *flexible production* (Hitomi, 1989) or *human-centred (anthropocentric) systems* (Cooley, 1987) are concerned with, as pointed out in Section 4.3.

—— **EXAMPLE 36.8** ——

Motorcar assembly factories (Volvo in Sweden, Honda in Japan) that do not use conveyors, industrial community systems by craftsmen in Italy, flexible manufacturing by workers' own networks in the USA, etc. □

Technology-centred full automation mainly comprising highly advanced production technologies, is not only insufficiently flexible but is also confronted with a lack of innovative flexibility, greater vulnerability to failure, and, above all, neglect of human skills, robbing workers of their pride and pleasure of work (Wobbe, 1990), in spite of an enormous amount of capital investment. J. Harrington, who coined the term *computer integrated manufacturing*, mentioned that 'CIM does not mean an automated factory; people are very much involved' (Harrington, 1973). This mode of production pays attention to a new metaphysical 'production philoso-technology' beyond the conventional subjective 'production skills/techniques' and objective 'production technology'.

(3) *High added-value production*. Production of sophisticated goods generate high added value.

As mentioned in Section 1.5, 'value added' or 'added value' is a value created purely during the process of production, and is calculated using equations (1.9)–(1.11). The situation of high-volume production, but 'dumping-like' sales activities, as Japanese manufacturers have often conducted, accompanies small unit profits, hence little added value is obtained. By developing highly original, quality products with high added values, manufacturers can assure themselves appropriate profits through better coordinated production.

In this respect cost reducing through the cost-control activity is especially important to decrease the total cost of producing a piece of product.

(4) *Manufacturing for customer satisfaction*. Mass production of a single product, as instanced by the Model T Ford, is over; today, manufacturing firms must supply what customers need quickly by minimising the production lead time. *Cellular manufacturing*, making a variety of products in the quick-response or just-in-time mode, has replaced traditional flow-like manufacturing. Assembly lines by a single worker or by a group of workers have emerged, as mentioned in Section 25.5. There is no limit to what products will be made in this way (*Business Week*, 19 December 1994, p. 28).

── **EXAMPLE 36.9** ──────────────────────────

Dell Computer, who received a customised product on Wednesday at 10.49 a.m., completed their computer at 8.37 p.m. on that day; the customer received the product on Friday at 10.31 a.m. (*Fortune*, 18 April 1994, pp. 73–76).

In this sense, manufacturing is not only a means to produce tangible goods, but it is 'service' (Warnecke, 1993). □

(5) *Green production*. Resource-saving and environment-preserving production is especially vital to reduce the environmental impact of manufacturing. This topic is discussed in the following section.

(6) *Socially appropriate production*. This is the ultimate manufacturing excellence to stop excessive production of industrial products. This topic is discussed in Section 36.5.

36.4 Green Production

36.4.1 Japan's Industrial Activity

Small Land, but Large Economy of Japan

Japan has very limited natural resources. In spite of this restriction Japan has been a great economic power producing 15.6% (3669 billion US dollars in 1992) of the world's GNP—the world's second largest economy following the United States (6020 billion US dollars or 25.6%).

Consumption of Natural Resources in Japan

Japan has built up an affluent society with full use of natural resources. As shown in Figure 36.1 based upon the research report on Japan's material balance (1992) in 1990 Japan used 1334 million tonnes of domestic natural resources and 711 million tonnes of imported resources (16% of the world's annual resources of around 14 billion tonnes (Honda, 1991)).

As a result 1246 million tonnes of buildings, structures and products were accumulated, 73 million tonnes of export products were manufactured and 84 million tonnes of foods (1.8 kg/day per capita) were consumed. In this process 368 million tonnes of energy resources were employed, whilst producing industrial wastes (19 kinds) of 222 million tonnes and general wastes (also generated from

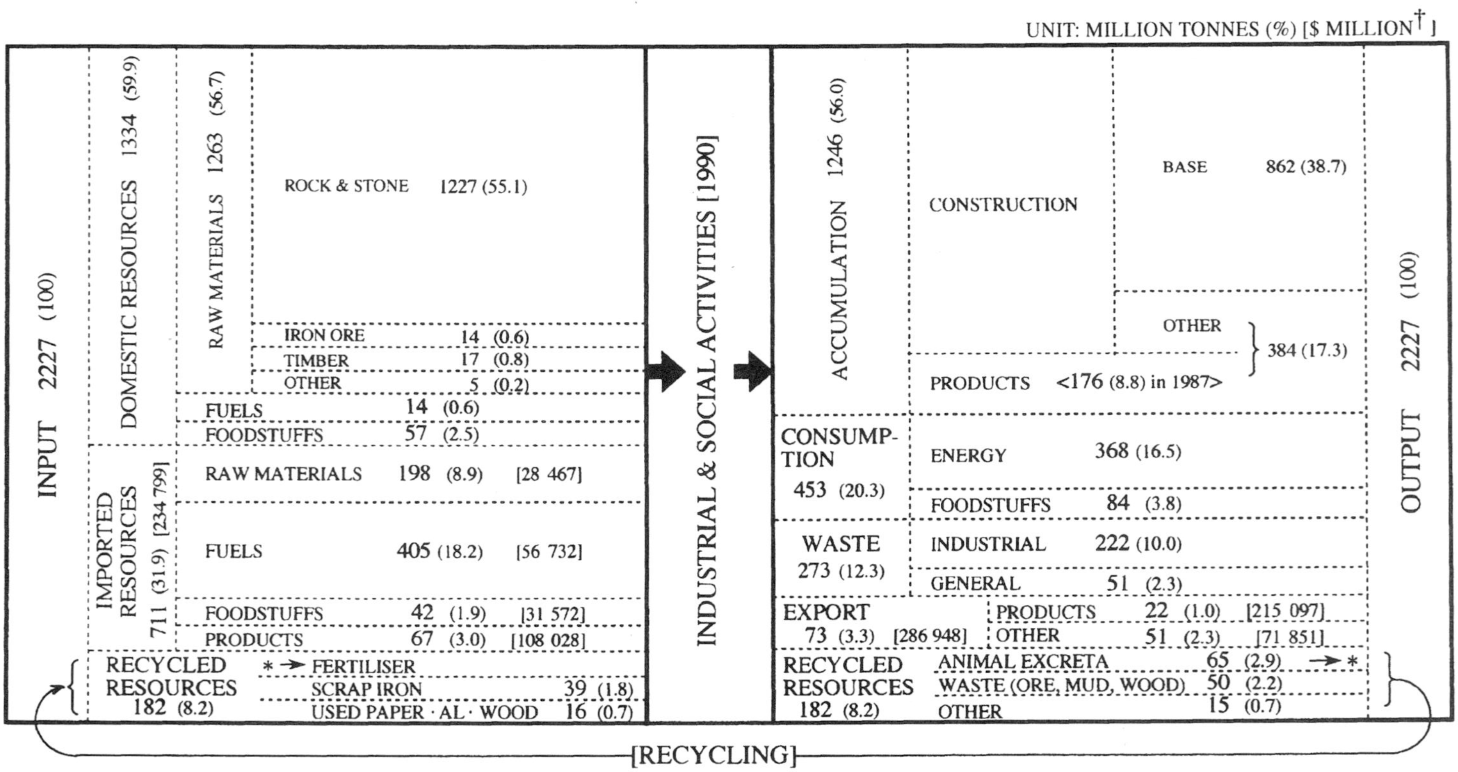

502 SOCIAL SYSTEMS FOR MANUFACTURING

UNIT: MILLION TONNES (%) [$ MILLION†]

INPUT 2227 (100)

IMPORTED DOMESTIC RESOURCES 1334 (59.9)
- ROCK & STONE 1227 (55.1)
- IRON ORE 14 (0.6)
- TIMBER 17 (0.8)
- OTHER 5 (0.2)
- FUELS 14 (0.6)
- FOODSTUFFS 57 (2.5)

IMPORTED RESOURCES 711 (31.9) [234 799]
- RAW MATERIALS 198 (8.9) [28 467]
- FUELS 405 (18.2) [56 732]
- FOODSTUFFS 42 (1.9) [31 572]
- PRODUCTS 67 (3.0) [108 028]

RAW MATERIALS 1263 (56.7)

RECYCLED RESOURCES 182 (8.2)
- *→ FERTILISER 39 (1.8)
- SCRAP IRON 16 (0.7)
- USED PAPER · AL · WOOD

↓↓ INDUSTRIAL & SOCIAL ACTIVITIES [1990] ↓↓

OUTPUT 2227 (100)

ACCUMULATION 1246 (56.0)
- CONSTRUCTION BASE 862 (38.7)
- OTHER } 384 (17.3)
- PRODUCTS <176 (8.8) in 1987>

CONSUMPTION 453 (20.3)
- ENERGY 368 (16.5)
- FOODSTUFFS 84 (3.8)
- INDUSTRIAL 222 (10.0)
- GENERAL 51 (2.3)

WASTE 273 (12.3)
- PRODUCTS 22 (1.0) [215 097]
- OTHER 51 (2.3) [71 851]

EXPORT 73 (3.3) [286 948]
- ANIMAL EXCRETA 65 (2.9) →*→
- WASTE (ORE, MUD, WOOD) 50 (2.2)
- OTHER 15 (0.7)

RECYCLED RESOURCES 182 (8.2)

[RECYCLING]

(Note) Components of recycled resources are indicated at both input and output sides. / → * → : represents a recycling example.

(*Source*: The Environment Agency of Japan: *Research Report on Japan's Material Balance* (March, 1992).)

† Japan Institute for Social and Economic Affairs: *Japan 1993 – An International Comparison*, pp. 40-41.

Figure 36.1 Material balance in Japan's industrial activity (1990) — total usage is about 16% of the world's usage; imported resources exceed 700 million tonnes, whilst those exported are only one tenth of this.

industrial activities) of 51 million tonnes, consequently amounting in total to 273 million tonnes. The resources recycled amounted to 182 million tonnes, approximately 8% of the overall resources consumed.

Material Imbalance

Though Japan is a leading export country in financial balance (export: 360.9 billion US dollars (third in the world) with surplus of 135 billion US dollars (world's top) in 1995), as mentioned in Section 33.3, it is an extremely conspicuous import country in terms of material resource balance. Viewing the import and export of materials and products to/from Japan in Figure 36.1, the import is 711 million tonnes whilst the export is only 73 million tonnes, which proves an extreme imbalance of materials.

In the case of Britain, an island country like Japan, both import and export respectively amount to approximately 150 million tonnes and are well balanced. The United States, which is a large country with hugh natural resources, imported 295 million tonnes and exported 410 million tonnes in 1986 (OECD Report) (Nakamura, 1989). On the other hand, the material balance in Japan is abnormal on a time series, as exhibited in Figure 36.2; hence the Japanese Islands are becoming garbage-dumping grounds.

Net National Welfare

Environmental pollution by the above garbage dumping and the urbanisation result in reducing the value of nominal GNP. The *green GNP* or *net national welfare (NNW)*, which preserves the environment in good order and enhances quality of human life, has been recently declining; in Japan NNW over GNP was 94% in 1975, 86% in 1980, 80% in 1985, and 75% in 1990.

36.4.2 Recycle-oriented Manufacturing Systems

Green Design and Production for Sustainable Development

Recognising the above actualities in consuming a huge amount of natural resources for industrial and manufacturing activities, design and production in the coming age should be directed to preservation of a better environment in view of the material flow of procurement–production–distribution–consumption–recycling–disposal. This must be achieved in the early stage of design (Overby, 1991), thereby resulting in permanently sustainable development by taking resource saving into account. This is *green design/green production*.

Recycle-oriented Society

The material flows in current society and in the recycle-oriented society are represented in Figure 36.3. In the current society the flow

(1) starts on extraction of natural resources from the earth;
(2) is followed by production and consumption; then
(3) the entire wastes are dumped into the environment.

504　　　　　　　　　SOCIAL SYSTEMS FOR MANUFACTURING

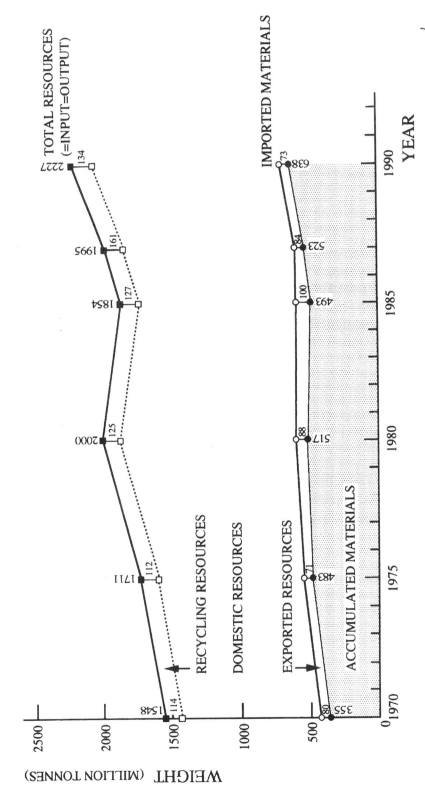

Figure 36.2 Trends of material balance and trade surplus in Japan, 1970–90: the difference between the imported and exported materials accumulates.

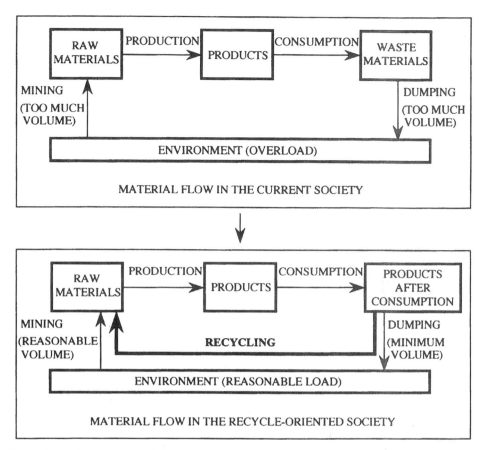

Figure 36.3 The current and the recycle-oriented societies.

On the other hand, in the recycle-oriented society the flow includes a bypass called *recycling*, to reduce the environmental burden to a reasonable level.

Recycle-oriented Manufacturing

The following activities are vital in the recycle-oriented manufacturing system.

(1) *Considering the recycling of used materials at the design stage.* The product design is made not only from the viewpoint of function/quality, but by taking efficient recycling and minimum harmful waste discharge (emission) into consideration.
(2) *Enhancing re-production and re-use of the parts.* After consumption of the products, parts/materials capable of re-production and re-use are collected.
(3) *Analysing the cost effectiveness of the recycling activity.* The recycling activity should be sustained economically. The cost-effectiveness analysis is useful for this purpose; that is, the expense required for the recycling should be shared in the society or covered by reasonable product pricing.

--- EXAMPLE 36.10 ---

A model was constructed for a firm with the recycle-oriented manufacturing system, which purchases necessary parts from the raw-material suppliers, and assembles a product item with the purchased parts together with the recyclable parts processed inside the firm's plant. The firm sells reproducible parts to the outside raw-material suppliers. The other wastes are dumped. This firm sets the following two kinds of goals:

(1) maximise the recycling rate for each part beyond a specified aspiration level;
(2) maximise the total profit.

Figure 36.4 shows a solution by taking account of re-use and re-production for the parts after the product's use. As seen in this figure, the total profit and recycling rates have a trade-off relationship (Hoshino et al., 1995). □

36.4.3 Product's Life-cycle Assessment

Life-cycle Analysis/Assessment (LCA)

Quantitative analysis and evaluation of the total environmental burden through a product's life—extraction of natural resources and material balance, energy consumption through production and use of the product, emission of air-contaminating materials and gases, effects on animals and plants, etc.—is called *life-cycle analysis*. *Life-cycle assessment* is to reduce, even minimise, this environmental burden based upon the total cost-minimum principle. Environmental management to prevent air pollution and earth destruction is presently being regulated by ISO 14000.

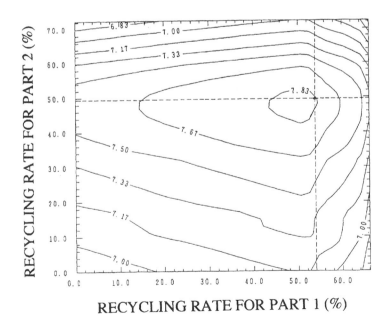

Figure 36.4 Iso-profit contours with respect to recycling rates for parts—figures indicate the amount of total profit in billion yen (the maximum profit obtained is 79 billion yen given at the point of intersection of the two dashed lines).

36.5 Socially Appropriate Production as Ultimate Manufacturing Excellence

Collapse of Production-firstism

Industrial, particularly manufacturing, activities commonly expect to consume huge amounts of natural resources. This has been accelerated by the rapid growth of human population—over 5.6 billion in 1995. Manufacturing, which has benefited human beings in the three aspects pointed out at the beginning of this chapter, has brought about mass (even 'excess') production, mass consumption, and mass disposal, thereby resulting in great fear of complete exhaustion of natural resources and energy and destruction of the Earth (geocatastrophe); human beings have been constructing the 'Tower of Babel'.

Socially Appropriate Manufacturing Through the Satisfaction-consciousness Principle

We human beings, who have been favoured with manufacturing/production, are now largely responsible for environmental issues of sustaining our Earth. Unless 'excessive' production of industrial products accompanying global destruction is now stopped, and the way for obtaining appropriate profits and long-term moderate growth capable of contributing to the public welfare is found, humankind will no longer be able to live.

This is the discipline of *socially appropriate manufacturing/production* (Hitomi, 1991). The essential spirit to support this discipline may be the 'satisfaction consciousness' based on Taoism and Buddhism that emerged in the Orient. This vital spirit says that 'the man who realises that he is satisfied is (spiritually) wealthy' (Chapter 33, the book of Laozi written in ancient China).

Satisfaction-conscious Production/Consumption and Minimum Disposal

Production based upon the spirit of satisfaction-consciousness is *satisfaction-conscious production*. The traditional mass (excessive) production should be replaced with this satisfaction-conscious production mode as early as possible. Under this production mode, 'socially useful production' (Cooley, 1987) is followed; only durable products are produced in amounts as small as society requires.

On the other hand, consumers must also follow 'satisfaction-conscious' consumption, i.e., refrain from 'mass' or 'excessive' consumption, and the society must proceed toward the goal of 'minimum disposal'.

The above concept of manufacturing/production is considered to be what we can call the ultimate 'manufacturing excellence' (production aesthetics), which should be the vision of the future management strategy of each individual manufacturing firm.

Manufacturing Ethics

Socially appropriate manufacturing the world over or satisfaction-conscious production by each individual manufacturing firm is not easy; it will be attained by recognition of 'manufacturing ethics'.

There exists:

- the spirit of capitalism based upon the ethics of Protestantism in the West, and
- the spirit of Buddhism/Confucianism in the Orient.

It is said that the latter idea has contributed to the economic and industrial developments of Asian countries in recent years. However, the rate of adopting the ethical codes is high in the West; it is rather low in Japan.

--- **EXAMPLE 36.11** ---

The rate of adopting the ethical codes by industrial firms: USA 85%, Germany 51%, EU 41%, and Japan 30% (*Japan Productivity Newspaper*, 21 February 1992, p. 1). ☐

Engineers' Ethics

Professional societies, which emerged in the United States at the end of the 19th century, declared their ethics (Martin and Schinzinger, 1989).

In 1970 the 'order of the engineer' was established in order to enhance the engineer's understanding of high professional ethics and contribution to public welfare. In 1974 ethical codes were set up for professional engineers; in the following year the American Society of Mechanical Engineers, Institute of Industrial Engineers, and other professional societies approved the codes.

Notes

1. Chronological figures are mostly based upon *The Shorter Oxford Dictionary of Historical Principles* (Oxford University Press, 1973).

References

BOURDIEU, P. (1992) *The Field of Cultural Production* (Cambridge: Polity Press).
COOLEY, M. (1987) *Architect or Bee?—The Human Price of Technology* (London: Hogarth Press), Chap. 8.
HALL, R.W. (1987) *Attaining Manufacturing Excellence* (Homewood, IL.: Dow Jones–Irwin).
HARRINGTON, J. Jr. (1973) *Computer Integrated Manufacturing* (New York: Industrial Press), p. 6.
HITOMI, K. (1989) Non-mass, multi-product, small-sized production: the state of the art, *Technovation*, **9** (4).
HITOMI, K. (1991) Strategic integrated manufacturing systems: the concept and structures, *International Journal of Production Economics*, **25** (1–3).
HITOMI, K. (1994) *Introduction to Today's Advanced Manufacturing* (in Japanese) (Tokyo: Dobunkan), p. 131.
HONDA, A. (1991) *Recycling Industrial Wastes* (in Japanese) (Tokyo: Energy-Saving Centre), p. 21/Private communication, 12 August 1993.
HOSHINO, T., YURA, K. and HITOMI, K. (1995) Optimization analysis of recycle-oriented manufacturing systems, *International Journal of Production Research*, **33** (8).
IDE, T. (1972) *Introduction to Aristotle's Philosophy* (in Japanese) (Tokyo: Iwanami Publishing), pp. 30–31.

KIMURA, S. (1992) *The Age of 'Cultivating Culture'* (in Japanese) (Tokyo: PHP Institute), p. 122.
MARTIN, M.W. and SCHINZINGER, R. (1989) *Ethics in Engineering* (2nd edn) (New York: McGraw-Hill), p. 5.
NAKAMURA, S. (1989) *Rich Asia, Poor Japan* (in Japanese) (Tokyo: Gakuyo Publishing), p. 49.
OVERBY, C. (1991) Design for environmental elegance, *Engineering Foundations/NSF Conference.*
Research Report on Japan's Material Balance (1992) (in Japanese) (Tokyo: The Environment Agency of Japan).
SAHLINS, M. (1976) *Culture and Practical Reason* (Chicago: University of Chicago Press), Chap. 4.
The Shorter Oxford Dictionary of Historical Principles (1973) (Oxford: OUP).
WARNECKE, H.J. (1993) *The Fractal Company* (Berlin: Springer), p. 93.
WILLIAMS, R. (1976) *Keywords—A Vocabulary of Culture and Society* (London: Collins).
WOBBE, W. (1990) Science, technology and society towards the 21st century: European choice (in Japanese), *Economic Review*, **39** (9).

Supplementary reading

HITOMI, K. (1996) Manufacturing excellence for 21st century production, *Technovation*, **16** (1).
KOBAYASHI, A. (1993) *Philosophy of Goods Production* (in Japanese) (Tokyo: Kogyo-chosa-kai).
PFEIFER, T. *et al.* (eds) (1994) *Manufacturing Excellence—The Competitive Edge* (London: Chapman & Hall).

Concluding Remarks

System Integration of Three Major Activities Concerning Manufacturing

This book has been intended to integrate three major concerns in manufacturing:

- *The flow of materials.* This forms the process or logistic system covering a range from the acquisition of raw materials, through their conversion, to the shipment of the finished products, constituting a chain of functions—procurement–production–inventory–distribution–sales.
- *The flow of information.* This deals with managing production logistics through the management cycle of planning and control, constituting strategic planning and operational production management—aggregate production planning–process design–scheduling–production control.
- *The flow of costs* (or *value*). This clarifies the product[ion] cost structure and contributes to the creation of the value added through manufacturing.

More specifically the 'flow of value' is viewed as the social aspect of manufacturing, including international production, industrial structure and input–output analysis, and socially appropriate production for manufacturing excellence.

Six Approaches to Manufacturing Systems Engineering

The unified, systematic study of manufacturing/production is termed 'manufacturing systems engineering', which emphasises six approaches:

(1) systems analysis and design for manufacturing;
(2) optimum decision-making in production;
(3) industrial automation and computer-integrated manufacturing (CIM);
(4) manufacturing information processing by computers;
(5) cost engineering and management for product and production process;
(6) social science for manufacturing issues.

The book hence deals with both hardware and software, such as:

- *technological aspect*—manufacturing technology;
- *managerial aspect*—production management;
- *economic aspect*—cost engineering;
- *social aspect*—industrial economy (or production economics).

Global View of Manufacturing

In social life 'production' is closely related to 'consumption' and 'inventory'. The balance of these three functions plays a vital role in developing as well as stabilising the economic trends in the long run.

Manufacturing of goods by private firms is no longer confined within their domestic country; they can establish production sites in any foreign countries where labour costs are low, even though de-industrialisation occurs in their own country. This international production is now common for world-class manufacturers in order to keep a competitive edge.

Recognising Production Objectives

The primary concern of production is to produce economic goods, thus creating utilities. In a capitalist society, needless to say, private firms seek 'profit' through this productive activity. This is the 'profit objective' of production.

However, the attitude of seeking only profit by any means ought to be avoided; at least a small portion of the revenue is to be distributed for the sake of social welfare. This is the 'social objective' of production. It is important that these two objectives be balanced within the activity of a company.

Human Problems in Production

Recently young people's feelings towards the value of labour have changed and there has been a strong tendency to hate repetitive and dirty jobs, especially in highly developed countries. One approach to cope with this trend is to establish a labour-saving organisation by installing the computer-integrated, fully automated manufacturing system, though this system needs a huge amount of capital investment.

The other approach is to design jobs for an individual worker in order to achieve an improved quality of working life. Human problems within man–machine manufacturing systems are everlasting problems which need adequate solutions. Establishing human-centred aspects in integrated manufacturing systems is of vital importance today.

Proposing 'Socially Appropriate Manufacturing'

There is no need to mention the importance of production resources, especially raw materials, which are being drained from the world. Excess production and consumption associated with a huge amount of waste is now going to result in destruction of our only Earth; hence, although human beings have the responsibility of producing goods, production is now dead! (J. Baudrillard, 1982).

Today we have to recognise that manufacturing/production is an activity that destroys the Earth. Accordingly manufacturing for the future should be friendly to the Earth, aiming at resource conservation by recycling of waste as much as possible and preventing the generation of environment-polluting 'public bads (or hazards)' (creation of 'negative' utility).

Excess production must be stopped; for this purpose I now propose 'socially appropriate manufacturing (or production)'. The 'satisfaction-consciousness' principle based upon Buddhism and Taoism in the Orient will give a hint to save the Earth from drainage of natural resources, the contamination of the atmosphere, and the extinction of all living creatures.

The present mass (excess) production and mass (excess) consumption associated with a huge amount of waste, which have formed a sort of 'manufacturing (or throw-away) culture', is to be replaced by 'satisfaction-concious' production and 'satisfaction-conscious' consumption associated with the minimum amount of disposal, which is an ultimate 'manufacturing excellence'.

Social Role of 'Manufacturing Systems Engineering'

The manufacturing ethic is of vital importance to achieve the ultimate goal of socially appropriate manufacturing. For this purpose the traditional and conventional production technique and technology are raised up to the level of the so-called 'philoso-technology' to be discussed as manufacturing philosophy.

Needless to say, manufacturing/production is an essential useful activity in human history; we can not live at all without this activity in our highly civilised societies. Manufacturing technologies, production management techniques, and industrial or production economics are the most powerful tools for both efficiently technical and effectively economical production, thereby generating a variety of useful goods for living, whilst creating the wealth and benefiting our social welfare. 'Manufacturing systems engineering (or [philoso-]technology)' is a vital tool for this purpose. This discipline searches for excellence in human [working] life, symbiosis of all life on Earth, and world peace.

This tool is only for human happiness; it is strictly prohibited to apply manufacturing systems engineering to the fabrication of any weapons, and any drugs that are contrary to human moral welfare and to any manufacturing aimed merely at profit-making.

In the end, keep in mind that technology, science, management, economics, sociology, philosophy, ... are all in one for manufacturing [systems engineering].

Review Questions and Problems

Note: Problems marked with an asterisk are difficult, and may be skipped.

CHAPTER 1

1.1   Explain three steps of development of the mode of production.
1.2   What is the difference between production and manufacturing?
1.3   Why does manufacturing matter?
1.4   Show and explain three major flows concerning manufacturing.
1.5   Examine Marx' capital circulation.
1.6   List kinds of utility.
1.7   Briefly explain inputs, conversion (or transformation), and outputs for manufacturing/production.
1.8   How is profit calculated and utilised?
1.9   By what criteria is product[ion] evaluated?
1.10 Define 'productivity', and explain kinds of productivity.
1.11 Explain typical 'production functions'.
1.12 Calculate the 'marginal labour productivity' for the Cobb–Douglas production function, and prove that it is proportional to 'average labour productivity'. (*Hint*: Partially differentiate equation (1.6) with respect to L; refer to Example 1.9.)
1.13* Prove that the CES production function coincides with the Cobb–Douglas function if the elasticity of substitution, $\sigma = 1$. (*Hint*: Set 1 to σ for equation (1.7).)
1.14 How is value added calculated? Show two ways.

CHAPTER 2

2.1   Explain three features concerning system integration.
2.2   Explain four essential attributes in defining systems, and give four essential definitions of systems.
2.3   List the kinds of system.
2.4   Why is the 'systems approach' followed these days?
2.5   Explain five basic problems relating to the 'input–output system'. Exemplify those problems as to production at a factory.

2.6 Depict a system as a 'module' and as a 'work system model'.
2.7 List and explain the major 'models' used in manufacturing systems analysis/design.
2.8 Demonstrate some optimisation techniques.
2.9 Explain decision-making criteria.
2.10 What are two basic design methodologies?
2.11 How can a large-scale system be treated?
2.12 How does rational decision-making proceed?
2.13 List and explain kinds of decision-making.
2.14 Describe the procedure of optimum decision-making.
2.15 Obtain graphically the optimal solution in the case of setting the weights of 'social and profit objectives' as 1 and 2 in Example 2.13.

CHAPTER 3

3.1 Explain three major aspects meant by 'manufacturing systems'.
3.2 Explain 'strategic production planning', and list the five steps of 'operational production management'.
3.3 Discuss functions involved in an integrated manufacturing system (IMS).
3.4 Point out six basic approaches to manufacturing systems engineering?

CHAPTER 4

4.1 Classify and explain types of production.
4.2 What principle has contributed to establish the mass-production mode?
4.3 Given that the fixed cost is $1 million and the unit variable cost $1000, calculate the unit production cost for producing only 1 piece, then 10, 100, and 1000 pieces. Based upon these data, draw the total cost and the unit production cost diagrams. (*Hint*: Use equations (4.1) and (4.2) and refer to Figure 4.1.)
4.4 Derive equation (4.7) from equation (4.6). (*Hint*: Differentiate equation (4.6) with respect to x and set it to 0.)
4.5 Using the data of Example 4.3, obtain the optimum production scale, if $k = 1.3$ for the case of nonlinear total cost expressed by equation (4.5). Draw the total cost and the unit production cost—Silberston—curves. (*Hint*: Use equations (4.5)–(4.7) and refer to Figure 4.2.)
4.6 Explain the characteristics of non-mass, jobbing—multi-product, small-batch—production, and describe effective approaches to this type of production.
4.7 Obtain the reasonable job sequences for Example 4.4. (*Hint*: Two other suitable sequences exist beside the solution shown in the example.)

CHAPTER 5

5.1 Explain two aspects of the structure of a management system.
5.2 Explain the basic framework of an integrated manufacturing management system.
5.3 Explain four major functions and the framework of an integrated manufacturing system.

CHAPTER 6

6.1 Explain material flow (logistic system) from the macro-economic and social viewpoints.

6.2   Describe briefly three basic activities inside the manufacturing system.
6.3   List a chain of functions involved in technological information flow.

CHAPTER 7

7.1   Explain the product life cycle [analysis/assessment].
7.2   Describe a procedure for planning and entry of a new product. (*Hint*: Refer to Figure 7.2.)
7.3   List three phases of product design, and important features to be considered in this process.
7.4   Briefly discuss ways to increase the reliability of a product.
7.5   What is the reliability of a serial system composed of 100 independent components, each having a reliability of 99%? (*Hint*: Use equation (7.1).)
7.6   What is the reliability of a redundant, parallel system composed of four independent components, each having reliability of 60%? (*Hint*: Use equation (7.2).)
7.7   Briefly mention methods of reducing design cost.
7.8   Describe five steps taken in value analysis (VA).
7.9   Explain how useful the application of group technology (GT) is in manufacturing.
7.10   Exemplify the classification and coding system.
7.11   Explain two ways of expressing the bill of material[s].

CHAPTER 8

8.1   What are two problems involved in process planning?
8.2   Explain two decision items in process design.
8.3   How is the precedence relationship between two successive operations expressed?
8.4   Describe three process patterns of work flow (a sequence of operations).
8.5   Analyse total production time and total production cost. (*Hint*: Refer to equations (8.1) and (8.2).)
8.6   How is an optimum production facility selected by break-even analysis?
8.7   List 'charts' useful to process planning.
8.8   Explain useful tools in conducting time and motion study/work simplification.
8.9   Carefully examine the principles of motion economy.
8.10   Explain the structure of standard time.
8.11   How can the repetitive inhumane work involved in technological division of labour be managed?
8.12   Explain methods of job design.
8.13   Explain how to use the 'learning effect' and the 'production progress function'.
8.14   Calculate the cumulative (total) production time for a production volume of 100, given the 'production-progress function' using equation (8.3'). (*Hint*: Use equation (8.5).)
8.15   Calculate the learning rate for the production-progress function using equation (8.3'). (*Hint*: Use equation (8.7).)
8.16   Explain the basic structure of dynamic programming (DP).
8.17   In Figure 8.13 hourly costs are required to utilise machine tools, such as $11 for lathe, $13 for milling machine, $18 for planer, $8 for shaper, $35 for machining centre, $9 for drilling machine $15 for boring machine, and $20 for grinding machine: (a) determine the optimal process route for minimising the total production cost; (b) determine the minimum cost. (*Hint*: Use dynamic programming or network techniques.)
8.18   Explain the difference between 'network' and 'graph'.
8.19   What are the objectives of 'line balancing'?
8.20   Ascertain the solution result of Figure 8.15 based upon Jackson's step-by-step enumeration.
8.21   Show that, if the number of workstations is five, the cycle time decreases, but the total delay time increases, compared to the line-balancing result obtained in Table 8.5 for Example 8.16. What if the number of workstations is three?

CHAPTER 9

9.1 What are the objectives of layout design and what effects are expected in optimum plant layout?
9.2 Explain three patterns of plant layout in relation to the $P-Q$ chart.
9.3 Briefly describe the procedure of systematic layout planning (SLP).
9.4 Explain how to evaluate alternatives.
9.5 Draw a layout plan, based upon the space relationship chart shown in Figure 9.5.
9.6 Explain a pattern search for the case of designing a plant layout.

CHAPTER 10

10.1 Explain three ways of obtaining an initial solution for transportation problems expressed by linear programming.
10.2 Obtain an initial solution for a transportation problem using Table 10.2 modified such that $a_1 = 5$ and $a_2 = 11$. How many [non-]basic solutions are there?
10.3 Explain two ways of solving travelling salesperson problems.

CHAPTER 11

11.1 What is meant by 'profit rate'?
11.2 Explain three basic criteria to be utilised as the production objectives in manufacturing optimisation.
11.3 Briefly explain mathematical models constructed for manufacturing optimisation.
11.4 State the [generalised] Taylor tool-life equation.
11.5 Machining a steel bar longitudinally on a lathe with cut depth of 1.00 mm and feed rate of 0.03 mm/rev needs 45 s for setup. The tool life for machining this material is 256 min for the case of a machining speed of 100 m/min, but only 16 min at 200 m/min. It takes 1 min to replace the worn cutting tool with a new edge. Obtain the Taylor tool-life equation for this case, and calculate the maximum production rate (or minimum production time) machining speed and tool life, and the minimum [unit] production time. (*Hint*: Determine n and C by using equation (11.16), and use equations (11.23) and (11.25), then equation (11.27).)
11.6 Formulate the lot production time and cost for producing a product item by a lot of l pieces.
11.7 Analyse the unit production time and cost for milling and drilling. (*Hint*: Refer to Figure 11.1 and supplementary references.)
11.8 Derive equations (11.23) to (11.27) and discuss optimal machining speeds.
11.9 What quantitative relationship exists among three optimal machining speeds for the maximum production rate (or minimum time), the minimum cost, and the maximum profit rate? *Mathematically prove this relationship. (*Hint*: Compare which is greater, v_t (minimum time speed) given by equation (11.23) or v_c (minimum cost speed) given by equation (11.25) for the case of ordinary machines; verify that the first derivative of equation (11.22) is positive at v_c, 0 at v_p (maximum profit-rate speed), and negative at v_t and that its second derivative is negative at v_p.)
11.10 What is meant by the '[high-]efficiency [speed] range'? Why is it important?
11.11* Prove that the efficiency-sensitivity function:

$$e(v) = dr(v)/dt(v) \tag{11.112}$$

is a monotone-increasing function on the efficiency speed range $[v_c, v_t]$; 0 at v_c, infinite at v_t and intersects at v_p with the profit-rate function $r(v)$ given by equation (11.22). (*Hint*: Formulate the above equation with equations (11.20) and (11.22), evaluate its values at v_c, v_p and v_t and show that the first derivative is positive.)
11.12 Explain why any machining speed on the efficiency speed range $[v_c, v_t]$ is the Pareto optimum solution and why the maximum profit-rate machining speed is recognised as a preferred solution.

11.13* Formulate a product-mix problem to select optimally proper items from among M items ($j = 1, 2, \ldots, M$) to be produced within an allowable time D of a machine tool such that the total profit is maximised, and examine which optimal machining speed should be employed for total optimality. (*Hint*: For the product-mix formulation, refer to Section 13.4.)

11.14 Why are the 'so-and-so' time (e.g. 4 h non-replacement [time]) of cutting tools and the 'in-life machine quantity' important practically?

11.15 What constraints are to be considered in practical machining?

11.16 Explain how to obtain both optimal feed rate and optimal machining speeds.

11.17* Solve Example 11.6 by means of geometric programming. (*Hint*: Refer to Duffin, R.J., Peterson, E.L., and Zener, C.M. (1967) *Geometric Programming* (New York: Wiley); Walvekar, A.G. and Lambert, B.K. (1970) An application of geometric programming to machining variable selection, *International Journal of Production Research*, **8** (No. 3).)

11.18 How are multistage manufacturing systems characterised?

11.19 Briefly discuss the optimality of a multistage manufacturing system.

11.20 Ascertain equations (11.81)–(11.83) and equations (11.96)–(11.99).

11.21* Construct the optimising algorithm for determining the optimal machining speeds to be utilised on the multistage manufacturing system under the maximum profit-rate criterion.

11.22 Ascertain and carefully examine optimal solutions given in Table 11.2 for the multistage manufacturing system.

CHAPTER 12

12.1 List a cycle of functions and decision problems involved in managerial information flow.

CHAPTER 13

13.1 What problems are involved in 'production planning'?

13.2 Draw the three-dimensional graph showing the optimal solution to Example 13.1 by setting the z-axis in addition to x_1 and x_2. (*Hint*: Refer to Figure 13.2.)

13.3 Consider application of the graphical solution to LP problems with three decision variables.

13.4 Formulate the maximum-production production planning for the resources given as in Example 13.1, and obtain the optimal solution—optimal production quantities for P_1 and P_2. (*Hint*: The objective function is the sum of production of two products; solve graphically.)

13.5 Compare the two optimal solutions for the maximum profit obtained in Example 13.2 and the maximum production obtained in the previous problem. (*Hint*: Refer to Example 2.13; the maximum production for the 'social objective' does not always ascertain the maximum profit—the 'profit objective' of manufacturers.)

13.6 Graphically solve a nonlinear program:

maximise $kx + y$ (13.101)

subject to $x^2 + y^2 \leq 16$ (13.102)

or $(x - 10)^2 + y^2 \leq 16$ (13.103)

$x, y \geq 0$ (13.104)

where k is a parameter. (*Hint*: Solve graphically.)

13.7 Classify the optimal solution obtained by linear programming.

13.8 Briefly explain the 'simplex method'.

13.9 How many numbers of basic solutions are there for the problem of Example 13.1? List all the basic solutions, and ascertain that the basic feasible solutions correspond one-to-one to the individual 'extreme points' in Figure 13.1.

13.10 Solve a linear program with an objective given by equation (13.7) and constraints by equations (13.5) and (13.6) by the simplex algorithm.
13.11 What are meant by 'shadow price' and 'reduced profit (cost)'?
13.12 Obtain the maximum profit by the simplex method for Example 13.1 excepting that the available material is 41, and ascertain that the shadow price of the material is what is obtained in Table 13.2(c).
13.13 Explain the relationship between the primal and dual problems.
13.14 Show the graphical solution for the problem expressed by equations (13.21) to (13.23). (*Hint*: Draw a three-dimensional graph or a two-dimensional one with $y_3 = 0$ as obtained in Example 13.9.)
13.15 Consider the significance of the shadow price being 0 for the production facility, as obtained in Example 13.9, for the production planning given by Example 13.1.
13.16 A special-steel manufacturer needs more than 36 kg and 48 kg for materials A and B, respectively. These materials can be extracted from three kinds of raw materials—M_1, M_2, and M_3. Per 1 tonne 4.0 kg of A and 1.0 kg of B from M_1, 1.8 kg of A and 2.0 kg of B from M_2, and 1.0 kg of A and 2.0 kg of B from M_3 are extracted. The purchase costs for these raw materials are \$760 for M_1, \$360 for M_2 and \$300 for M_3, per tonne. How much raw materials should the firm buy to get necessary amounts for A and B with a minimum costs? (*Hint*: Formulate this problem by linear programming, make the dual problem and solve graphically.)
13.17 Briefly mention the limitation of application of linear programming techniques to production planning problems.
13.18 Solve the problem indicated in Example 13.10 by the big-M (or penalty) method. (*Hint*: Set *M* at 100 and use the simplex method.)
13.19 Apply the two-phase method to Example 13.1 with another condition that the total production required is at least 7, and show that the 'feasible region' does not exist graphically and the 'optimal solution' is never derived, as the process stops at phase I. (*Hint*: Refer to Example 13.4.)
13.20 Why is multiple-objective optimisation hard to solve?
13.21 List and briefly explain methods for multiple-objective optimisation.
13.22 Solve graphically multi-objective production planning in the case where the second and third priorities are exchanged in Example 13.14.
13.23 How can the 'profit rate' be utilised to solve 'knapsack problems' heuristically?
13.24 Enumerate all the solutions for Example 13.16 by complete enumeration, and obtain the optimal solution.
13.25 Formulate the minimum unit cost for lot production. Calculate it for the case of Example 13.18. (*Hint*: Substitute equation (13.68) into equation (13.67).)
13.26 Draw curves of setup, inventory-holding, and total costs with lot size as abscissa for Example 13.18. (*Hint*: Refer to Figure 13.6.)
13.27 Explain the procedure of MRP.
13.28 If the present stock on hand is 4 and the production lead time is two weeks for subassembly A in Example 13.20, calculate the scheduled requirements and orders by the MRP procedure.
13.29 List several policies for production smoothing.
13.30 Obtain Table 13.12(a) and then obtain (b) from Table 13.11 by means of the transportation algorithm. (*Hint*: Refer to Section 10.1.)
13.31 What is a special feature of long-term production planning?
13.32 What is the difference between prediction and forecasting?
13.33 Explain types of time series.
13.34 List and briefly explain methods of forecasting.
13.35 Determine the forecast sales for October, if the actual sales amount in September is 208 for Table 13.15 in Example 13.25.
13.36 What characteristic is involved in the Gomperz and logistic models?

CHAPTER 14

14.1 Briefly describe three intrinsic problems involved in production scheduling.

14.2 Explain the structure and the role of the Gantt chart.
14.3 List problems to be solved in operations scheduling.
14.4 Explain the 'scheduling criteria'.
14.5 Why is the mean flow time important? To what criteria is it related?
14.6 Construct an algorithm for obtaining an optimal schedule that minimises the maximum tardiness, and, in addition, that minimises the mean flow time, if there exists a sequence with zero tardiness. (*Hint*: Refer to Theorem 14.3.)
14.7 Calculate the minimum mean flow time obtained by the SPT rule for Example 14.3. Show that the maximum tardiness for this sequence is not zero. (*Note*: There rarely exists an optimal sequence that has not only zero tardiness but the minimum mean flow time.)
14.8 Explain briefly Johnson's algorithm and its extensions to three-stage scheduling, Petrov's method, group scheduling technique, etc. (*Hint*: Refer to Section 28.1.)
14.9 Given a five-job, three-machine flow-shop scheduling problem for minimising the makespan, as represented in the table below, (a) check whether or not the extended Johnson's algorithm can be applicable to this problem. (b) Determine an optimal sequence. (*Hint*: Calculate equation (14.18) and apply Johnson's algorithm.) (c) Obtain the minimum makespan by means of Gantt charting, direct calculation by equation (14.17), and table calculation by Table 14.3.

Job number	Processing time		
	Stage 1	Stage 2	Stage 3
1	4	5	8
2	9	6	10
3	8	2	7
4	7	3	9
5	4	4	11

14.10 Given that the transfer of workparts between two stages takes 0.1 h for the data of two-stage flow-shop scheduling given by Table 14.2, (a) determine the optimal job sequence. (*Hint*: Apply the three-stage flow-shop scheduling technique by considering the transfer as a virtual second stage and the actual second stage as a third stage.) (b) Draw a Gantt chart, indicating the critical path. (c) Obtain the makespan.
14.11 Explain the procedure of the branch-and-bound method.
14.12 Calculate the lower bound at node $\{J_2J_1\}$ in Figure 14.6. (*Hint*: Calculate equation (14.20) as in Example 14.7.)
14.13 Draw the Gantt chart for the 'exact'-optimal solution obtained in Example 14.8 for a large-scale flow-shop scheduling problem.
14.14 Draw the Gantt chart for the 'near'-optimal solution obtained by Petrov's method in Example 14.9 for large-scale flow-shop scheduling and obtain the makespan. Also obtain the makespan by table calculation.
14.15 Compare the minimum mean flow time for the optimal solution obtained in Example 14.8 and that for the 'near'-optimal solution obtained by Petrov's method in Example 14.9 for large-scale flow-shop scheduling.
14.16 Briefly explain the graphical procedure to solve a minimum-makespan two-job, job-shop scheduling problem.
14.17 In the graphical solution for two-job, job-shop scheduling, show that the makespan is equal to the sum of the length of perpendicular (horizontal) broken lines showing the optimal path and the abscissa (ordinate) of the end point. (*Hint*: Refer to Figure 14.9.)
14.18 Draw the Gantt chart for the optimal solution obtained graphically in Example 14.11 for a two-job, job-shop scheduling problem. (*Hint*: Refer to Figure 14.9.)

14.19 Draw the Gantt chart for the optimal solution obtained by Jackson's method in Example 14.12 for a two-machine, job-shop scheduling problem.
14.20 Why is job-shop (or scheduling) simulation needed?
14.21 List and explain typical dispatching (or priority) rules.
14.22 Explain the term 'simulation languages' and list some.
14.23 Briefly explain the difference between PERT and CPM.
14.24 Why is the critical path important? How can it be found?
14.25 Explain the structure of an arrow diagram.
14.26 Ascertain the earliest and latest node times for the PERT diagram indicated in Figure 14.12(b).
14.27 Calculate floats on slack paths for the PERT diagram indicated in Figure 14.12(b).
14.28 Show that, if the deadline of the project is given as 45 weeks for the problem indicated in Example 14.19, it is almost impossible to complete this project.
14.29 Explain the procedure of CPM.
14.30 Draw the project cost curve based upon the result obtained in Example 14.20.

CHAPTER 15

15.1 What are aims and kinds of inventory?
15.2 What are basic decision factors for inventory control?
15.3 Briefly explain typical inventory models.
15.4 Explain typical ways to practical inventory.
15.5 Draw the inventory cost curve for the data given in Example 15.1. (*Hint*: Use equation (15.1); also refer to Figure 13.6.)
15.6 Determine the reorder point with the confidence level of 95% for the data given in Example 15.1 in the case of a lead time of 9 days and daily demand deviation of 1 piece.
15.7 Examine three patterns of inventory fluctuation caused by fixed-order quantity, replenishment, and (S, s) inventory models.
15.8 What are main restrictions for multiple-product inventory?
15.9 Describe the optimum order policy.

CHAPTER 16

16.1 Why is production control needed?
16.2 State the major functions involved in production control.
16.3 Explain the production control procedure.
16.4 Explain the major process of implementing JIT production.
16.5 How is JIT production evaluated and criticised?
16.6 Explain two types of failure.
16.7 Explain three types of productive maintenance.
16.8 By what is the concept of reliability characterised?
16.9 List and explain indices that express reliability.
16.10 How are failure rate, failure density function, and reliability function related?
16.11 How is the failure rate related to types of failure?
16.12 Show typical reliability functions.
16.13* Prove that the failure rate for the normal-type failure density function,

$$f_N(t) = \frac{1}{\sqrt{2\pi}\sigma} e^{-(t-\mu)^2/2\sigma^2} \qquad (16.42)$$

where μ is the mean and σ is the standard deviation of lifetime, approaches an asymptote $(t-\mu)/\sigma^2$ as time elapses.
16.14 Express the relation between the Poisson and exponential distributions.
16.15 Derive the 'availability' in the steady state given by equation (16.34).

REVIEW QUESTIONS AND PROBLEMS   523

16.16   Why is no preventive maintenance needed for a machine with exponential-type failure?

CHAPTER 17

17.1   What are the roles of TQM and QC circle activities?
17.2   List and explain control charts that are frequently employed in quality control.
17.3   How is sequential sampling conducted?
17.4   Explain the role of the operating characteristic (OC) curve?
17.5   List useful tools for quality control activities.
17.6   What sort of organisation should be established in manufacturing firms for quality assurance for products?
17.7   Briefly mention the procedure of quality function deployment (QFD).
17.8   How is quality defined in 'quality engineering' or the 'Taguchi method'?
17.9   List three phases of design in quality engineering.
17.10 Express an example of loss function.
17.11 Why is the signal-to-noise (S/N) ratio regarded as a measure of robustness?

CHAPTER 18

18.1   Why is the flow of costs important in manufacturing? List similar terms.
18.2   Examine the cash flow in and outside a manufacturing firm. (*Hint*: Refer to Figure 18.1.)
18.3   Derive basic formulae for the time-series value of money.
18.4   How much do you get after 1, 5, 10 and 100 years if $100 is invested with an interest rate of 6%? (*Hint*: Use equation (18.1).)
18.5   Intending installation of a modern production facility of US$100 million after 5 years, how much money should be saved at the end of each year if the rate of interest is 5%? (*Hint*: Use equation (18.5).)
18.6* Prove that the capital recovery factor converges to r after $h \to \infty$.

CHAPTER 19

19.1   Classify costs from the morphological and economical standpoints.
19.2   What is the total cost of a product made up of?
19.3   How is the product price set?
19.4   Show the relationship between 'markup rate' and 'margin rate'. (*Hint*: Equalise equations (19.1) and (19.3).)
19.5   Explain two forms of depreciation.
19.6   How is the social equilibrium price established?
19.7   Explain how to estimate the components of manufacturing cost.
19.8   List kinds of cost accounting.

CHAPTER 20

20.1   Why is the profit-making (profit objective) of a firm justified?
20.2   Mention the difference of the meanings involved in equations (20.1) and (20.2).
20.3   How are profits made by a firm calculated? (*Hint*: Refer to equations (20.3)–(20.8).)
20.4   Examine the DuPont chart for the ROI structure carefully. (*Hint*: Refer to Figure 20.1.)
20.5   What effects do the fixed cost, the variable cost, and the selling price give to the break-even point?
20.6* Prove Proposition 20.2.
20.7* Given a production function, (a) show that the maximum profit is obtained if the marginal value product—the product of the marginal productivity of a certain input and

the unit revenue of the output—is equal to the price of that input. (b) Show that maximum profit is obtained if all of the ratio of the marginal productivity to the price is equal for every input.

CHAPTER 21

21.1 What is the present worth of profits which are obtained at the end of each period over the economic life of a production facility? (*Hint*: Refer to equation (21.1).)
21.2 List and explain methods of evaluating the effectiveness of capital investment.
21.3 Compare the following machines—the existing and the new automatic—and determine which alternative should be selected by the minimum-cost approach. Use an interest rate of 8%.

	Existing machine	New automatic machine
Initial investment ($)	25 000	45 000
Annual operating cost ($)	10 000	8 000
Annual repair cost ($)	400	250
Overhaul cost ($)	2 000 (every 2 years)	3 000 (every 5 years)
Life (years)	10	15

21.4 Compare the following flexible manufacturing systems—A and B—and select an alternative with a larger value for the rate of discount.

	A	B
Initial investment ($)	100 000	150 000
Annual operating cost ($)	30 000	25 000
Annual income ($)	45 000	47 000
Salvage value ($)	0	15 000
Life (years)	8	10

CHAPTER 22

22.1 List three steps in the development of manufacturing efficiency towards automation.
22.2 What are three basic components of a machine?
22.3 Explain three basic concepts concerning automation.
22.4 What is the difference between mechanisation and automation?
22.5 List kinds of automation.
22.6 Describe the development of automatic machines.
22.7 Discuss the influence of automation on social life.

CHAPTER 23

23.1 Explain three basic computer aids to manufacturing.
23.2 Define computer-integrated manufacturing (CIM) and consider CIM in practice.
23.3 Discuss the merits and effects of CIM.

CHAPTER 24

24.1 How has computer-aided design/drafting (CAD[D]) developed?
24.2 List typical geometric models used in CAD.

24.3 Explain two different types of solid model for CAD.
24.4 What effective methods are used for optimum design or computer-aided engineering (CAE)?
24.5 Explain two basic methods of computer-automated process planning (CAPP).
24.6 Describe typical algorithms for CAPP.
24.7 How is concurrent (or simultaneous) engineering useful?
24.8 Explain three principal functions involved in an autoprogramming system.
24.9 Briefly explain typical autoprogramming languages.
24.10 State two types of computerised layout planning.
24.11 Describe typical algorithms for computerised layout planning.

CHAPTER 25

25.1 List and explain types of automatic machine tools for mass production.
25.2 What type of construction is employed in transfer machines?
25.3 Explain important time factors for the efficient use of transfer machines.
25.4 Describe basic types of numerical control (NC) systems.
25.5 How are machining/turning centres characterised?
25.6 How is adaptive control applied to machining operation?
25.7 Briefly explain CNC, DNC, FMS and FAS.
25.8 List the kinds of FMS.
25.9 List design points for components and products for automated assembly.
25.10 List four basic configurations and the motion patterns of industrial robots.
25.11 What are the major functions of an automated warehouse?
25.12 Briefly mention how to make automatic inspection.
25.13 How does the three-coordinate measuring machine act in factory automation?
25.14 Exemplify the unmanned factory.

CHAPTER 26

26.1 How is information produced and used in decision-making?
26.2 Point out important points information must possess.
26.3 Describe the structure and functions of information systems.
26.4 What types of computation or data processing can the computer system conduct?
26.5 Explain the functions of a management information system (MIS) and a strategic information system (SIS) in a firm's activities.
26.6 Discuss the use of today's information networking in manufacturing information systems.

CHAPTER 27

27.1 Explain the major structure of parts-oriented production and its information systems.
27.2 Discuss the advantages of parts-oriented production.

CHAPTER 28

28.1 Explain the characteristics of group scheduling.
28.2 How does interactive group scheduling work?
28.3 What is computer-aided line balancing (CALB)?

CHAPTER 29

29.1 What do on-line production control and its information systems do in the changing manufacturing environment?

29.2 Briefly explain the procedure of optimum-seeking machining (OSM).
29.3 Ascertain optimal machining speeds demonstrated in Table 29.1(c) by the optimum seeking machining (OSM) algorithm. Set the initial machining speed as 120 m/min, and the initial slope constant as 0.10. Other data needed are: work diameter, 40.00 mm; work length, 30.00 mm; preparation time, 7.00 min/pc; tool-replacement time, 3.00 min/edge; direct labour cost and overhead, $1.00/min machining cost, ¢25/min; tool cost, $9.00/edge; and net revenue, $33.50/pc. (*Hint*: Refer to Section 11.2.)
29.4* Construct the optimum-seeking machining algorithm for determining both optimal feed rate and optimal machining speed.

CHAPTER 30

30.1 Mention the characteristics of computer-based production management systems.
30.2 Briefly explain the structure and functions of COPICS and BAMCS.

CHAPTER 31

31.1 How is production [system/structure] concerned with other activities in society?
31.2 How has the production mode (power/relation) changed historically?
31.3 Describe the historical development of production modes.
31.4 Trace the historical development of management.
31.5 What is the difference between the Taylor[ism] system and the Ford[ism] system?

CHAPTER 32

32.1 What is the difference between strategy and tactics?
32.2 List the major features of strategy and tactics.
32.3 Explain two aspects of corporate strategy.
32.4 Why is manufacturing strategy important?
32.5 Describe strategic aspects of a computer-integrated manufacturing (CIM).

CHAPTER 33

33.1 Why does de-industrialisation occur in developed countries?
33.2 List and explain theories of globalisation.
33.3 Explain the necessity and the trend of global manufacturing.
33.4 What is the situation of the export-driven economy?

CHAPTER 34

34.1 Classify industries.
34.2 How has industrial structure changed?
34.3 Review the development and growth of Japan's manufacturing and industry.
34.4 Explain two kinds of measure for evaluating industrial efficiency.
34.5 In 1994 Japan's industry: agriculture, forestry and fishing, mining, manufacturing, construction, electric, gas, and water, wholesale and retail trade, finance and insurance, real estate, transportation and communication, services, and government and non-profit services held labour forces of 4919, 86, 15424, 6937, 433, 11103, 2093 1020, 3665, 15138, and 5765 thousand persons, produced output of 17530.5; 1951.4; 314202.2; 100286.2; 21602.4; 91376.7; 35974.6; 67843.7; 46572.8; 141086.8; and 71138.4 billion yen, and created value added (GDP) of 10149.1; 1027.2; 117151.3; 51643.5;

13355.5; 60770.3; 24778.6; 60868.9; 30655.6; 80451.7; and 48161.6, where $1 = 102.23 yen. Calculate and compare the yield rates and efficiency indices of Japan's industrial sectors. (*Hint*: Make a table like Table 34.1.)

34.6 Discuss labour efficiency in Japan, the United States, Britain, Germany, Korea, Thailand, and China.
34.7 Compare the industrial efficiency internationally among Japan, the United States, and Germany.
34.8 Discuss the economic growth in Asia.
34.9 Compare manufacturing efficiency internationally.

CHAPTER 35

35.1 Exemplify the ripple effects among industrial sectors.
35.2 Explain the procedure of input–output analysis and list tools used in this method.
35.3 How is the demand–supply equilibrium formulated in input–output analysis?
35.4 What conditions are needed to obtain the required production in the industrial sectors?
35.5 Explain the requirements involved in input–output analysis.
35.6 Formulate the Leontief model and the Leontief inverse matrix for two industrial sectors.
35.7 Calculate the input coefficient and inverse matrix coefficient tables for Tables 35.2 and 35.3. (*Hint*: Calculate equations (35.3) and (35.7).)

CHAPTER 36

36.1 Given the importance of manufacturing, what issues is today's manufacturing faced with?
36.2 Define manufacturing excellence and explain how it differs from manufacturing culture.
36.3 What is meant by cultural/civilised production?
36.4 List and explain methods of manufacturing excellence.
36.5 Outline the trend of consumption of natural resources for industrial activities in Japan.
36.6 What effect does environmental pollution have on 'green GNP' or 'net national welfare'?
36.7 Explain the structure and functions of the recycle-oriented manufacturing system.
36.8 How can socially appropriate manufacturing—an ultimate manufacturing excellence—be achieved?
36.9 What attitudes are vital in manufacturing ethics?

Index

Aachen system 101–3
ABC inventory classification 271
acceptance quality level (AQL) 305–6
activity analysis 191
activity duration 262–3
activity relationship (REL) diagram 139–42
   layout planning 374–6
adaptability of systems 26–7, 41
adaptive control 388–9
aggregate production planning 53, 187, 188, 190–234, 456
   production control 282, 286
AGV (automated guided vehicles) 346, 393, 400
American system of manufacture 62
AMP (automated manufacturing planning) 358, 365–6
APT (automatically programmed tools) 358, 372–3
AQL (acceptance quality level) 305–6
arrow diagram 258–61, 265
Asia 484–5, 494
assembly 25, 359, 501
   automation 339, 347, 395–8
   POPIS 420–4
automated factory 346–7
automated guided vehicle (AGV) 346, 393, 400
automated manufacturing planning (AMP) 358, 365–6
automatic carousel 404
automatically programmed tools (APT) 358, 372–3
automatisation 339, 342
autoprogramming systems 339, 370–2
availability of machines 299–300

Balas' additive algorithm 211–12
BAMCS (business and manufacturing control system) 409, 448, 449
batch production 62, 69, 213, 237
big-M (penalty) method 200, 203
bill of materials 68, 103–5, 218
bottlenecks 177, 178, 179, 180, 181

Boundary representation (B-rep) 360–1
branch-and-bound technique 152, 248–52, 425
break-even analysis 113–14, 328–33
British efficiency 480, 482
business and manufacturing control system (BAMCS) 409, 448, 449

CAD see computer-aided design
CAD/CAM 369–70
CADD (computer-aided design and drawing) 339, 359
CAE (computer-aided engineering) 359, 363, 462
CALB (computer-aided line balancing) 409, 430–1
CALS (computer-aided acquisition and logistic support) 56
CAM see computer-aided manufacturing
CAP (computer-aided planning) 80, 339, 353–5, 461, 462
capacity requirements planning (CRP) 69, 218, 220–1
   BAMCS 449
   COPICS 443, 444
   CAPIS (computer-aided production information system) 432–9
capital
   circulation 10, 11
   efficiency 474, 482, 483
   investment 188, 334–8, 347, 381, 386–7, 406
   return 326–8, 335–6
capitalism 456
CAPP see computer-automated process planning
CAR (computer-aided routing) 367–9
cash flow 316
CAT (computer-aided testing) 405–6
CIM-related factories 187
chance-constrained programming 200, 204–5
chaos 24–5, 28
Chinese efficiency 488, 489
CIM see computer-integrated manufacturing
classification and coding system 67, 100–3

529

530 INDEX

CMM (coordinate measuring machine) 405
CMS (computerised manufacturing system) 70, 347
CNC (computer numerical control) 386
communications-oriented production information and control system (COPICS) 409, 441–7, 449
computer-aided acquisition and logistic support (CALS) 56, 462
computer-aided design (CAD) 80, 339, 353–5, 358–80, 417
　FMS 391
　strategy 461, 462
computer-aided design and drawing (CADD) 339, 359
computer-aided engineering (CAE) 359, 363, 462
computer-aided line balancing (CALB) 409, 430–1
computer-aided manufacturing (CAM) 80, 339, 353–5, 461
　FMS 390
computer-aided planning (CAP) 80, 339, 353–5, 461, 462
computer-aided production information system (CAPIS) 432–9
computer-aided routing (CAR) 367–9
computer-aided testing (CAT) 405–6
computer-automated process planning (CAPP) 339, 353, 364–70, 461
　CAD 358
　FMS 391
computer-controlled automation 339, 346
computer-integrated automation 347
computer-integrated manufacturing (CIM) 8, 29, 57, 81–2, 339, 347, 353–7, 441, 510–11
　capital investment 334
　databases 412
　DNC 391
　excellence 500
　FMS 390–1
　strategy 460, 461–2
　unmanned factory 406–7
computerised relationship layout planning (CORELAP) 145, 374–6
computerised relative allocation of facilities technique (CRAFT) 145, 374, 375–9
computerised manufacturing system (CMS) 70, 347
constructive solid geometry (CSG) 360–1
continuous production 61–2
conversion 85–6
coordinate measuring machine (CMM) 405
COPICS (communications-oriented production information and control system) 409, 441–7, 449
copying machine tools 382
CORELAP (computerised relationship layout planning) 145, 374–6
corporate strategy 460, 461–2
corrective maintenance 289
cost 83, 317, 319–24
　break-even analysis 329, 330–3
　classification 319
　comparisons 336
　computerisation 322–4, 357
　control 81, 189, 283, 288, 289–90, 324

COPICS 447
CRAFT 376–80
group technology 99–100
information technology 413
inventory management 271–2, 274–80
labour 319, 322–3, 498
layout planning and design 135, 138, 143–5
lot size 214, 215, 216
machine assignment 143–4
manufacturing 319–20, 322–4
mass production 63
material 319, 322
materials handling 376–8
process design 111, 113
product design 94–5, 98–100
project scheduling 264–8
transportation 147–51
see also production cost
cost accounting 319–30, 323–4
cost flow 1, 5–6, 8, 13–15, 56–7, 187, 315–18
　system integration 510
cost-over-time (COVERT) 257
cost-volume-profit (CVP) analysis 329
CRAFT (computerised relative allocation of facilities technique) 145, 374, 375–80
critical path 427, 429–30
　method (CPM) 35, 236, 258, 261–3, 264–8
CRP see capacity requirements planning
CSG (constructive solid geometry) 360–1
customer order servicing 441–3
CVP (cost-volume-profit) analysis 329
cycle time 286, 288, 384
　line balancing 130–4
　lot size 214
　minimum 132–4
　optimisation 174, 176–82

databases 355, 411–12
data processing 413–17
demand–supply equilibrium 492
decision-making 35–6, 38–45
　information technology 411–13
　optimum 57, 66, 510
decision support system (DSS) 440–1
delay in process design 110–11
delivery 13–14
depreciation 321, 323
design 30–8, 57, 88, 339, 358–80
　automation 344, 353, 356–7
　fail-safe 97
　foolproof 97
　green 98, 503
　IMMS 79–81
　IMS 55
　large-scale system 36–7
　operations 114–25
　optimum 40–5, 363–4
　parametric 364
　process 55, 83, 107–14
　product 55, 56, 81, 83, 93–103
　quality 307–9
　simplification 98
　work 36–7
　see also computer-aided design

deterioration 288
diminishing returns 331
direct numerical control (DNC) 389–91, 393, 406–7
   COPICS 445
   information networking 417
disassembly 94
discounted cash earnings 335, 336, 337
discrete-part production 62
dispatching rules 256–7
distribution problems 152–3
diversification 72–3, 90
division of labour 122
DNC *see* direct numerical control
DP (dynamic programming) 126–7, 295
DSS (decision support system) 440–1
duality 198–200
DuPont chart 326–7
dynamic programming (DP) 126–7, 295
DYNAMO (dynamic models) 257

earliest due date (EDD) rule 242, 243, 257
eclectic paradigm 465
eco-design 98
economic order quantity (EOQ) 218–19, 271–3, 275, 277–9
economic production 315
economies of network 465
economies of scale 63, 381
economies of scope 72
econometric models 232–3
EDD (earliest due date) rule 242, 243, 257
EDPS (electronic data processing system) 413
efficiency 22, 41, 135, 384
   international comparison 489
   machine speed 162, 164, 177, 181
   manufacturing 332, 469–90
efficiency index 474, 476–82, 484–9
electronic data processing system (EDPS) 413
environmental concerns 498, 501–7, 511–12
EOQ (economic order quantity) 218–19, 271–3, 275, 277–9
equilibrium price 321–2
ethics of production 507–8, 512
European Community input–output 494
expense 317
exponential distribution 239, 297–8, 299, 300–1
exports 466–7, 494, 502–4
extended APT (EXAPT) 358, 372–3

factor analysis method 140–1, 142
factory automation (FA) 187, 339, 344, 346, 381
   manufacturing excellence 500
   strategy 460, 462
   see also unmanned factories
failure of machines 288–301
   catastrophic 289, 292
   initial 292
   wear-out 288–93
fail-safe design 97
FAS (flexible assembly system) 339, 397–8
feasible solutions 149, 191–6, 200–2
files 411–12
finance management system 76, 79
first-come, first-served (FCFS) rule 257

fixed-order quantity 271, 273, 275
flexible assembly system (FAS) 339, 397–8
flexible machining cell (FMC) 392
flexible manufacturing system (FMS) 70, 339, 346–7, 355, 390–5, 400
   flow-type 392
   random-access type 392, 393, 394
   strategy 461
flexible transfer line (FTL) 392
flexibility 41, 390, 500
   automation 70, 339, 345, 390–5
   layout planning 135, 138
flow-shop 99–100
   group scheduling 425–6, 429
   production scheduling 235, 237, 243–52
flow time 239–42, 245–7
flow-type FMS 392
FMC (flexible machining cell) 392
FMS *see* flexible manufacturing systems
forecasting 188, 228–33, 421–2, 441, 443
foreign production rate 465
form utility 11, 15

Gantt charts 235–6, 245, 247, 251, 253–4, 257, 284
   group scheduling 427, 429–30
GDP *see* gross domestic product
geometric modelling 360–3
Germany 465, 466–7
   efficiency 481, 482, 483–4
GERT (graphical evaluation and review technique) 35, 236, 258
global manufacturing 464–8, 511
   input–output 494
GNP (gross national product) 470, 479, 485, 501, 503
goal programming 26, 207–9
graphical evaluation and review technique (GERT) 35, 236, 258
green production 13, 501–6, 508
gross domestic product (GDP) 18, 19, 467
   efficiency 470–3, 476–7, 479–82, 484–9
gross national product (GNP) 470, 479, 485, 501, 503
group scheduling technique (GST) 68, 409, 425–30
group technology (GT) 67–8, 99–103, 364, 393, 425–30
   layout planning 136–7, 145–6
   product design 99–103

heuristic approach 35–6, 69–70, 144–5, 152–3
hierarchical system 37, 38
human-factor analysis 115–20

ICAM (integrated computer-aided manufacturing) 30, 354
IGS (interactive group scheduling) 409, 425, 427–30
IMMS (integrated manufacturing management systems) 74–82
imports 466–7, 493–4, 502–4
IMS *see* integrated manufacturing systems
industrial engineering 67, 98, 316
industrial property 92

industrial structure 469–90
information flow 1, 5–6, 8, 28–9, 54, 56–7, 315
   automation 339, 344, 353–4
   computers 411, 413, 415
   FMS 393
   IMMS 76, 80–2
   managerial 187–9, 409, 411, 413, 415, 457
   system integration 510
   technical 83, 88
   unmanned factory 406
information networking 409, 417–18
information systems 413–15
information technology 411–19
initial feasible solution 149
input coefficient 493
input–output relations 491–6
inspection 110–11, 339, 404–5
   COPICS 445
   quality 304, 306, 307
integer programming 200, 203–4
integer-valued production 203–4
integrated computer-aided manufacturing (ICAM) 30, 354
integrated manufacturing systems (IMS) 1, 55–6, 353–4
   FMS 391
   framework 79–82
integrated manufacturing management systems (IMMS) 74–82
interactive group scheduling (IGS) 409, 425, 427–30
intermittent production 61–2, 65, 69, 213
international manufacturing 188, 464–8, 494, 511
inventory 87–8, 135, 214, 357
   ABC classification 271
   control 189, 283
   COPICS 443–4, 446
   IMMS 76
   just-in-time 270, 288
   management 270–81
   operations scheduling 239–41
   POPIS 420, 422, 424
   social production 453–5

Jackson's method 254
Japan
   efficiency 470–9, 482–4
   global manufacturing 465–7
   green production 501–4, 508
   input–output 493, 494
JIT *see* just-in-time
jobbing production 61–2, 65, 237, 432
job satisfaction 286
job shops 66, 236, 237, 252–6
Johnson's method 244, 247, 252, 254, 426
just-in-time production 8, 14, 71–2, 187, 285–8, 501
   computers 354–5
   IMS 55–6
   inventory 270, 288
   storage 88
   strategy 460, 461, 462

kanban 14, 72, 285–8
knapsack problem 210

Korean efficiency 485, 486
Kuhn–Tucker conditions 178, 179, 181, 182–3, 277

labour 9, 10
   cost 319, 322–3, 498
   division 122
   efficiency 470–1, 475–7, 479–89
   future trends 498
   management 76, 79
   productivity 17, 18
large-scale system design 36–7
lateness (tardiness) 239, 241–3
layout planning and design 1, 48–9, 81, 83, 135–8, 143–6
   computerised 339, 353, 373–80
   CORELAP 145, 374–6
   process design 108
   production line 136–7
   systematic (SLP) 138–43
   technological information flow 88
   transportation 87, 144–5, 376, 380
lead times 384, 462
   inventory 272–4
   POPIS 420, 423
learning curves 123–5
learning effect 67, 122–4, 385
least unit transportation 149–50
least-work-remaining rule 257
Leontief models 16, 493, 495
life cycles of products 90–1
   assessment (LCA) 98, 506
line balancing 130–4, 284
   computer-aided (CALB) 409, 430–1
   mixed-model 69
linear programming (LP) model
   product mix 210, 211
   short-term production planning 190–203
   transportation 147–50, 223, 225, 226
loading and unloading 217, 220–2, 223–4
   automation 382, 392–3, 399–400
logistic systems 50, 76–8
   inventory 270
   material flow 85, 86
   planning and design 83, 147–53
long-term profit planning 188
loss function 309, 310
lot scheduling 69
lot-size analysis 188, 213–17, 218–20
lot tolerance percentage defective (LTPD) 305–6
low-cost automation (LCA) 344

machines and machining
   assignment 143–4
   availability 299–300
   conditions 83
   coordinate measuring (CMM) 405
   depth of cut 155, 158
   failure 288–301
   feed rate 155, 158, 162, 167–74, 175
   layout 285
   loading 217, 220–2, 223–4, 382, 392–3, 399–400
   maintenance 283–301

NC (numerically controlled) 114, 339, 342, 345, 384–9, 406
  optimisation 154–83
  optimum-seeking (OSM) 433–4, 436–9
  replacement 294–6
  speeds 155, 158–74, 175–81, 436–9
machining centres 386–7, 392
makespan 240, 244–56, 426–7, 429
management information systems (MIS) 409, 415, 417–18
  computerised production management 440–9
management systems 1, 50–3, 74–82, 456–7
  automation 339, 353
  by exception 283
  finance 76, 79
  information flow 187–9, 409, 411, 413, 415, 457
  integrated manufacturing (IMMS) 74–82
  inventory 270–81
  planning 47, 53, 74, 76, 77–81
  production 53–5, 75–9, 346, 409, 440–9
  strategy 460–1
man–machine system 114–15
manufacturing 1, 3–23, 47–60, 85
  automated planning (AMP) 358, 365–6
  computerised system (CMS) 70, 347
  cost 319–20, 322–4
  defined 4, 8
  efficiency 332, 469–90
  excellence 499–501, 507–8, 510
  future production 497–509
  information systems 55, 188
  integrated computer-aided (ICAM) 30, 354
  integrated management systems (IMMS) 74–82
  international 188, 464–8, 494, 511
  optimisation 83, 154–84
  socially appropriate 507–8, 510, 511–12
  strategy 459–63
  see also computer integrated manufacturing (CIM); flexible manufacturing system (FMS); integrated manufacturing systems (IMS)
Manufacturing Automation Protocol (MAP) 417–18
manufacturing resource planning (MRPII) 68, 217
  COPICS 443, 444
MAPI (Machinery and Allied Products Institute) method 335, 338
marketing see sales
market segmentation 94
markup rate 320–1
mass production 1, 62–5, 287
  automation 339, 342, 344–5, 381, 384
  future trends 501, 507, 512
  group technology 99
  layout planning 136
  materials handling 399
  operations design 122
  scheduling 237
master production schedule (MPS) 68, 217, 218, 443, 449
material cost 319, 322
material flow 1, 5–6, 8, 15, 56–7, 83, 85–8, 187, 315
  automation 339, 353–4, 403
  computers 411, 415, 416

FMS 393
group scheduling 425
IMMS 76, 77, 78, 80–2
inventory 270
layout planning 139–40, 376–7, 380
manufacturing systems 50, 85
production modes 66, 68
system integration 510
unmanned factory 406
material requirements planning (MRP) 68–9, 105, 188, 217–20
  BAMCS 449
  COPICS 443, 444
materials handling 87, 376–8, 399–404
mean time between failures (MTBF) 291, 293, 298
mean time to failure (MTTF) 291
mechanical hands 397, 399
mechanisation 339, 341–3, 348–51
mechatronics 342
MIS see management information systems
models and modelling 31–3, 40–2, 464
  CAD 360–3
  econometric 232–3
  geometric 360–3
  inventory 271, 275–6, 279–80
  layout planning 143
  Leontief 16, 493, 495
  long-term production planning 225–7
  mixed line-balancing 69
  see also linear programming (LP) model
modules and modular systems 30, 36–7, 69–70
  reliability 96–7
  transfer machines 384
modular production 69
motion (or method) study 115–20
moving-average method 229
MPS (master production schedule) 68, 217, 218, 443, 449
MRP see material requirements planning
MRPII see manufacturing resource planning
multinational enterprises see global manufacturing
multiple-objective production planning 200, 205–9
multiple-product inventory system 277–9
multi-product, small-batch production 1, 65–72, 286–7
  computerisation 355
  FMS 393
  group technology 99
  information systems 409
  inventory 270
  life cycle 90
  on-line control 432
multistage manufacturing systems 83, 174–83

network techniques 126, 128–9, 409, 417–18
  economies 465
northwest corner rule 149, 151–2
numerical control (NC) 461
  CAD 358, 369–73
  machine tools 114, 339, 342, 345, 384–9, 406

office automation (OA) 339, 343, 344, 355
  IMMS 80
  strategy 461, 462

on-line production control 409, 432–9
operating characteristic (OC) curve 305–7
operations
  control system 78
  conversion 86
  design 114–25
  planning 76–8, 370–3
  process design 107, 108, 110
  scheduling 188, 235–6, 237–58
  sheets 120, 121, 427, 429
opportunity cost (loss) 335
optimisation 26, 31, 33–5
  analysis 140
  CAPP 369
  cycle time 174, 176–82
  inventory systems 279–80
  manufacturing 83, 154–84
  multiple-objective 205–8
optimised production technology (OPT) 258
optimum design 40–5, 363–4
optimum plant size 332
optimum production scale 332–3
optimum routing analysis 125–9
optimum-seeking machining (OSM) system 433–4, 436–9
order release 445
overheads 319, 322–3

parametric design 364
Pareto optimum 44–5, 164–5, 205–7, 307
parts 62, 103, 105, 397, 406
  machines 297, 298
  recycling 506
parts-oriented production information system (POPIS) 68, 69, 88, 409, 420–4
  inventory 270
payoff period 334–5
PDM (product data management) systems 105
penalty rule 257
PERT *see* program evaluation and review technique
Petri net 258
Petrov's method 248, 252, 425–6
Petty–Clark law 93
place utility 11, 15
planning 8, 190–203
  automated manufacturing (AMP) 358, 365–6
  capacity requirements (CRP) 69, 218, 220–1, 443–4, 449
  computer-aided (CAP) 80, 339, 353–5, 461, 462
  management systems 47, 53, 74, 76, 77–81
  operations 76–8, 370–3
  process 56, 81, 83, 107, 302, 364–70
  product 83, 90–3, 302
  production control 282, 284
  profit 43–4, 188, 325–8
  resource 53
  see also computer-automated process planning (CAPP); layout planning and design; material requirements planning (MRP); production planning
plant monitoring and control 445, 446
Poisson distribution 297
pollution 12–13, 512

possession utility 13, 93
power constraint on machining speed 170, 173
P–Q analysis 135–6
precedence relationship 108–9
preventive maintenance 289–90, 293
  optimum 299–301
  tools 290–4
price 197, 198, 320–2
primary industry 3, 93
  efficiency 469–73, 475, 480–9
  input–output 493
principle of optimality 127
probabilistic inventory models 279–80
procedural aspect manufacturing system 48, 50–5
process 62
  charts 110–11, 112
  control 189, 283–4
  design 55, 83, 107–14
  layout 136–7
  planning 56, 81, 83, 107, 302, 364–70
  production 1, 14–22, 50, 51, 187, 456
  see also computer-automated process planning (CAPP)
product
  cost structure 319–20
  cycle model 464
  data management (PDM) 105
  design 55, 56, 81, 83, 93–103
  differentiation 93–4
  liability 92–3
  life cycle 90–1, 98, 506
  mix 188, 209–12
  planning 83, 90–3, 302
  structure 103–5
production
  control 53–5, 81, 187, 188, 189, 282–301, 346, 457
  cycle 214–17
  defined 3, 4, 6, 8
  design 94
  ethics 507–8, 512
  forecasting 188, 228–33
  future trends 497–509
  green 13, 501–6, 508
  implementation 53–4, 55, 56, 187
  information 9, 11, 432–9
  integer-valued 203–4
  intermittent 61–2, 65, 69, 213
  international 465–6
  jobbing 61–2, 65, 237, 432
  management systems 53–5, 75–9, 346, 409, 440–9
  means 9, 10
  modes 3–4, 61–73, 455–6
  on-line control 409, 432–9
  optimised technology (OPT) 258
  optimum scale 332–3
  outputs 11–14
  process 1, 14–22, 50, 51, 187, 456
  rate 154, 156–7, 161, 166, 177–9, 465
  resources 8–11
  scheduling 235, 237, 243–52
  smoothing 188, 221–2, 286
  social structure 453–8
  stochastic programming 200, 204–5

stock replenishment 61, 66, 271, 273–5, 420
structure 47
systems 47, 48
to order 61
see also just-in-time production; mass production; master production schedule (MPS); multi-product, small batch production
production cost
  automation 381–2
  CAPP 365
  manufacturing optimisation 154–8, 161–7
  multistage manufacturing 176, 178, 179–82
production flow analysis 145–6
production-line layout design 136–7
production planning 42–4, 53–5, 61, 66, 77, 81, 105, 346
  aggregate 53, 187, 188, 190–234, 282, 286, 456
  long-term 188, 190, 225–8
  multiple-objective 200, 205–9
  POPIS 421–2
  process 187, 456
  short-term 188, 190–205
  strategic 53, 78, 187–8
production-progress function 123–4
production scheduling 53, 66, 187, 188, 235–69, 457
  computerised 409, 425–31
  IMMS 81
production time 83, 138, 239, 384
  CAPP 365, 368
  group scheduling 427
  manufacturing optimisation 154–6, 158, 161–5, 167
  multistage manufacturing 175, 177–82
  operations design 120–5
  project scheduling 264–6, 268
productive maintenance 283, 288–94
productivity 15–17, 18, 87, 357
profit 11, 13, 83, 315, 320, 511
  break-even analysis 330–2
  calculation 326
  per machine hour 224
  planning 43–4, 188, 325–8
  rate 154–8, 161–4, 166, 177, 180–2, 210
  reduced 197
  sharing 330
program evaluation and review technique (PERT) 35, 129
  scheduling 236, 258, 259–64, 265
projects
  probability of completion 263–4
  scheduling 188, 236, 258–68
pull-through systems 286
purchasing and receiving 446

QCD 13–14
quality 13–14, 122, 187, 357
  circles 14, 287, 302, 460
  control 81, 189, 283, 287, 302–4
  engineering 302–11
  function deployment (QFD) 307–9
  product design 94–6
quick response *see* just-in-time production

random-access-type FMS 392, 393, 394
rapid prototyping technology 370
raw materials 5–6, 8–9
  COPICS 445
  green production 505–6
  IMMS 76–81
  logistic system 50
  quality 302, 306
  see also material flow
R-charts 303–4
recursive functional equation 127
recycling 13, 94, 512
  green production 502–6
re-design of systems 42, 45
redundant systems 96
regression model 232
relationships of systems 25–6
REL (activity relationship) diagrams 139–42, 374–6
reliability 41, 94–5, 96–7
  preventive maintenance 290–3
renewal function 296, 298–300
reorder point 271, 272–3, 275
replenishment of stock 61, 66, 271, 273–5, 420
research and development (R&D) 78, 91–2, 258, 353
return on capital 326–8, 335–6
return on equity (ROE) 328
return on investment (ROI) 326–8, 331, 335–6, 338
right- and left-hand chart 116, 117
ripple effect 491
risk 305
robots 339, 392, 397, 399, 400–2
robustness 41, 311
roughness constraint on machine speed 170
routing 125–9, 367–9
Russian efficiency 482

SADT (structured analysis and design technique) 30, 31
safety of product design 94
sales 13, 354
  break-even analysis 329, 330
  management systems 76–7, 81
  product life cycle 90–1
scheduling
  group 68, 409, 425–30
  on-line 432–5
  operations 188, 235–6, 237–58
  project 188, 236, 258–68
  see also master production schedule (MPS); production scheduling
scheduling and control (SAC) system 433–5, 438, 439
scientific management 67, 457
seasonal variation 230–1
secondary industry 3, 93
  efficiency 469–75, 478, 480–9
  input–output 493
sequencing 188, 235, 237–8, 241–3, 252
  COPICS 445
  group scheduling 426–7
sequential sampling 304–5
service level 274, 276
  POPIS 420, 423–4

setup time 99, 286
shadow price 197, 198
shortage loss 275–6
shortest path algorithm 129
shortest processing time (SPT) rule 242, 257
signal-to-noise ratio 310–11
Silberston curve 64–5
Simplex method 193–7
simplification design 98
simulation 32, 256–8
single-stage manufacturing 83, 156–74
SIS (strategic information system) 409, 415–16, 460
slack 199, 245, 257, 427
 project scheduling 261–2, 266–8
SLAM-FACTOR 258
SLP (systematic layout planning) 138–43
social manufacturing systems 453–5
social production structure 453–8
socially appropriate manufacturing 507–8, 510, 511–12
space relationship diagrams 140, 142
special-purpose machine tools 382–3
SPT (shortest processing time) rule 242, 257
(S,s) system 271, 274, 275
stability 41, 135
standard time 120, 121
standardisation 423–4
stepping-stone path method 150, 151–2
stochastic programming 200, 204–5
stockless production 72
storage 85–6, 87–8, 110–11, 399, 403–4
strategic information systems (SIS) 409, 415–16, 460
strategic production planning 53, 78, 187–8
strategy 459–63
structural aspect manufacturing system 48
structured analysis and design technique (SADT) 30, 31
synergy 24, 25
system integration 440, 510
systematic layout planning (SLP) 138–43

tactics 459
Taguchi method 309–11
target costing 325
Taylor system 457
Taylor tool-life equation 159–60, 167, 175
TCR (total closeness rating) 374–6
Technical and Office Protocol (TOP) 417–18
technical production 315
technological information flow 83, 88
tertiary industry 3–4, 93, 415
 efficiency 469–73, 475, 480–9
 input–output 493
testing 339, 404–6, 445
Thai efficiency 485, 487
therbligs 116, 119
time study 115–20
time-series 228–9, 317–18
time utility 11, 15

tools 339, 341, 382–3
 automatically programmed (APT) 358, 372–3
 life 159–60, 167, 175, 436, 439
 manufacturing optimisation 156, 159–62, 164, 176
 NC (numerically controlled) 114, 339, 342, 345, 384–9, 406
 preventive maintenance 290–4
TOP (Technical and Office Protocol) 417–18
total closeness rating (TCR) 374–6
total cost 319–20
total productive maintenance (TPM) 289
total quality control (TQC) 287, 302
total systems 25, 37–8, 440
trade balance 466–7
transfer 339, 344–5, 346, 383–4, 399–400
transformational aspect manufacturing system 48–52
transportation 85–6, 87, 110–11
 layout planning 87, 144–5, 376, 379
 least unit 149–50
 linear programming 147–50, 223, 225, 226
 problems 147–52
travelling salesperson problem 83, 152–3
turning centres 386–7, 392
two-bin system 271
two-phase method 200–3

unmanned factories 339, 347, 406–7, 411
USA 465–7, 494
 efficiency 478, 480, 483
utilities 3–4, 11, 15, 93

value 98–9, 413
 engineering 316
 flow 1, 5–6, 8, 15, 315–18, 510
value added 17, 20, 321, 417, 500
Vogel's approximation method (VAM) 150

Wagner–Whitin method 219
Walras-type pricing 321–2
warehouses
 automated 339, 393, 403–4, 407, 423
 storage 87–8
waste disposal 12–13, 503, 505, 511–12
Weibull distribution 293–4, 300
wild-geese flying model 465
Wilson model 185, 271
Winters' method for seasonal fluctuations 230–1
work design 36
work flow 107–13, 145–6
 patterns 109–10
workstations 107, 113–14, 174, 432–3, 436
 CALB 430–1
 line-balancing 130–4
 optimal number 131–2

$\bar{X}$-charts 303–4

zero defect concept 306, 307